Genetics and Genomics of Pulmonary Arterial Hypertension

Genetics and Genomics of Pulmonary Arterial Hypertension

Editors

Laura Southgate
Rajiv D. Machado

MDPI • Basel • Beijing • Wuhan • Barcelona • Belgrade • Manchester • Tokyo • Cluj • Tianjin

Editors
Laura Southgate
St George's University of
London
UK

Rajiv D. Machado
St George's University of
London
UK

Editorial Office
MDPI
St. Alban-Anlage 66
4052 Basel, Switzerland

This is a reprint of articles from the Special Issue published online in the open access journal *Genes* (ISSN 2073-4425) (available at: https://www.mdpi.com/journal/genes/special_issues/Arterial_Hypertension).

For citation purposes, cite each article independently as indicated on the article page online and as indicated below:

LastName, A.A.; LastName, B.B.; LastName, C.C. Article Title. *Journal Name* **Year**, *Volume Number*, Page Range.

ISBN 978-3-0365-4829-6 (Hbk)
ISBN 978-3-0365-4830-2 (PDF)

Contents

About the Editors

Laura Southgate

Dr. Southgate is a Senior Lecturer (Associate Professor) in Genetics at St George's University of London, UK. Her research is focused on rare disease genetics, specifically understanding the molecular mechanisms underlying inherited vascular disorders, including pulmonary arterial hypertension (PAH) and the developmental disorder Adams–Oliver syndrome (AOS). Through an investigation of the genetic risk factors, she aims to identify molecular pathways to improve the diagnosis and clinical management of these conditions. Dr. Southgate gained her Ph.D. in Medical & Molecular Genetics from King's College London, UK and completed a post-doctoral Research Fellow at Queen Mary University of London, UK. She is a current member of the International Consortium for Genetic Studies in PAH (PAH-ICON), the Pulmonary Hypertension Gene and Variant Curation Expert Panels (ClinGen), and the Genomics England Clinical Interpretation Partnership (Cardiovascular GeCIP).

Rajiv D. Machado

Dr. Rajiv Machado has had a long-standing interest in the genetics and genomics of PAH. He was central to one of the teams that determined the mutation of the BMPR2 gene as the major risk factor for PAH. Since, he has worked to define additional genetic risk factors in this fatal disease and use these insights to uncover molecular pathways that might be amenable to intervention and treatment. Having previously held an Intermediate Basic Science Research Fellowship from the British Heart Foundation, he is currently a researcher and tutor at St George's University of London, UK. As an active member of the ClinGen Gene and Variant Curation Expert Panels for Pulmonary Hypertension and the International Consortium for Genetic Studies in PAH (PAH-ICON), Dr. Machado seeks to further improve the understanding of the genetic landscape of PAH to aid both research and patient care.

 *genes*

Editorial

Pulmonary Arterial Hypertension: A Deeper Evaluation of Genetic Risk in the -Omics Era

Rajiv D. Machado [1,2] **and Laura Southgate** [2,*]

1   Institute of Medical and Biomedical Education, St George's University of London, London SW17 0RE, UK; rmachado@sgul.ac.uk
2   Molecular and Clinical Sciences Research Institute, St George's University of London, London SW17 0RE, UK
*   Correspondence: lasouthg@sgul.ac.uk

Citation: Machado, R.D.; Southgate, L. Pulmonary Arterial Hypertension: A Deeper Evaluation of Genetic Risk in the -Omics Era. *Genes* **2021**, *12*, 1798. https://doi.org/10.3390/genes12111798

Received: 19 October 2021
Accepted: 3 November 2021
Published: 16 November 2021

Publisher's Note: MDPI stays neutral with regard to jurisdictional claims in published maps and institutional affiliations.

Pulmonary arterial hypertension (PAH) is a highly heterogeneous disorder with a complex, multifactorial aetiology. PAH is characterized by a sustained elevation of mean pulmonary arterial pressure resulting from the occlusion of distal pulmonary arterioles due to the uncontrolled proliferation of endothelial and smooth muscle cell populations. Of interest, pulmonary artery endothelial cells display monoclonal expansion suggestive of a neoplastic, quasi-cancerous phenotype. While the majority of PAH cases present with idiopathic disease (IPAH), approximately 10% of patients report a positive family history wherein the mode of transmission is predominantly autosomal dominant. More recently, examples of autosomal recessive inheritance have been recorded, albeit in rare instances. PAH may also be associated with co-morbidities, including congenital heart disease, connective tissue disease, thromboembolism and exposure to anorexigens (APAH). Of note, PAH displays a marked sex bias towards females (4.3:1) and reduced penetrance both within and between families. These features strongly suggest the presence of modifying factors that may be genetic and/or environmental in origin.

The identification of heterozygous, germline mutations in the bone morphogenetic protein receptor 2 (*BMPR2*) gene, encoding a type-II receptor of the TGF-β signalling superfamily, represented a critical breakthrough in defining the pathogenesis of PAH [1]. In the last decade, next-generation sequencing methodologies have been successfully harnessed in the detection of novel causative genes that, together, have significantly expanded the genetic landscape of this disease. However, several key clinical and molecular aspects of PAH remain enigmatic. First, missing heritability in both familial and idiopathic disease is yet to be fully explored and, indeed, solved. Historically, gene identification and variant analyses have largely targeted Caucasian, adult populations to the exclusion of paediatric cohorts and populations of different genetic ancestry that may provide important insights. A strong focus has been placed on the analysis of the exome in both gene identification and molecular diagnostic efforts. The comprehensive interrogation of non-coding regions of the genome may represent a fruitful means of accessing variation that explains the apparent mutation shortfall observed thus far. Modifying factors of PAH, for example, expression quantitative trait loci is a field that requires further investigation. Moreover, the preponderance of females afflicted with PAH remains a vexed question demanding deeper analysis. The cellular phenotype of PAH is considered to be largely similar to the Warburg phenomenon of cancer cell progression, namely monoclonal expansion of key cell populations, altered glycolysis and mitochondrial dysfunction [2]. However, this research avenue remains to be fully elucidated. Finally, although major breakthroughs have been made in the management and treatment of this devastating disease, clinical intervention remains ameliorative, and PAH typically continues to be a fatal outcome.

In this *Genes* Special Issue, a number of these outstanding questions are addressed. Welch et al. distinguish between the clinical and molecular genetic features of paediatric and adult PAH to draw distinct parallels between these disease subtypes [3]. Specifically, the authors indicate a significantly higher mutation burden in early onset disease (~42% by comparison to 12.5% in adults). Gelinas et al. performed whole-exome sequencing on a panel of paediatric patients to explore the genetic background to this specific and pernicious form of PAH [4]. This study provides novel insights into the wider mutation burden in childhood disease and also establishes the first independent validation of *BMP10* and *PDGFD* as genetic risk factors for PAH. Both reports emphasise the need to consider childhood-onset PAH as a genetically and clinically discrete entity to better facilitate improved diagnostic surveillance and novel gene identification.

The granular assessment of patient groups based on clearly defined ethnicity has furthered awareness of divergent genetic architecture in both PAH and other inherited diseases. To this end, Tenorio Castaño et al. surveyed a cohort of three hundred Spanish patients in a customised panel that included 21 genes associated with PAH [5]. Likely deleterious variants were detected in 15% of the patients analysed, including those with APAH. Moreover, the study recorded a sub-set of variants of uncertain significance (VUS) that require further annotation to reach clinical utility. The value of incorporating emerging genes, likely to underpin PAH, into existing screening protocols is exemplified by van der Heuvel et al., who added 17 newly identified causal factors to the previous diagnostic panel, which initially consisted of just *BMPR2* and *SMAD9*, in the Netherlands [6]. The diagnostic yield in a cohort comprising 28 previously tested patients and an additional 56 patients with IPAH or pulmonary veno-occlusive disease (PVOD) was subsequently augmented, thereby affording improved prospects for early diagnosis, screening of at-risk relatives and personalised management of patients. As detailed in some depth by Swietlik et al., genetic screening now plays a pivotal role in the diagnosis of PAH; yet, there is more work to be done to fully elucidate the genetic architecture, particularly in idiopathic disease. An intriguing approach is the assessment of stable intermediate phenotypes that, in turn, leads to a fluid interplay between forward and reverse genetics and reverse phenotyping. Whilst conceptual in nature, this study nonetheless suggests multi-omics approaches to otherwise intractable genetic disorders may yield subtle and/or masked insights into PAH pathogenesis [7].

With the exception of reports limited to single genes and small study panels, the non-coding regions of putative causative genes and the wider genome remain insufficiently investigated. An early example of the value of analysing these regions is provided by the identification of a deleterious variant in the 5′ untranslated region of the *BMPR2* gene consolidated by functional evidence suggesting loss of transcript by nonsense-mediated decay [8]. To assess the role of non-coding sequence defects leading to the promotion of the disease phenotype, Song et al. examined nine different *BMPR2* promoter variants previously identified in seven PAH families and three IPAH patients by in vitro over-expression luciferase assays. In human pulmonary artery smooth muscle cells, the majority of these variants (c.-575A>T, c.-586dupT, c.-910C>T, c.-930_-928dupGGC, c.-933_-928dupGGCGGC, c.-930_-928delGGC and c.-1141C>T) significantly depressed activation of the reporter gene [9]. This preliminary examination of regulatory elements in known PAH risk factors draws much-needed attention to a hitherto poorly explored avenue of research into disease manifestation and modulation. Expansion of this proof-of-principle study into readily available whole-genome sequence data will establish a framework for the comprehensive determination of causal variation.

A powerful analytic technique for investigating the genetic triggers underlying PAH is driven by the assessment of expression quantitative trait loci (eQTL), a nexus between genetic association study signals and fundamental biology. Herein, Ulrich et al. describe transcriptome- and genome-wide eQTL mapping in an idiopathic and hereditary PAH cohort employing RNA sequencing (RNAseq) methodology [10]. By comparison to at least one published study, the authors confirm approximately 75% eQTLs from an identified total of 2314, of which 90% were cis-acting. These data implicate genes involved in immune related processes, an acknowledged association of PAH pathogenesis. This insight has wider relevance to diseases that share a molecular signature with PAH.

Genomic instability in PAH has been a long-standing mechanism of the pathobiological process. To assess the importance of DNA repair in the nuclear and mitochondrial genome, Sharma et al. analysed aberration in damage response pathways in both the nuclear and mitochondrial genomes. Apoptotic escape associated with concomitant pro-proliferation processes provides a temporal sequence that may aid in early diagnosis and therapeutic intervention [11]. Of interest, expression profiles of key cellular sites of disease in PAH patients indicated that 586 genes were up-regulated and 372 down-regulated. Moreover, 35 of these genes were involved in DNA repair [12]. The authors report an intriguing link between *BRCA1* and BMP signalling, further providing an interface between PAH and neoplastic expansion.

Cirulis et al. posit the 'oestrogen puzzle' as a confounding factor in the development of PAH [13]. Typically, oestrogen and its metabolites protect organisms from developing indicative symptoms of pulmonary hypertension. However, in human disease, female carriers of causal variants are significantly more likely to develop the disease than male variant carriers. The authors in this review examine the interplay between oestrogen and the BMP signalling pathway in regard to the role of gender and PAH susceptibility. In relevant cell lines, the most prominent genetic risk factor, namely *BMPR2*, was significantly repressed by the administration of estradiol (E2) and estriol (E3). Indeed, over-expression in a cell line lacking native oestrogen receptors with elevated concentrations of ERα generated de-repression of the *BMPR2* promoter. Of interest, these findings have been reproduced in murine models; in particular *Bmpr2* gene expression was depressed in ovariectomized females. These studies indicate a multi-layered nexus between oestrogen signalling and BMP activity. The dysregulation of these pathways might promote a PAH phenotype that signals a shift toward mitogenic elements, for example, 16α-hydroxyestrone (16α-OHE$_1$).

Treatment options in PAH have focussed on addressing pathobiological processes, namely endothelial dysfunction, excessive cell proliferation and vasoconstriction. Although efficacious in prolonging 5-year survival metrics, notably an improvement of 34% to 60%, these approaches are not curative, and the disease remains fatal often due to resistance to medication. Dannewitz Prosseda et al. propose employing the fundamental genetic and molecular findings key to disease pathogenesis as complementary avenues to the development of targeted, effective measures in combating this disease [14]. In particular, they note the fundamental importance of the BMPR2 pathway as a precision tool to the development of therapeutic modalities. Specifically, the authors present alternatives to current therapeutic strategies, which include enhancing BMPR2 availability at the cell membrane, relieving receptor inhibition, driving transcription of BMPR2 target genes or gene therapy approaches.

In conclusion, the series of articles presented in this Special Issue highlight that PAH genetics have now moved beyond a simple Mendelian 'one gene = one disease' model to the recognition of PAH as a multigenic and multifactorial disorder. Of note, independent validation of newly described genes by expert review has become an increasingly important consideration in the establishment of high-throughput diagnostic testing panels. By expanding contemporary analyses into considerations of the wider genomic architecture, these reports demonstrate that the examination of the non-coding genome, eQTL mapping, DNA damage repair and gender-specific bias will provide fundamental insights into the pathways and molecular mechanisms critical to the development and maintenance of the pulmonary vasculature. In the near future, further focussed investigations based on these findings will afford novel diagnostic and clinical avenues to improve outcomes for patients and their families.

Funding: This research received no external funding.

Conflicts of Interest: The authors declare no conflict of interest.

References

1. Southgate, L.; Machado, R.D.; Gräf, S.; Morrell, N.W. Molecular genetic framework underlying pulmonary arterial hypertension. *Nat. Rev. Cardiol.* **2020**, *17*, 85–95. [CrossRef] [PubMed]
2. Archer, S.L. Pyruvate kinase and Warburg metabolism in pulmonary arterial hypertension: Uncoupled glycolysis and the cancer-like phenotype of pulmonary arterial hypertension. *Circulation* **2017**, *136*, 2486–2490. [CrossRef] [PubMed]
3. Welch, C.L.; Chung, W.K. Genetics and genomics of pediatric pulmonary arterial hypertension. *Genes* **2020**, *11*, 1213. [CrossRef] [PubMed]
4. Gelinas, S.M.; Benson, C.E.; Khan, M.A.; Berger, R.M.F.; Trembath, R.C.; Machado, R.D.; Southgate, L. Whole exome sequence analysis provides novel insights into the genetic framework of childhood-onset pulmonary arterial hypertension. *Genes* **2020**, *11*, 1328. [CrossRef] [PubMed]
5. Tenorio Castaño, J.A.; Hernández-Gonzalez, I.; Gallego, N.; Pérez-Olivares, C.; Ochoa Parra, N.; Arias, P.; Granda, E.; Acebo, G.G.; Lago-Docampo, M.; Palomino-Doza, J.; et al. Customized massive parallel sequencing panel for diagnosis of pulmonary arterial hypertension. *Genes* **2020**, *11*, 1158. [CrossRef] [PubMed]
6. Van den Heuvel, L.M.; Jansen, S.M.A.; Alsters, S.I.M.; Post, M.C.; van der Smagt, J.J.; Handoko-De Man, F.S.; van Tintelen, J.P.; Gille, H.; Christiaans, I.; Vonk Noordegraaf, A.; et al. Genetic evaluation in a cohort of 126 Dutch pulmonary arterial hypertension patients. *Genes* **2020**, *11*, 1191. [CrossRef] [PubMed]
7. Swietlik, E.M.; Prapa, M.; Martin, J.M.; Pandya, D.; Auckland, K.; Morrell, N.W.; Gräf, S. 'There and back again'—Forward genetics and reverse phenotyping in pulmonary arterial hypertension. *Genes* **2020**, *11*, 1408. [CrossRef] [PubMed]
8. Aldred, M.A.; Machado, R.D.; James, V.; Morrell, N.W.; Trembath, R.C. Characterization of the BMPR2 5′-untranslated region and a novel mutation in pulmonary hypertension. *Am. J. Respir. Crit. Care Med.* **2007**, *176*, 819–824. [CrossRef] [PubMed]
9. Song, J.; Hinderhofer, K.; Kaufmann, L.T.; Benjamin, N.; Fischer, C.; Grünig, E.; Eichstaedt, C.A. BMPR2 promoter variants effect gene expression in pulmonary Arterial hypertension patients. *Genes* **2020**, *11*, 1168. [CrossRef] [PubMed]
10. Ulrich, A.; Otero-Núñez, P.; Wharton, J.; Swietlik, E.M.; Gräf, S.; Morrell, N.W.; Wang, D.; Lawrie, A.; Wilkins, M.R.; Prokopenko, I.; et al. Expression quantitative trait locus mapping in pulmonary arterial hypertension. *Genes* **2020**, *11*, 1247. [CrossRef] [PubMed]
11. Sharma, S.; Aldred, M.A. DNA Damage and repair in pulmonary arterial hypertension. *Genes* **2020**, *11*, 1224. [CrossRef] [PubMed]
12. Li, M.; Vattulainen, S.; Aho, J.; Orcholski, M.; Rojas, V.; Yuan, K.; Helenius, M.; Taimen, P.; Myllykangas, S.; De Jesus Perez, V.; et al. Loss of bone morphogenetic protein receptor 2 is associated with abnormal DNA repair in pulmonary arterial hypertension. *Am. J. Respir. Cell Mol. Biol.* **2014**, *50*, 1118–1128. [CrossRef] [PubMed]
13. Cirulis, M.M.; Dodson, M.W.; Brown, L.M.; Brown, S.M.; Lahm, T.; Elliott, G. At the X-roads of sex and genetics in pulmonary arterial hypertension. *Genes* **2020**, *11*, 1371. [CrossRef] [PubMed]
14. Dannewitz Prosseda, S.; Ali, M.K.; Spiekerkoetter, E. Novel advances in modifying BMPR2 signaling in PAH. *Genes* **2020**, *12*, [CrossRef] [PubMed]

Review

Genetics and Genomics of Pediatric Pulmonary Arterial Hypertension

Carrie L. Welch [1] and Wendy K. Chung [1,2,*]

[1] Department of Pediatrics, Columbia University Irving Medical Center, 1150 St. Nicholas Avenue, New York, NY 10032, USA; cbw13@columbia.edu

[2] Department of Medicine, Columbia University Irving Medical Center, 622 W 168th St, New York, NY 10032, USA

* Correspondence: wkc15@columbia.edu; Tel.: +212-851-5313

Received: 2 September 2020; Accepted: 13 October 2020; Published: 16 October 2020

Abstract: Pulmonary arterial hypertension (PAH) is a rare disease with high mortality despite recent therapeutic advances. The disease is caused by both genetic and environmental factors and likely gene–environment interactions. While PAH can manifest across the lifespan, pediatric-onset disease is particularly challenging because it is frequently associated with a more severe clinical course and comorbidities including lung/heart developmental anomalies. In light of these differences, it is perhaps not surprising that emerging data from genetic studies of pediatric-onset PAH indicate that the genetic basis is different than that of adults. There is a greater genetic burden in children, with rare genetic factors contributing to ~42% of pediatric-onset PAH compared to ~12.5% of adult-onset PAH. De novo variants are frequently associated with PAH in children and contribute to at least 15% of all pediatric cases. The standard of medical care for pediatric PAH patients is based on extrapolations from adult data. However, increased etiologic heterogeneity, poorer prognosis, and increased genetic burden for pediatric-onset PAH calls for a dedicated pediatric research agenda to improve molecular diagnosis and clinical management. A genomics-first approach will improve the understanding of pediatric PAH and how it is related to other rare pediatric genetic disorders.

Keywords: genomics; pediatrics; lung disease; pulmonary arterial hypertension

---

1. Introduction

Pulmonary arterial hypertension (PAH) is a rare disease with an estimated prevalence of 4.8–8.1 cases/million for pediatric-onset [1] and 15–50 cases/million for adult-onset disease [2]. Pathogenic changes in the pulmonary vasculature—including endothelial dysfunction, aberrant cell proliferation, and vasoconstriction—give rise to the clinical consequences of increased pulmonary vascular pressures, increased vascular resistance, heart failure, and premature death [3]. The disease is caused by genetic, epigenetic, and environmental factors, as well as gene–environment interactions wherein genetic contributions to disease risk are modified by environmental exposures. Causal genetic factors for PAH are typically autosomal dominantly inherited for genes such as *BMPR2*, the major gene causing familial forms of PAH (FPAH) [4,5]. Environmental risk factors include hypoxia and exposure to drugs and toxins [4]. Epigenetic factors include active histone mark H3K27ac [6]. Most of our understanding of PAH etiology and treatment is based upon studies in adults [7,8]. However, emerging clinical and genetic data indicate that there are fundamental differences between pediatric- and adult-onset disease.

Pediatric PAH differs from the adult-onset disease in several important aspects, including sex bias, clinical presentation, etiology, and response to therapy [7–9]. The frequency of PAH is ~3–4-fold higher in females relative to males for adult-onset disease. However, data from the National Biological Sample and Data Repository for PAH (aka PAH Biobank, Table 1) [10,11] and other studies [12,13] indicate that the frequency of pediatric-onset PAH is similar for females and males, suggesting less dependence on sex-specific factors in children. Children present with increased severity of disease, e.g., elevated mean pulmonary artery pressure (mPAP), decreased cardiac output, and increased pulmonary vascular resistance, compared to adults at diagnosis (Table 1) [10,11]. The clinical manifestations likely reflect the complex etiology of disease in children. While prenatal and early postnatal influences on lung growth and development can contribute to the development of PAH across the lifespan, early developmental influences play a particularly prominent role in pediatric-onset PAH in which patients frequently have complex comorbidities such as congenital heart disease (APAH-CHD), Down syndrome, congenital diaphragmatic hernia (CDH), and other developmental lung diseases, including persistent pulmonary hypertension of the newborn (PPHN) [7,14,15]. Histopathological studies have identified abnormal lung development and lung hypoplasia as common features of PAH, CHD, CDH, and Down syndrome [14,16]. While the mechanisms for impaired lung development are not known, altered expression of angiogenic and anti-angiogenic genes likely contribute [17–20]. Decreased lung vascular and alveolar growth predispose one to vascular injury during susceptible periods of growth and adaptation. The presentation of pediatric PAH with developmental comorbidities contributes to poor outcomes in these children [7,15]. Few pharmaceutical therapies are approved for use in children due to the lack of safety and efficacy data [8]. However, a retrospective study of pediatric PAH patients suggested that PAH patients with Down syndrome may be less responsive to PAH treatments than non-Down syndrome patients [21]. Clearly, pediatric-focused studies are needed to increase our understanding about the natural history, the pathogenic mechanisms, and the treatment of PAH in children.

Table 1. Clinical characteristics and hemodynamic parameters of child- vs. adult-onset pulmonary arterial hypertension (PAH) cases at diagnosis. Data are from the PAH Biobank ($n = 2572$). Child-onset, <18 years of age at diagnosis. Mean ± SD.

Group (n)	Age at dx (y)	F:M Ratio	mPAP (mm Hg)	mPCWP (mm Hg)	CO Fisk (L/min)	PVR (Woods Units)	Common Comorbidities
Child (226)	7.7 ± 5.4	1.65:1	55.1 ± 18.6	9.0 ± 3.0	3.2 ± 1.6	18.1 ± 11.7	CHD, CDH, DS, lung growth/development
Adult (2345)	51.6 ± 14.7	4.02:1	49.6 ± 13.9	10.2 ± 4.2	4.6 ± 1.7	10.0 ± 5.9	HTN, hypothyroidism, other pulmonary & metabolic diseases
p-value	<0.0001 *	<0.0001 **	<0.0001 *	<0.0001 *	<0.0001 *	<0.0001 *	

Abbreviations: dx, diagnosis; F:M, female:male; mPAP, mean pulmonary artery pressure; mCWP, mean pulmonary capillary wedge pressure; CO, cardiac output; PVR, pulmonary vascular resistance; CHD, congenital heart disease; CDH, congenital diaphragmatic hernia; DS, Down syndrome; HTN, systemic hypertension. * Student's t-test, 2-tailed. ** Fisher exact test.

2. Genetics of Pediatric PAH—Current Knowledge

Emerging data from genetic studies of pediatric-onset PAH indicate that the genetic basis in children is different from that of adults [10,11,13]. There is a greater genetic burden in children, with rare genetic factors contributing to ~42% of pediatric-onset PAH compared to ~12.5% of adult-onset PAH (Figure 1). De novo variants are frequent in children, likely contributing to ~15% of pediatric PAH [10,13]. Among rare inherited variants, variants in *BMPR2* are causal in ~6.5–7% of both pediatric- and adult-onset PAH; most of the cases are FPAH or idiopathic PAH (IPAH), rarely PAH associated with other diseases (APAH) [11,22], and no occurrences in PPHN have been reported to date [23]. Notably, two of the other known causal genes with the highest frequencies of rare deleterious variants among pediatric cases—*TBX4* and *SOX17*—are highly expressed in embryonic tissues and have prominent roles in lung and vasculature development [24–26]. The mean age of PAH onset by risk gene is shown for twelve of the genes in Figure 2.

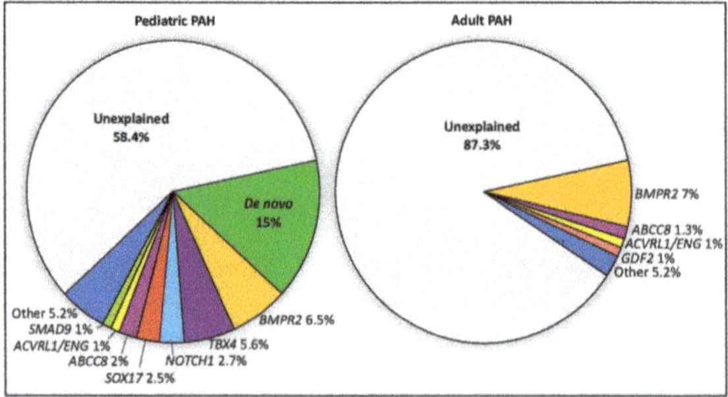

Figure 1. Relative contributions of *de novo* mutations and 18 PAH risk genes in a cohort of 443 pediatric and 2628 adult cases from CUIMC and the PAH Biobank. Risk genes include *BMPR2, ABCC8, ACVRL1, ATP13A3, BMPR1B, CAV1, EIF2AK4, ENG, GDF2, KCNA5, KCNK3, KDR, NOTCH1, SMAD1, SMAD4, SMAD9*, and *TBX4*. PAH cases include IPAH, APAH, FPAH and other rarer cases.

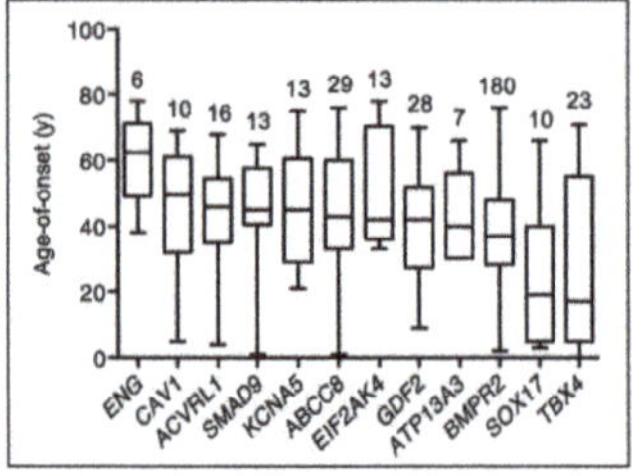

Figure 2. Age-of-disease onset for all PAH Biobank cases with rare deleterious variants in known PAH risk genes. Box plots showing median, interquartile range and min/max values for age-of-disease onset (i.e., age at diagnostic right heart catheterization). The number of cases carrying variants for each gene is given above each box plot. Genes represented by less than four cases are not shown.

2.1. TBX4

Unlike *BMPR2* and other known causal PAH genes, *TBX4* is not expressed in pulmonary arterial endothelial cells or smooth muscle cells. *TBX4* is a transcription factor in the T-box gene family that is co-expressed with *TBX5* throughout the mesenchyme of developing lung and trachea [24]. Lung-specific *Tbx4/Tbx5* deficient mice exhibit impaired lung branching and hypoplasia during gestation as well as early postnatal death due to severe respiratory disease [24]. *TBX4* is also expressed in the developing atrium of the heart and the limb buds [27]. In humans, rare but recurrent microdeletions of chromosome 17q23, including *TBX4*, have been observed in children with complex phenotypes including PAH, heart and skeletal defects, and neurodevelopmental delay [28–30]. More recently, *TBX4*-specific likely gene-disrupting (LGD) and damaging missense variants have been associated with PAH with or without small patella syndrome (OMIM #147891), most frequently in pediatric cases [11,13,22,31,32]. In two independent cohorts [11,13], rare deleterious variants in *TBX4* showed significant enrichment among pediatric- compared to adult-onset IPAH cases (Columbia University Irving Medical Center, CUIMC, cohort: 10/130 vs. 0/178; PAH Biobank: 12/155 vs. 1/257, respectively). In the PAH Biobank, ten additional *TBX4* variants were identified for other PAH subtypes, including three APAH-CHD cases with heart defects. In a cohort of 256 APAH-CHD cases (144 pediatric- and 112 adult-onset), we identified *TBX4* variants in seven cases with age-of-onset from newborn to 11 years, one associated with alveolar hypoplasia [22]. Together, the data suggest that rare *TBX4* variants contribute to 7.7% of pediatric IPAH and 4.9% of pediatric APAH-CHD cases. Notably, *TBX4* variants have not been observed in CHD alone [33].

Skeletal and other developmental defects are not routinely assessed as part of a PAH diagnosis and, as such, some dual diagnoses may have been missed for PAH cases. To this end, Galambos et al. [34] carried out detailed clinical and histopathologic characterization of 19 pediatric PAH cases with *TBX4* variants: 6 microdeletions, 12 LGD, and 1 missense. Seven infants had evidence of abnormal distal lung development, and there was a high frequency of heart and skeletal developmental anomalies; neurodevelopmental delay was also observed among those patients with microdeletions, likely due to haploinsufficiency of other adjacent genes. Ten newborns presented with PPHN which resolved but recurred later in infancy or childhood [34]. A report from the National French Registry [35] concurred these findings of skeletal, heart, and lung developmental anomalies in PAH cases. Why some patients present with PAH alone, small patella syndrome alone, PAH with small patella syndrome, or PAH with other developmental defects is not understood at this time but may depend on the variant type or the protein location of gene variants, other genetic or epigenetic factors, or other environmental factors affecting the specific transcriptional pathways regulated by TBX4. It *is* clear that genetic diagnosis of a rare deleterious *TBX4* variant or *TBX4*-containing microdeletion in pediatric PAH predicts a more complex developmental phenotype (*TBX4* syndrome [36]). Chest imaging for severe and diffuse features of pulmonary growth arrest, assessment for congenital heart defects, physical examination of hands and feet, and radiological assessment of pelvic areas are recommended. In addition, a *TBX4* diagnosis predicts potential recurrence of PAH following neonatal PPHN suggesting that annual screening by echocardiography may be useful.

2.2. SOX17

SOX17 is a member of the conserved family of SRY-related HMG box transcription factors, originally identified as key regulators of male sex determination but now recognized to have critical roles in embryogenesis [25,26]. *SOX17* is specifically required for endoderm formation and vascular morphogenesis [25,37,38], and germline deletion of *Sox17* results in embryonic lethality by E10.5 [25]. In the developing murine lung, *Sox17* is expressed in mesenchymal progenitor cells and is then restricted to endothelial cells of the pulmonary vasculature [39]. Conditional deletion of *Sox17* in mesenchymal progenitor cells causes abnormal pulmonary vascular morphogenesis, resulting in postnatal cardiopulmonary dysfunction and juvenile death [39]. Endothelial-specific inactivation of *Sox17* in mice leads to impaired arterial specification and embryonic death or, with conditional postnatal inactivation, arterial-venous malformations [37]. Transcriptional activation of *Sox17* via hypoxia-induced factor 1α leads to upregulation of cyclin-E1 and endothelial regeneration in response to lung injury [40]. We identified *SOX17* as a candidate risk gene for PAH using exome sequencing data in a cohort of 256 APAH-CHD patients [22]. Thirteen cases with rare predicted deleterious *SOX17* variants were identified, seven LGD and six missense variants located primarily within the conserved HMG box domain (Figure 2). Fifty-six percent of the overall cohort were pediatric cases, but nine of thirteen cases with rare deleterious variants in *SOX17* were pediatric cases with mean age-of-onset of 14 years. A recurrent frameshift variant, p.(Leu167Trpfs*213), was identified in three APAH-CHD cases with age-of-onset ranging from 7 months to 5 years. We [11,22] and others [41,42] have identified *SOX17* variants in IPAH cases but with lower frequency in adults. Combined data from five cohorts ([11,13,22,41,42] indicate that *SOX17* variants contribute to 7% of all pediatric-onset PAH cases compared to 0.4% of adult-onset cases (Figure 3). Protein modeling indicates that at least three of the APAH-CHD case missense variants localize to the transcription factor DNA binding pocket [22], and missense variants in this region have been shown to impair both direct DNA binding and SOX17/β-catenin nucleoprotein complexes at target gene promoters [43,44]. These data suggest that haploinsufficiency with complete or partial loss of function alleles is the likely mechanism of *SOX17* risk in PAH.

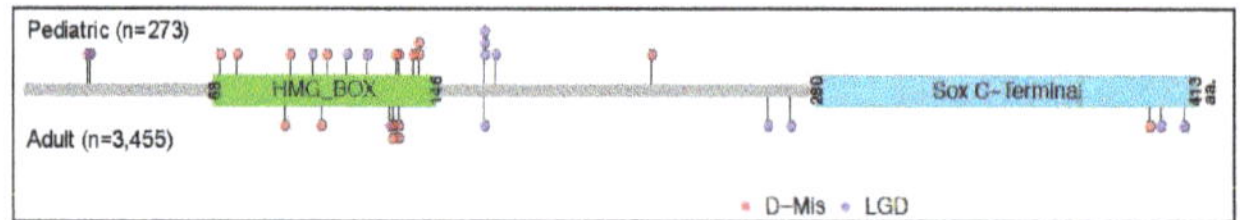

Figure 3. Locations of SOX17 likely gene disrupting (LGD) and rare predicted deleterious missense (D-Mis) variants carried by PAH cases from five cohorts from the US, UK and Japan. Variants carried by pediatric cases (*n* = 19) are shown above the protein schematic and variants carried by adult cases (*n* = 13) below the schematic. The combined datasets include 273 pediatric and 3455 adult cases [11,13,22,41,42].

Some variants in SOX17 downstream target genes may be predicted to mimic the consequences of *SOX17* LGD variants or haploinsufficiency. We identified 163 rare predicted deleterious variants in 149 putative SOX17 target genes, most with prominent expression in pulmonary artery endothelial cells and/or developing heart [22]. For the 32 LGD and the 131 missense variants, we observed a moderate but significant enrichment of rare missense variants in cases compared to controls. Approximately one-third of these genes had top quartile gene expression in both pulmonary artery endothelial cells and developing heart. Pathway analysis indicated that the genes have likely roles in developmental biology, small molecule transport/homeostasis, and extracellular matrix interactions (Table 2). While these results are intriguing, they require confirmation in larger cohorts to determine which specific SOX17-regulated genes/pathways contribute to PAH risk.

Table 2. Biological pathway analysis of SOX17 target genes harboring PAH-CHD patient rare deleterious variants. Data obtained using Reactome 2016. Pathways with false discovery rate (FDR)-adjusted *p*-value ≤ 0.05 are listed.

Term	Reactome ID	# Genes in Overlap	*p*-Value	Adjusted *p*-Value	Genes
Developmental biology	R-HAS-1266738	16/786	6.8×10^{-5}	0.03	*KLB, ROBO2, LAMA1, EGF, ANK3, LAMC, SLC2A4, MED6, SPRED2, MEIS1, NRFA2, PCMC4, NF1, EP300, TCF4, EPHB4*
Transmembrane transport of small molecules	R-HAS-382561	13/594	1.7×10^{-4}	0.03	*RYR2, ABCC4, ABCC1, SLC1A3, SL#3A4, SLC8A1, CLCN5, SLCA9, ATPB7, ASPH, WNK1, NUP35, EMB*
Non-integrin membrane extracellular matrix interactions	R-HAS-3000171	4/42	1.7×10^{-4}	0.03	*LAMA1, LAMA4, LAMC1, THBS1*
Ion homeostasis	R-HAS-5578775	4/51	1.7×10^{-4}	0.03	*RYR2, ASPH, TPR3, SLC8A1*

2.3. The Relative Contribution of Other Known PAH Risk Genes

Among the combined CUIMC and PAH Biobank pediatric cohort of 443 non-overlapping cases, rare deleterious variants in other known PAH risk genes altogether account for ~12% of cases (27 IPAH, 26 APAH-CHD, 2 APAH-HHT). Variants in *NOTCH1* account for 2.7% of pediatric PAH cases (2 IPAH, 10 APAH-CHD). *NOTCH1* encodes a transmembrane receptor that facilitates intercellular interactions and signaling with known roles in development and is a known CHD risk gene. Variants in *ABCC8* and *SMAD9* account for 2% and 1% of cases, respectively, including both IPAH and APAH-CHD. Variants in *ACVRL1/ENG* account for 1% of cases, including two APAH-HHT cases. *BMPR1B*, *CAV1*, *GDF2*, *KCNK3* and *KDR/BMP9* were identified in one to four cases each, accounting for <1% for each gene. In addition to autosomal dominant inheritance, recessively inherited *EIF2AK4* variants have been identified in 1–3% of children in European and Chinese cohorts [31,45,46]. In addition, a rare occurrence of recessively inherited *GDF2* variants has been reported for a 3-year-old boy with right heart failure [47]. Autosomal recessive inheritance of other risk variants may cause very early-onset severe PAH, and additional pediatric studies are necessary to evaluate rare recessive genetic etiologies.

2.4. De Novo Variants

De novo variants have emerged as an important class of genetic factors underlying rare diseases, especially early-onset severe conditions [15,33,48–50], due to strong negative selection decreasing reproductive fitness [51]. We recently assessed the role of rare deleterious de novo variants in pediatric PAH using a cohort of 124 parent-child trios (56% IPAH, 38% APAH-CHD, 6% other PAH) [10]. We observed a 2.5-fold enrichment of de novo variants among all PAH cases compared to the expected rate, almost entirely due to genes that are highly expressed in developing lung or heart (Table 3). Among the PAH cases identified with de novo variants, 54% were IPAH, 32% were APAH-CHD, and 14% other PAH; at least 20% of the de novo variant carriers had additional diagnoses of other congenital anomalies. De novo variants were identified in three known PAH risk genes (four variants in *TBX4*, two in *BMPR2*, one in *ACVRL1*) and 23 additional genes with high expression in developing lung and/or heart but little to no previous association with PAH. Based on the enrichment rate, we estimate that ~18 of the identified variants are likely to be implicated in pediatric PAH. The identified genes fit a general pattern for developmental disorders—genes intolerant to LGD variants (pLI >0.5 for 40% of the PAH genes) and with known functions important for coordinated organogenesis, including transcription factors, RNA binding proteins, protein kinases, and chromatin modification. Three of the genes are known CHD risk genes (*NOTCH1*, *PTPN11*, and *RAF1*). *NOTCH1* is the most commonly associated gene for the congenital heart defect of tetralogy of Fallot, [52] and the *NOTCH1* de novo variant carrier had a diagnosis of APAH-CHD with tetralogy of Fallot. Rare variants in *PTPN11* and *RAF1* are causal for Noonan syndrome, which has a high frequency of congenital heart defects. The de novo variants identified in both of these genes are known causal Noonan syndrome variants [53], and three cases of fatal pediatric PAH with Noonan syndrome have been previously reported [54,55]. We previously reported rare inherited variants in *NOTCH1* (*n* = 5), *PTPN11* (*n* = 1), and *RAF1* (*n* = 2) carried by APAH-CHD cases [22]. Aside from known PAH and CHD genes, at least eight of the other genes with identified de novo variants have known or plausible roles in lung/vascular development (Table 4). For example, *AMOT* (angiomotin) encodes an angiostatin-binding protein involved in embryonic endothelial cell migration and tube formation as well as endothelial cell tight junctions and angiogenesis [56–58]. *HSPA4* (heat shock protein A4) encodes a chaperone that, together with HSPA4L, functions in embryonic lung maturation and dual deletion of *Hspa4/Hspa4l*, which results in intrauterine pulmonary hypoplasia and early neonatal death [59]. *KEAP1* (Kelch-like ECH associated protein 1) regulates oxidative stress and apoptosis through interactions with NRF2 in murine vascular cells [60], and endothelial-specific deletion of *NRF2* reduces endothelial sprouting in vivo [61] and increases susceptibility to bronchopulmonary dysplasia and other respiratory diseases [62]. An NRF2 activator is currently being investigated in a phase 2 clinical trial for PAH (ichgcp.net/clinical-trials-registry/NCT02036970). Finally, one third of all of the de novo variants identified in the trio analysis are in causal genes for developmental syndromes, consistent with the enrichment of developmental phenotypes among the variant carriers [10]. The genes identified in this study require replication in a larger pediatric cohort. In addition, genes with rare variants can be entered into GeneMatcher to identify other cases with rare variants in the same gene and compare genotypes and phenotypes. Due to the low background rate of rare deleterious de novo variants [63], the statistical evidence for a candidate risk gene is effectively equivalent to multiplicity. That is, genes with ≥2 rare deleterious de novo variants are unlikely to be mutated by chance and should be considered candidate risk genes. The genes and the variants identified in the pediatric trio analysis have not been observed in adult-onset cases and likely will be specific to pediatric PAH. Thus, it is imperative that larger pediatric-focused PAH cohorts are studied to advance our knowledge of the causal genes specific to pediatric-onset PAH.

Table 3. Burden of de novo variants in 5756 genes highly expressed in developing lung (murine E16.5 lung stromal cells) and/or developing heart (murine E14.5 heart) in pediatric-onset PAH ($n = 124$ child/parent trios).

Variant Type *	Observed in Trios ($n = 124$)	Expected by Chance	Enrichment	p-Value	Estimated # of True Risk Variants
SYN	18	14.0	1.3	0.28	—
LGD	11	4.7	2.4	0.06	—
MIS	40	31.7	1.3	0.15	—
D-MIS	19	7.2	2.6	2.0×10^{-4}	12
LGD + D-MIS	30	11.8	2.5	7.0×10^{-6}	18

* SYN, synonymous; LGD, likely gene disrupting; MIS, missense; D-MIS, predicted deleterious missense based on REVEL score > 0.5.

Table 4. Novel genes with rare deleterious de novo variants in pediatric-onset PAH ($n = 124$ trios).

Gene Symbol	Variant Type	Protein Change	REVEL Score	CADD Score	Allele Frequency (gnomAD)	E16.5 Lung Expression Rank	E14.5 Heart Expression Rank	Variant Carrier PAH Subtype
AMOT	LGD	p.(Leu320Cysfs*55)	.	31	.	68	95	IPAH
CSNK2A2	D-MIS	p.(His184Leu)	0.50	25	.	55	77	IPAH
HNRNPF	LGD	p.(Tyr210Leufs*14)	.	29	.	85	98	PPHN, PAH
HSPA4	D-MIS	p.(pro684Arg)	0.62	30	4.1×10^{-6}	43	96	PAH-CHD
KDM3B	D-MIS	p.(Pro1100Ser)	0.66	29	.	89	87	IPAH
KEAP1	LGD	p.(Tyr584*)	.	35	.	79	82	IPAH with dev delay
MECOM	D-MIS	p.(Phe762Ser)	0.76	32	.	82	60	IPAH
ZMYM2	LGD	p.(Arg540*)	.	36	.	93	77	IPAH with skeletal anomalies

AMOT, angiomotin; *CSKN2A2*, casein kinase II, α 2; *HNRNPF*, heterogeneous nuclear ribonucleoprotein F; *HSPA4*, heat shock protein A (HSP70), member 4; *KDM3B*, lysine demethylase 3B; *KEAP1*, Kelch-like ECH-associated protein 1; *MECOM*, MDS1 and EVI1 complex locus; *ZMYM2*, zinc finger protein 620. LGD, likely gene disrupting; MIS, missense; D-MIS, predicted deleterious missense based on REVEL score > 0.5. Allele frequency "." absent from gnomAD.

2.5. Genetic Ancestry

Most of the large genetic studies conducted to date have utilized cohorts of predominantly European ancestry. However, the role of specific genes in PAH may be heterogeneous across genetic ancestries, and the results of these studies may not be generalizable to all other populations. For example, the frequency of *ACVRL1* and *ENG* variants combined is ~1% among pediatric IPAH cases of European ancestry [11,13], but the frequency of *ACVRL1* alone may be closer to 13% among Asian children [64].

GDF2/BMP9 was recently identified as a novel PAH risk gene with genome-wide significance in both European [41] and Asian [65] cohorts with replication in the PAH Biobank cohort [11]. Similar to other PAH risk genes, the mode of inheritance was autosomal dominant. The frequency of *GDF2/BMP9* variants among children was 2.1% (2/94 cases) in the PAH Biobank and 5.2% (3/57 cases) in the Asian cohort, suggesting that *GDF2/BMP9* variants might be a more frequent cause of PAH among Asian children. Further study is required to determine whether this difference is a true genetic ancestry effect or random variation due to relatively small sample size or differences in bioinformatic pipelines. A PAH case study of a five-year-old boy of Hispanic ancestry identified a homozygous *GDF2/BMP9* LGD variant, NM_016204.1:c.76C > T; p.(Gln26Ter) [47]. The unaffected parents were heterozygous for the variant. Interestingly, the gnomAD population database (gnomADv2.1.1, n = 141,456 samples) [66] contains only two heterozygous counts of this allele, both of Latino ancestry, suggesting that this might be an ancestry-specific allele. Clearly, larger studies of children with greater diversity are needed to define population-specific risk gene allele frequencies as well as ancestral-specific genetic factors.

2.6. The Role of Other "Omics" in PAH

In addition to DNA sequencing to identify genetic etiologies of PAH, other "omics", including RNA sequencing, metabolomics, and proteomics, can provide valuable predictions of who is at risk for disease, define endophenotypes, and guide effective therapies [67,68]. For example, West and colleagues performed RNA sequencing of blood lymphocytes derived from *BMPR2* variant carriers with and without PAH to identify transcriptional patterns relevant to disease penetrance [69]. More recently, *FHIT* was identified as a potentially clinically relevant BMPR2 modifier gene through an siRNA screen of BMPR2 signaling regulatory genes combined with publicly available PAH RNA expression data. Subsequently, the authors showed that pharmaceutical upregulation of FHIT prevented and reversed experimental pulmonary hypertension in a rat model [70]. Rhodes and colleagues utilized metabolomics to identify circulating metabolites that distinguish PAH cases from healthy controls, to predict outcomes among PAH cases, and to monitor metabolite levels over time to determine whether correction could affect outcomes [71]. Stearman et al. combined gene expression data with pathway analyses to identify a transcriptional framework for PAH-affected lungs [72]. Similarly, Hemnes and colleagues used transcriptomics to identify RNA expression patterns predictive of vasodilator responsiveness among PAH patients [73]. These studies highlight the promise of other omics in predictions of PAH risk, diagnosis, classification, drug responsiveness, and prognosis. However, such studies have not been conducted in children. Detailed omic phenotyping requires biologic sampling, which can be difficult in pediatric patients, especially for the very young or those with complex medical conditions. We propose a pilot genomics-first approach followed by detailed phenotyping of patients grouped by genetic diagnosis to enrich the biologic sampling and assess utility before performing larger studies across all pediatric PAH patients.

3. A Genomics First Approach towards Better Understanding of Pediatric PAH

Identification of molecular subtypes of PAH has been proposed as a means to improve risk stratification, treatment, and outcomes. Obtaining a genetic diagnosis in children requires more extensive genetic testing than in adults (Figure 4). If testing for a panel of genes known to be associated with PAH is not diagnostic, children should be evaluated genome-wide for rare de novo and inherited variants with trio (parental and child) exome sequencing/chromosome microarray or genome sequencing. With knowledge of the causal gene, natural history, penetrance, and response to treatment can be refined for that specific genetic subtype of PAH to allow for more precise care for each genetically defined group. Individuals across genes in the same biological pathway can then be compared to assess similarities and differences.

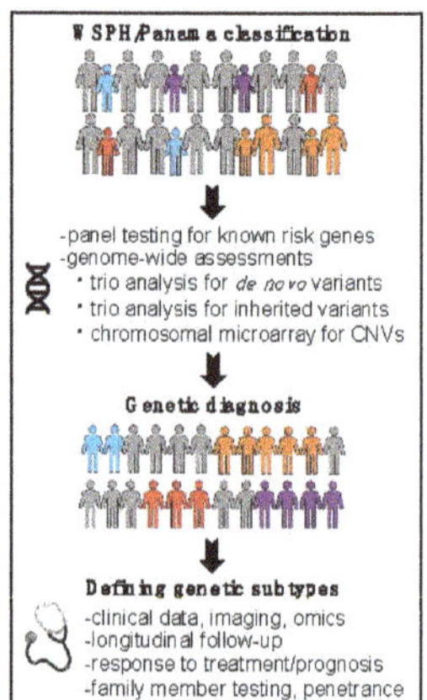

Figure 4. Genomic approach to improve understanding of pediatric PAH.

Collaboration across national and international clinical PH sites will be necessary to yield sufficient sample sizes due to the extremely small number of pediatric PAH patients at single PH sites, heterogeneity of risk genes for PAH, and need for ancestral diversity. PPHNet is an example of a pediatric-specific PAH consortium with ongoing recruitment across 13 North American clinical sites [74–78]. PVDomics [79], a US multicenter study launched in 2014, and PAH-ICON (pahicon.com), a new international effort, represent additional large-scale PAH cohorts. PAH was defined as mPAP $\geq$25 mmHg but was recently updated to include mPAP 20–25 mmHg in both children and adults [8]. Due to variability in pulmonary hemodynamics during the post-natal transition period, pediatric PAH is defined by elevated mPAP after 3 months of age in combination with pulmonary vascular resistance as indexed to body surface area, PVRI $\geq$3 Woods units/m^2 [8]. Clinical classification of PAH subtypes aims to improve clinical management and enhance research efforts and is typically based on the World Symposium on Pulmonary Hypertension (WSPH) system, updated during the 2018 Nice session [80]. In children, use of a pediatric-specific classification system developed in Panama by the Pulmonary Vascular Research Institute (PVRI) Pediatric Task Force [81] provides more definitive classification of developmental and complex phenotypes.

Biological trios composed of two unaffected parents and an affected child are preferred over singleton cases for pediatric studies of PAH in order to identify both inherited and de novo variants causal for disease. However, as the number of ethnically matched genomic data in public databases increases, the need for trios will decrease. DNA can be reliably obtained from small samples of blood or saliva. We have developed methods in a large national autism study to collect saliva from pediatric patients and their parents in their homes, with instructional videos to support clinical sites with remote biospecimen collection [82,83]. Biological samples can be shipped to a central biorepository for DNA extraction and then processed and sequenced using a single genetic platform. Since annotation tools for predicting relative pathogenicity of noncoding variants are still under development, and because the incremental yield of structural variants identified from genome sequencing is low, there is currently little added value in analysis of genome sequencing compared to exome sequencing data. Following extensive quality control, filtering, and annotation, sequencing data are screened for rare deleterious variants in known candidate genes or undergo trio analysis for de novo variants or association analysis for inherited variants [10,11]. Candidate genes identified by these methods are further assessed by mapping the locations of variants to protein structures, assessing expression in PAH-relevant tissues and cell types, assessing variant function in vitro, and assessing the impact of pathogenic variants in vivo in model organisms. Human mutations can be introduced into cells and PAH-sensitized mice using CRISPR technology. Phenotypic "rescue" by exogenous delivery of normal proteins can add evidence to support causation.

Once genetic subtypes are defined, demographic data, clinical phenotypes, and imaging data (pulmonary vascular angiography, chest X-ray or CT scans, chest/cardiac MRI, and lung biopsies) can be compared among cases with variants in the same gene/pathway. Relevant cases can be recalled or remotely interviewed for targeted clinical assessments to determine if there are similarities among cases with rare variants in the same gene/pathway.

To increase rigor and assess the full phenotypic spectrum of the new genetic subtypes, additional cases can be identified using GeneMatcher, clinical diagnostic laboratories, and large sequencing centers. Longitudinal phenotypes of the genetic subtypes can be assessed retrospectively and prospectively, including death/transplant, response to medication, other medical diagnoses, and changes in lung function. For inherited variants, cascade genetic testing of family members with clinical evaluation of "unaffected" individuals who carry the relevant genetic variant can inform penetrance by age and sex. To support families with genetic diagnoses, "virtual" family meetings can be organized to update families on new findings related to their conditions and build communities for each of the rare endophenotypes.

The value of a genetic diagnosis to families is three-fold: (**1**) identification of other associated features, (**2**) identification of family members at risk for developing PAH, and (**3**) clarification of reproductive risks and provide family planning options. Biallelic mutations in *EIF2AK4* are diagnostic for pulmonary veno-occlusive disease/pulmonary capillary hemangiomatosis [84], which can be difficult to diagnose clinically without a lung biopsy, and patients can be listed for transplant earlier in the course of disease, which may improve outcomes. As mentioned, *TBX4* variant carriers should be assessed for associated developmental (lung, heart) and orthopedic (hips, knees, feet) issues. *ENG/ACVRL1* variant carriers are prone to arteriovenous malformations in brain, intestine, lung, and liver [85]; these patients require periodic MRI surveillance. After making a genetic diagnosis in a PAH patient, additional family members can be screened for the family variant to identify those at risk who may benefit from annual surveillance and early diagnosis/treatment. Furthermore, diagnosed young adults can make informed decisions regarding family planning.

4. Conclusions

Pediatric PAH differs from adult-onset PAH in many important aspects, including clinical presentation, etiology, genetic burden, and specific genes involved. In many young children, PAH is a developmental disease with a complex phenotype. *TBX4* and *SOX17* are examples of developmental genes in which rare deleterious variants occur much more frequently in pediatric- compared to adult-onset PAH. De novo variants likely contribute to at least 15% of pediatric-onset PAH, but the specific genes require confirmation in larger pediatric cohorts. Many genes with de novo variants likely contribute to developmental phenotypes and complex medical conditions. A genomics-first approach to pediatric PAH starts with a genetic diagnosis followed by phenotypic characterization of cases with variants in the same genes/pathways. Large, diverse pediatric populations are needed to confirm the candidate genes identified thus far, identify new genes, characterize each rare endophenotype and natural history, and assess the efficacy of therapies to inform more precise clinical management. In addition, questions related to which children are at risk for developing PAH—especially children with CHD, CDH, bronchopulmonary dysplasia, and Down syndrome—may be answered. The yield of genetic diagnoses in pediatric-onset PAH cohorts is significantly greater than the yield in adult-onset cohorts. However, identification of genes, pathways, and networks in children could provide novel targets for therapy not only for children but for all patients with and at high risk for PAH.

Author Contributions: Writing—original draft preparation, C.L.W.; writing—review and editing, W.K.C.; funding acquisition, W.K.C. All authors have read and agreed to the published version of the manuscript.

Funding: This research was funded by NIH U01HL125218 and the JPB Foundation (W.K.C.).

Acknowledgments: We are grateful to the patients who have contributed to these studies.

Conflicts of Interest: The authors declare no conflict of interest.

References

1. Li, L.; Jick, S.S.; Breitenstein, S.; Hernandez, G.; Michel, A.; Vizcaya, D. Pulmonary arterial hypertension in the USA: An epidemiological study in a large insured pediatric population. *Pulm. Circ.* **2017**, *7*, 126–136. [CrossRef] [PubMed]

2. Humbert, M.; Sitbon, O.; Chaouat, A.; Bertocchi, M.; Habib, G.; Gressin, V.; Yaici, A.; Weitzenblum, E.; Cordier, J.F.; Chabot, F.; et al. Pulmonary arterial hypertension in France. *Am. J. Respir. Crit. Care Med.* **2006**, *173*, 1023–1030. [CrossRef]

3. Ryan, J.J.; Archer, S.L. The right ventricle in pulmonary arterial hypertension: Disorders of metabolism, angiogenesis and adrenergic signaling in right ventricular failure. *Circ. Res.* **2014**, *115*, 176–188. [CrossRef] [PubMed]

4. Morrell, N.W.; Aldred, M.A.; Chung, W.K.; Elliott, C.G.; Nichols, W.C.; Soubrier, F.; Trembath, R.C.; Loyd, J.E. Genetics and genomics of pulmonary arterial hypertension. *Eur. Respir. J.* **2019**, *53*, 1801899. [CrossRef] [PubMed]

5. Southgate, L.; Machado, R.D.; Gräf, S.; Morrell, N.W. Molecular genetic framework underlying pulmonary arterial hypertension. *Nat. Rev. Cardiol.* **2020**, *17*, 85–95. [CrossRef] [PubMed]

6. Reyes-Palomares, A.; Gu, M.; Grubert, F.; Berest, I.; Sa, S.; Kasowski, M.; Arnold, C.; Shuai, M.; Srivas, R.; Miao, S.; et al. Remodeling of active endothelial enhancers is associated with aberrant gene-regulatory networks in pulmonary arterial hypertension. *Nat. Commun.* **2020**, *11*, 1673. [CrossRef]

7. Hansmann, G. Pulmonary hypertension in infants, children, and young adults. *J. Am. Coll. Cardiol.* **2017**, *69*, 2551–2569. [CrossRef] [PubMed]

8. Rosenzweig, E.B.; Abman, S.H.; Adatia, I.; Beghetti, M.; Bonnet, D.; Haworth, S.; Ivy, D.D.; Berger, R.M. Paediatric pulmonary arterial hypertension: Updates on definition, classification, diagnostics and management. *Eur. Respir. J.* **2019**, *53*, 1801916. [CrossRef]

9. Saji, T. Update on pediatric pulmonary arterial hypertension. Differences and similarities to adult disease. *Circ. J.* **2013**, *77*, 2639–2650. [CrossRef]

10. Zhu, N.; Swietlik, E.M.; Welch, C.L.; Pauciulo, M.W.; Hagen, J.J.; Zhou, X.; Guo, Y.; Karten, J.; Pandya, D.; Tilly, T.; et al. Rare variant analysis of 4241 pulmonary arterial hypertension cases from an international consortium implicate FBLN2, PDGFD and rare de novo variants in PAH. *bioRxiv* **2020**, *2029*, 124255.

11. Zhu, N.; PAH Biobank Enrolling Centers' Investigators; Pauciulo, M.W.; Welch, C.L.; Lutz, K.A.; Coleman, A.W.; Gonzaga-Jauregui, C.; Wang, J.; Grimes, J.M.; Martin, L.J.; et al. Novel risk genes and mechanisms implicated by exome sequencing of 2572 individuals with pulmonary arterial hypertension. *Genome Med.* **2019**, *11*, 69. [CrossRef] [PubMed]

12. Barst, R.J.; McGoon, M.D.; Elliott, C.G.; Foreman, A.J.; Miller, D.P.; Ivy, D.D. Survival in childhood pulmonary arterial hypertension: Insights from the registry to evaluate early and long-term pulmonary arterial hypertension disease management. *Circulation* **2012**, *125*, 113–122. [CrossRef]

13. Zhu, N.; Gonzaga-Jauregui, C.; Welch, C.L.; Ma, L.; Qi, H.; King, A.K.; Krishnan, U.; Rosenzweig, E.B.; Ivy, D.D.; Austin, E.D.; et al. Exome sequencing in children with pulmonary arterial hypertension demonstrates differences compared with adults. *Circ. Genom. Precis. Med.* **2018**, *11*, e001887. [CrossRef] [PubMed]

14. Abman, S.H.; Baker, C.; Gien, J.; Mourani, P.; Galambos, C. The Robyn Barst memorial lecture: Differences between the fetal, newborn, and adult pulmonary circulations: Relevance for age-specific therapies (2013 grover conference series). *Pulm. Circ.* **2014**, *4*, 424–440. [CrossRef] [PubMed]

15. Qiao, L.; Wynn, J.; Yu, L.; Ms, R.H.; Zhou, X.; Duron, V.; Aspelund, G.; Farkouh-Karoleski, C.; Zygumunt, A.; Krishnan, U.S.; et al. Likely damaging de novo variants in congenital diaphragmatic hernia patients are associated with worse clinical outcomes. *Genet. Med.* **2020**, 1–9. [CrossRef]

16. Bush, D.; Abman, S.H.; Galambos, C. Prominent intrapulmonary bronchopulmonary anastomoses and abnormal lung development in infants and children with down syndrome. *J. Pediatr.* **2017**, *180*, 156–162.e1. [CrossRef]

17. Sánchez, O.; Dominguez, C.; Ruiz, A.; Ribera, I.; Alijotas-Reig, J.; Roura, L.C.; Carreras, E.; Llurba, E. Angiogenic gene expression in down syndrome fetal hearts. *Fetal Diagn. Ther.* **2016**, *40*, 21–27. [CrossRef]

18. Galambos, C.; Minic, A.D.; Bush, D.; Nguyen, D.; Dodson, B.; Seedorf, G.; Abman, S.H. Increased lung expression of anti-angiogenic factors in down syndrome: Potential role in abnormal lung vascular growth and the risk for pulmonary hypertension. *PLoS ONE* **2016**, *11*, e0159005. [CrossRef]

19. Reuter, M.S.; Jobling, R.; Chaturvedi, R.R.; Manshaei, R.; Costain, G.; Heung, T.; Curtis, M.; Hosseini, S.M.; Liston, E.; Lowther, C.; et al. Haploinsufficiency of vascular endothelial growth factor related signaling genes is associated with tetralogy of Fallot. *Genet. Med.* **2018**, *21*, 1001–1007. [CrossRef]

20. Zhao, W.; Wang, J.; Shen, J.; Sun, K.; Zhu, J.; Yu, T.; Ji, W.; Chen, Y.; Fu, Q.; Li, F. Mutations in VEGFA are associated with congenital left ventricular outflow tract obstruction. *Biochem. Biophys. Res. Commun.* **2010**, *396*, 483–488. [CrossRef]

21. Beghetti, M.; Rudzinski, A.; Zhang, M. Efficacy and safety of oral sildenafil in children with Down syndrome and pulmonary hypertension. *BMC Cardiovasc. Disord.* **2017**, *17*, 177. [CrossRef]

22. Zhu, N.; Welch, C.L.; Wang, J.; Allen, P.M.; Gonzaga-Jauregui, C.; Ma, L.; King, A.K.; Krishnan, U.; Rosenzweig, E.B.; Ivy, D.D.; et al. Rare variants in SOX17 are associated with pulmonary arterial hypertension with congenital heart disease. *Genome Med.* **2018**, *10*, 56. [CrossRef] [PubMed]

23. Byers, H.M.; Dagle, J.M.; Klein, J.M.; Ryckman, K.K.; McDonald, E.L.; Murray, J.C.; Borowski, K.S. Variations in CRHR1 are associated with persistent pulmonary hypertension of the newborn. *Pediatr. Res.* **2012**, *71*, 162–167. [CrossRef] [PubMed]

24. Arora, R.; Metzger, R.J.; Papaioannou, V.E. Multiple roles and interactions of Tbx4 and Tbx5 in development of the respiratory system. *PLoS Genet.* **2012**, *8*, e1002866. [CrossRef]

25. Kanai-Azuma, M.; Kanai, Y.; Gad, J.M.; Tajima, Y.; Taya, C.; Kurohmaru, M.; Sanai, Y.; Yonekawa, H.; Yazaki, K.; Tam, P.P.; et al. Depletion of definitive gut endoderm in Sox17-null mutant mice. *Development* **2002**, *129*, 2367–2379.

26. Lilly, A.J.; Lacaud, G.; Kouskoff, V. SOXF transcription factors in cardiovascular development. *Semin. Cell Dev. Biol.* **2017**, *63*, 50–57. [CrossRef]

27. Sheeba, C.J.; Logan, M.P. The roles of T-box genes in vertebrate limb development. *Curr. Top Dev. Biol.* **2017**, *122*, 355–381. [PubMed]

28. Ballif, B.C.; Theisen, A.; Rosenfeld, J.A.; Traylor, R.N.; Gastier-Foster, J.; Thrush, D.L.; Astbury, C.; Bartholomew, D.; McBride, K.L.; Pyatt, R.E.; et al. Identification of a recurrent microdeletion at 17q23.1q23.2 flanked by segmental duplications associated with heart defects and limb abnormalities. *Am. J. Hum. Genet.* **2010**, *86*, 454–461. [CrossRef]

29. Nimmakayalu, M.; Sheffield, V.C.; Solomon, D.H.; Patil, S.R.; Shchelochkov, O.A.; Major, H.; Smith, R.J. Microdeletion of 17q22q23.2 encompassing TBX2 and TBX4 in a patient with congenital microcephaly, thyroid duct cyst, sensorineural hearing loss, and pulmonary hypertension. *Am. J. Med. Genet. Part A* **2011**, *155A*, 418–423. [CrossRef]

30. Kerstjens-Frederikse, W.S.; Bongers, E.M.H.F.; Roofthooft, M.T.R.; Leter, E.M.; Douwes, J.M.; Van Dijk, A.; Vonk-Noordegraaf, A.; Dijk-Bos, K.K.; Hoefsloot, L.H.; Hoendermis, E.S.; et al. TBX4mutations (small patella syndrome) are associated with childhood-onset pulmonary arterial hypertension. *J. Med. Genet.* **2013**, *50*, 500–506. [CrossRef]

31. Levy, M.; Eyries, M.; Szezepanski, I.; Ladouceur, M.; Nadaud, S.; Bonnet, D.; Soubrier, F. Genetic analyses in a cohort of children with pulmonary hypertension. *Eur. Respir. J.* **2016**, *48*, 1118–1126. [CrossRef] [PubMed]

32. Navas, P.; Tenorio, J.; Quezada, C.A.; Barrios, E.; Gordo, G.; Arias, P.; Messeguer, M.L.; Santos-Lozano, A.; Doza, J.P.; Lapunzina, P.; et al. Molecular analysis of BMPR2, TBX4, and KCNK3 and genotype-phenotype correlations in Spanish patients and families with idiopathic and hereditary pulmonary arterial hypertension. *Rev. Esp. Cardiol.* **2016**, *69*, 1011–1019. [CrossRef] [PubMed]

33. Homsy, J.; Zaidi, S.; Shen, Y.; Ware, J.S.; Samocha, K.E.; Karczewski, K.J.; DePalma, S.R.; Mckean, D.; Wakimoto, H.; Gorham, J.; et al. De novo mutations in congenital heart disease with neurodevelopmental and other congenital anomalies. *Science* **2015**, *350*, 1262–1266. [CrossRef] [PubMed]

34. Galambos, C.; Mullen, M.P.; Shieh, J.T.; Schwerk, N.; Kielt, M.J.; Ullmann, N.; Boldrini, R.; Stucin-Gantar, I.; Haass, C.; Bansal, M.; et al. Phenotype characterisation of TBX4 mutation and deletion carriers with neonatal and pediatric pulmonary hypertension. *Eur. Respir. J.* **2019**, *54*, 1801965. [CrossRef] [PubMed]

35. Thoré, P.; Girerd, B.; Jaïs, X.; Savale, L.; Ghigna, M.R.; Eyries, M.; Levy, M.; Ovaert, C.; Servettaz, A.; Guillaumot, A.; et al. Phenotype and outcome of pulmonary arterial hypertension patients carrying a TBX4 mutation. *Eur. Respir. J.* **2020**, *55*, 1902340. [CrossRef]

36. Austin, E.D.; Elliott, C.G. TBX4 syndrome: A systemic disease highlighted by pulmonary arterial hypertension in its most severe form. *Eur. Respir. J.* **2020**, *55*, 2000585. [CrossRef]

37. Corada, M.; Orsenigo, F.; Morini, M.F.; Pitulescu, M.E.; Bhat, G.; Nyqvist, D.; Breviario, F.; Conti, V.; Briot, A.; Iruela-Arispe, M.L.; et al. Sox17 is indispensable for acquisition and maintenance of arterial identity. *Nat. Commun.* **2013**, *4*, 2609. [CrossRef]

38. Sakamoto, Y.; Hara, K.; Kanai-Azuma, M.; Matsui, T.; Miura, Y.; Tsunekawa, N.; Kurohmaru, M.; Saijoh, Y.; Koopman, P.; Kanai, Y. Redundant roles of Sox17 and Sox18 in early cardiovascular development of mouse embryos. *Biochem. Biophys. Res. Commun.* **2007**, *360*, 539–544. [CrossRef] [PubMed]

39. Lange, A.W.; Haitchi, H.M.; LeCras, T.D.; Sridharan, A.; Xu, Y.; Wert, S.E.; James, J.; Udell, N.; Thurner, P.J.; Whitsett, J.A. Sox17 is required for normal pulmonary vascular morphogenesis. *Dev. Biol.* **2014**, *387*, 109–120. [CrossRef]

40. Liu, M.; Zhang, L.; Marsboom, G.; Jambusaria, A.; Xiong, S.; Toth, P.T.; Benevolenskaya, E.V.; Rehman, J.; Malik, A.B. Sox17 is required for endothelial regeneration following inflammation-induced vascular injury. *Nat. Commun.* **2019**, *10*, 2126. [CrossRef]

41. Gräf, S.; Haimel, M.; Bleda, M.; Hadinnapola, C.; Southgate, L.; Li, W.; Hodgson, J.; Liu, B.; Salmon, R.M.; Southwood, M.; et al. Identification of rare sequence variation underlying heritable pulmonary arterial hypertension. *Nat. Commun.* **2018**, *9*, 1416. [CrossRef] [PubMed]

42. Hiraide, T.; Kataoka, M.; Suzuki, H.; Aimi, Y.; Chiba, T.; Kanekura, K.; Satoh, T.; Fukuda, K.; Gamou, S.; Kosaki, K. SOX17 Mutations in Japanese patients with pulmonary arterial hypertension. *Am. J. Respir. Crit. Care Med.* **2018**, *198*, 1231–1233. [CrossRef]

43. Liu, X.; Luo, M.; Xie, W.; Wells, J.M.; Goodheart, M.J.; Engelhardt, J.F. Sox17 modulates Wnt3A/β-catenin-mediated transcriptional activation of the Lef-1 promoter. *Am. J. Physiol. Lung Cell. Mol. Physiol.* **2010**, *299*, L694–L710. [CrossRef]

44. Banerjee, A.; Ray, S. Structural insight, mutation and interactions in human β-catenin and SOX17 protein: A molecular-level outlook for organogenesis. *Gene* **2017**, *610*, 118–126. [CrossRef]

45. Zhang, H.S.; Liu, Q.; Piao, C.M.; Zhu, Y.; Li, Q.Q.; Du, J.; Gu, H. Genotypes and phenotypes of Chinese pediatric patients with idiopathic and heritable pulmonary arterial hypertension—A single-center study. *Can. J. Cardiol.* **2019**, *35*, 1851–1856. [CrossRef]

46. Haarman, M.G.; Kerstjens-Frederikse, W.S.; Vissia-Kazemier, T.R.; Breeman, K.T.; Timens, W.; Vos, Y.J.; Roofthooft, M.T.; Hillege, H.L.; Berger, R.M. The genetic epidemiology of pediatric pulmonary arterial hypertension. *J. Pediatr.* **2020**, *225*, 65–73. [CrossRef]

47. Wang, G.; Fan, R.; Ji, R.; Zou, W.; Penny, D.J.; Varghese, N.P.; Fan, Y. Novel homozygous BMP9 nonsense mutation causes pulmonary arterial hypertension: A case report. *BMC Pulm. Med.* **2016**, *16*, 17. [CrossRef]

48. Qi, H.; Yu, L.; Zhou, X.; Wynn, J.; Zhao, H.; Guo, Y.; Zhu, N.; Kitaygorodsky, A.; Hernan, R.; Aspelund, G.; et al. De novo variants in congenital diaphragmatic hernia identify MYRF as a new syndrome and reveal genetic overlaps with other developmental disorders. *PLoS Genet.* **2018**, *14*, e1007822. [CrossRef] [PubMed]

49. Jin, S.C.; Homsy, J.; Zaidi, S.; Lu, Q.; Morton, S.; DePalma, S.R.; Zeng, X.; Qi, H.; Chang, W.; Sierant, M.C.; et al. Contribution of rare inherited and de novo variants in 2871 congenital heart disease probands. *Nat. Genet.* **2017**, *49*, 1593–1601. [CrossRef]

50. Epi, K.C.; Allen, A.S.; Berkovic, S.F.; Cossette, P.; Delanty, N.; Dlugos, D.; Eichler, E.E.; Epstein, M.P.; Glauser, T.; Goldstein, D.B.; et al. De novo mutations in epileptic encephalopathies. *Nature* **2013**, *501*, 217–221.

51. Veltman, J.A.; Brunner, H.G. De novo mutations in human genetic disease. *Nat. Rev. Genet.* **2012**, *13*, 565–575. [CrossRef]

52. Page, D.J.; Miossec, M.J.; Williams, S.G.; Monaghan, R.M.; Fotiou, E.; Cordell, H.J.; Sutcliffe, L.; Topf, A.; Bourgey, M.; Bourque, G.; et al. Whole exome sequencing reveals the major genetic contributors to nonsyndromic tetralogy of fallot. *Circ. Res.* **2019**, *124*, 553–563. [CrossRef]

53. Tartaglia, M.; Martinẽlli, S.; Stella, L.; Bocchinfuso, G.; Flex, E.; Cordeddu, V.; Zampino, G.; Van Der Burgt, I.; Palleschi, A.; Petrucci, T.C.; et al. Diversity and functional consequences of germline and somatic PTPN11 mutations in human disease. *Am. J. Hum. Genet.* **2006**, *78*, 279–290. [CrossRef]

54. Hopper, R.K.; Feinstein, J.A.; Manning, M.A.; Benitz, W.; Hudgins, L. Neonatal pulmonary arterial hypertension and Noonan syndrome: Two fatal cases with a specific RAF1 mutation. *Am. J. Med. Genet. Part. A* **2015**, *167A*, 882–885. [CrossRef] [PubMed]

55. Tinker, A.; Uren, N.; Schofield, J. Severe pulmonary hypertension in Ullrich-Noonan syndrome. *Br. Heart J.* **1989**, *62*, 74–77. [CrossRef] [PubMed]

56. Troyanovsky, B.; Levchenko, T.; Mansson, G.; Matvijenko, O.; Holmgren, L. Angiomotin: An angiostatin binding protein that regulates endothelial cell migration and tube formation. *J. Cell Biol.* **2001**, *152*, 1247–1254. [CrossRef] [PubMed]

57. Holmgren, L.; Ambrosino, E.; Birot, O.; Tullus, C.; Veitonmaki, N.; Levchenko, T.; Carlson, L.M.; Musiani, P.; Iezzi, M.; Curcio, C.; et al. A DNA vaccine targeting angiomotin inhibits angiogenesis and suppresses tumor growth. *Proc. Natl. Acad. Sci. USA* **2006**, *103*, 9208–9213. [CrossRef]

58. Zheng, Y.; Vertuani, S.; Nyström, S.; Audebert, S.; Meijer, I.; Tegnebratt, T.; Borg, J.P.; Uhlén, P.; Majumdar, A.; Holmgren, L. Angiomotin-like protein 1 controls endothelial polarity and junction stability during sprouting angiogenesis. *Circ. Res.* **2009**, *105*, 260–270. [CrossRef]

59. Mohamed, B.A.; Barakat, A.Z.; Held, T.; Elkenani, M.; Mühlfeld, C.; Männer, J.; Adham, I.M. Respiratory distress and early neonatal lethality in Hspa4l/Hspa4 double-mutant mice. *Am. J. Respir. Cell Mol. Biol.* **2014**, *50*, 817–824. [CrossRef]

60. Ungvari, Z.; Bagi, Z.; Feher, A.; Recchia, F.A.; Sonntag, W.E.; Pearson, K.; De Cabo, R.; Csiszar, A. Resveratrol confers endothelial protection via activation of the antioxidant transcription factor Nrf2. *Am. J. Physiol. Heart Circ. Physiol.* **2010**, *299*, H18–H24. [CrossRef] [PubMed]

61. Wei, Y.; Gong, J.; Thimmulappa, R.K.; Kosmider, B.; Biswal, S.; Duh, E.J. Nrf2 acts cell-autonomously in endothelium to regulate tip cell formation and vascular branching. *Proc. Natl. Acad. Sci. USA* **2013**, *110*, E3910–E3918. [CrossRef]

62. Liu, Q.; Gao, Y.; Ci, X. Role of Nrf2 and its activators in respiratory diseases. *Oxid. Med. Cell. Longev.* **2019**, *2019*, 7090534. [CrossRef]

63. Samocha, K.E.; Robinson, E.B.; Sanders, S.J.; Stevens, C.; Sabo, A.; McGrath, L.M.; Kosmicki, J.A.; Rehnström, K.; Mallick, S.; Kirby, A.; et al. A framework for the interpretation of de novo mutation in human disease. *Nat. Genet.* **2014**, *46*, 944–950. [CrossRef]

64. Chida, A.; Shintani, M.; Yagi, H.; Fujiwara, M.; Kojima, Y.; Sato, H.; Imamura, S.; Yokozawa, M.; Onodera, N.; Horigome, H.; et al. Outcomes of childhood pulmonary arterial hypertension in BMPR2 and ALK1 mutation carriers. *Am. J. Cardiol.* **2012**, *110*, 586–593. [CrossRef]

65. Wang, X.J.; Lian, T.Y.; Jiang, X.; Liu, S.F.; Li, S.Q.; Jiang, R.; Wu, W.H.; Ye, J.; Cheng, C.Y.; Du, Y.; et al. Germline BMP9 mutation causes idiopathic pulmonary arterial hypertension. *Eur. Respir. J.* **2019**, *53*, 1801609. [CrossRef]

66. Karczewski, K.J.; Francioli, L.C.; Tiao, G.; Cummings, B.B.; Alföldi, J.; Wang, Q.; Collins, R.L.; Laricchia, K.M.; Ganna, A.; Birnbaum, D.P.; et al. The mutational constraint spectrum quantified from variation in 141,456 humans. *Nat. Cell Biol.* **2020**, *581*, 434–443. [CrossRef]

67. Hemnes, A.R. Using omics to understand and treat pulmonary vascular disease. *Front. Med.* **2018**, *5*, 157. [CrossRef]

68. Harbaum, L.; Rhodes, C.J.; Otero-Núñez, P.; Wharton, J.; Wilkins, M.R. The application of 'omics' to pulmonary arterial hypertension. *Br. J. Pharmacol.* **2020**. [CrossRef] [PubMed]

69. West, J.; Cogan, J.; Geraci, M.W.; Robinson, L.; Newman, J.H.; Phillips, J.A.; Lane, K.B.; Meyrick, B.; Loyd, J.E. Gene expression in BMPR2 mutation carriers with and without evidence of Pulmonary Arterial Hypertension suggests pathways relevant to disease penetrance. *BMC Med. Genom.* **2008**, *1*, 45. [CrossRef] [PubMed]

70. Prosseda, S.D.; Tian, X.; Kuramoto, K.; Boehm, M.; Sudheendra, D.; Miyagawa, K.; Zhang, F.; Solow-Cordero, D.; Saldivar, J.C.; Austin, E.D.; et al. FHIT, a novel modifier gene in pulmonary arterial hypertension. *Am. J. Respir. Crit. Care Med.* **2019**, *199*, 83–98. [CrossRef]

71. Rhodes, C.J.; Ghataorhe, P.; Wharton, J.; Rue-Albrecht, K.C.; Hadinnapola, C.; Watson, G.; Bleda, M.; Haimel, M.; Coghlan, G.; Corris, P.A.; et al. Plasma metabolomics implicates modified transfer RNAs and altered bioenergetics in the outcomes of pulmonary arterial hypertension. *Circulation* **2017**, *135*, 460–475. [CrossRef]

72. Stearman, R.S.; Bui, Q.M.; Speyer, G.; Handen, A.; Cornelius, A.R.; Graham, B.B.; Kim, S.; Mickler, E.A.; Tuder, R.M.; Chan, S.Y.; et al. Systems analysis of the human pulmonary arterial hypertension lung transcriptome. *Am. J. Respir. Cell Mol. Biol.* **2019**, *60*, 637–649. [CrossRef]

73. Hemnes, A.R.; Trammell, A.W.; Archer, S.L.; Rich, S.; Yu, C.; Nian, H.; Penner, N.; Funke, M.; Wheeler, L.; Robbins, I.M.; et al. Peripheral blood signature of vasodilator-responsive pulmonary arterial hypertension. *Circulation* **2014**, *131*, 401–409. [CrossRef]

74. Abman, S.H.; Raj, U. Towards improving the care of children with pulmonary hypertension: The rationale for developing a pediatric pulmonary hypertension network. *Prog. Pediatr. Cardiol.* **2009**, *27*, 3–6. [CrossRef] [PubMed]

75. Abman, S.H.; Kinsella, J.P.; Rosenzweig, E.B.; Krishnan, U.; Kulik, T.; Mullen, M.; Wessel, D.L.; Steinhorn, R.; Adatia, I.; Hanna, B.D.; et al. Implications of the U.S. Food and drug administration warning against the use of sildenafil for the treatment of pediatric pulmonary hypertension. *Am. J. Respir. Crit. Care Med.* **2013**, *187*, 572–575. [CrossRef] [PubMed]

76. Krishnan, U.; Feinstein, J.A.; Adatia, I.; Austin, E.D.; Mullen, M.P.; Hopper, R.K.; Hanna, B.D.; Romer, L.; Keller, R.L.; Fineman, J.; et al. Evaluation and management of pulmonary hypertension in children with bronchopulmonary dysplasia. *J. Pediatr.* **2017**, *188*, 24–34.E1. [CrossRef] [PubMed]

77. Kinsella, J.P.; Steinhorn, R.; Mullen, M.P.; Hopper, R.K.; Keller, R.L.; Ivy, D.D.; Austin, E.D.; Krishnan, U.; Rosenzweig, E.B.; Fineman, J.R.; et al. The left ventricle in congenital diaphragmatic hernia: Implications for the management of pulmonary hypertension. *J. Pediatr.* **2018**, *197*, 17–22. [CrossRef]

78. Levy, P.T.; Jain, A.; Nawaytou, H.; Teitel, D.; Keller, R.; Fineman, J.; Steinhorn, R.; Abman, S.H.; McNamara, P.J. Risk assessment and monitoring of chronic pulmonary hypertension in premature infants. *J. Pediatr.* **2020**, *217*, 199–209.e4. [CrossRef]

79. Hemnes, A.R.; Beck, G.J.; Newman, J.H.; Abidov, A.; Aldred, M.A.; Barnard, J.; Berman, R.E.; Borlaug, B.A.; Chung, W.K.; Comhair, S.A.A.; et al. PVDOMICS: A multi-center study to improve understanding of pulmonary vascular disease through phenomics. *Circ. Res.* **2017**, *121*, 1136–1139. [CrossRef]

80. Simonneau, G.; Montani, D.; Celermajer, D.S.; Denton, C.P.; Gatzoulis, M.A.; Krowka, M.; Williams, P.G.; Souza, R. Haemodynamic definitions and updated clinical classification of pulmonary hypertension. *Eur. Respir. J.* **2019**, *53*, 1801913. [CrossRef]

81. Del Cerro, M.J.; Abman, S.; Diaz, G.; Freudenthal, A.H.; Freudenthal, F.; Harikrishnan, S.; Haworth, S.G.; Ivy, D.; Lopes, A.A.; Raj, J.U.; et al. A consensus approach to the classification of pediatric pulmonary hypertensive vascular disease: Report from the PVRI pediatric taskforce, Panama 2011. *Pulm. Circ.* **2011**, *1*, 286–298. [CrossRef] [PubMed]

82. Feliciano, P.; Daniels, A.M.; Snyder, L.G.; Beaumont, A.; Camba, A.; Esler, A.; Gulsrud, A.; Mason, A.; Gutierrez, A.; Nicholson, A.; et al. SPARK: A US cohort of 50,000 families to accelerate autism research. *Neuron* **2018**, *97*, 488–493. [CrossRef] [PubMed]

83. Feliciano, P.; The SPARK Consortium; Zhou, X.; Astrovskaya, I.; Turner, T.N.; Wang, T.; Brueggeman, L.; Barnard, R.; Hsieh, A.; Snyder, L.G.; et al. Exome sequencing of 457 autism families recruited online provides evidence for autism risk genes. *NPJ Genom. Med.* **2019**, *4*, 19. [CrossRef]

84. Chaisson, N.F.; Dodson, M.W.; Elliott, C.G. Pulmonary capillary hemangiomatosis and pulmonary veno-occlusive disease. *Clin. Chest Med.* **2016**, *37*, 523–534. [CrossRef]

85. Plauchu, H.; De Chadarévian, J.P.; Bideau, A.; Robert, J.-M. Age-related clinical profile of hereditary hemorrhagic telangiectasia in an epidemiologically recruited population. *Am. J. Med. Genet.* **1989**, *32*, 291–297. [CrossRef]

Publisher's Note: MDPI stays neutral with regard to jurisdictional claims in published maps and institutional affiliations.

 *genes* 

Article

Whole Exome Sequence Analysis Provides Novel Insights into the Genetic Framework of Childhood-Onset Pulmonary Arterial Hypertension

Simone M. Gelinas [1], Clare E. Benson [1], Mohammed A. Khan [1], Rolf M. F. Berger [2], Richard C. Trembath [3], Rajiv D. Machado [1,4,*] and Laura Southgate [1,3,*]

[1] Genetics Research Centre, Molecular and Clinical Sciences Research Institute, St George's University of London, London SW17 0RE, UK; simone.gelinas@gstt.nhs.uk (S.M.G.); p1805584@sgul.ac.uk (C.E.B.); m1500901@sgul.ac.uk (M.A.K.)

[2] Center for Congenital Heart Diseases, Department of Pediatric Cardiology, Beatrix Children's Hospital, University Medical Center Groningen, 9700 RB Groningen, The Netherlands; r.m.f.berger@umcg.nl

[3] Department of Medical & Molecular Genetics, Faculty of Life Sciences & Medicine, King's College London, London SE1 9RT, UK; richard.trembath@kcl.ac.uk

[4] Institute of Medical and Biomedical Education, St George's University of London, London SW17 0RE, UK

* Correspondence: rmachado@sgul.ac.uk (R.D.M.); lasouthg@sgul.ac.uk (L.S.)

Received: 22 October 2020; Accepted: 9 November 2020; Published: 11 November 2020

Abstract: Pulmonary arterial hypertension (PAH) describes a rare, progressive vascular disease caused by the obstruction of pulmonary arterioles, typically resulting in right heart failure. Whilst PAH most often manifests in adulthood, paediatric disease is considered to be a distinct entity with increased morbidity and often an unexplained resistance to current therapies. Recent genetic studies have substantially increased our understanding of PAH pathogenesis, providing opportunities for molecular diagnosis and presymptomatic genetic testing in families. However, the genetic architecture of childhood-onset PAH remains relatively poorly characterised. We sought to investigate a previously unsolved paediatric cohort ($n = 18$) using whole exome sequencing to improve the molecular diagnosis of childhood-onset PAH. Through a targeted investigation of 26 candidate genes, we applied a rigorous variant filtering methodology to enrich for rare, likely pathogenic variants. This analysis led to the detection of novel PAH risk alleles in five genes, including the first identification of a heterozygous *ATP13A3* mutation in childhood-onset disease. In addition, we provide the first independent validation of *BMP10* and *PDGFD* as genetic risk factors for PAH. These data provide a molecular diagnosis in 28% of paediatric cases, reflecting the increased genetic burden in childhood-onset disease and highlighting the importance of next-generation sequencing approaches to diagnostic surveillance.

Keywords: exome sequencing; molecular genetics; lung disease; paediatrics; pulmonary arterial hypertension

---

1. Introduction

Pulmonary arterial hypertension (PAH) is an uncommon vascular disorder that remains incurable despite significant advances in current treatment regimens. PAH is characterised by obstruction and occlusion of the pulmonary arterioles leading to progressive pulmonary arterial pressure overload, right ventricular hypertrophy and failure of the right side of the heart [1]. Typical histopathological features of PAH include marked vascular remodelling of the pulmonary arterioles as a consequence of exuberant proliferation of pulmonary artery endothelial (PAEC) and smooth muscle (PASMC) cells. In families, PAH predominantly segregates as an autosomal dominant trait displaying features of complex disease, namely variable expressivity, reduced penetrance and a gender bias favouring females [2]. Despite major advances in delineating PAH aetiology, the associated morbidity and early mortality burden of this disease remains perniciously high, with an average life expectancy of 3–5 years from diagnosis [3].

Paediatric PAH represents a clinically distinct form of disease. Although less prevalent, at an estimated 4.8–8.1 cases/million [4,5], than the adult form (15–50 cases/million) [6], childhood-onset PAH (cPAH) is marked by greater morbidity, depressed responses to therapy and poor survival metrics [7,8]. Taken together, these features suggest a specific genetic background. Putatively causal genetic variants have been described in at least 26 PAH risk genes [9–14], of which deleterious variation in *BMPR2*, encoding a type II receptor of the transforming growth factor beta (TGF-beta) superfamily, was firmly established as the major risk factor in adult PAH [13]. In paediatric cases, rare deleterious mutations of *BMPR2* and other key components of the bone morphogenetic protein (BMP) signalling pathway have been determined to be rare causal factors. These include the receptor proteins activin A receptor-like type 1 (*ACVRL1*, encoding ALK1), *BMPR1B* and endoglin (*ENG*), as well as *GDF2* (encoding the ligand BMP9) and the cytoplasmic signalling mediator *SMAD9* [8,15–20].

Whole exome sequencing (WES) has been instrumental in the accelerated gene discovery in PAH, notably in genetic risk unrelated to the canonical BMP pathway. Moreover, these findings both expanded the mutation spectrum in cPAH and shed light on the epidemiology of this discrete disease entity. For example, mutations within the transcription factor genes T-box 4 (*TBX4*) and SRY-box 17 (*SOX17*) show enriched contribution to paediatric PAH compared to adult-onset, explaining up to 7.7% and 3.3% of paediatric cases, respectively [8,21,22]. Of note, *SOX17* mutations show a strong correlation to PAH associated with congenital heart disease in both adult and paediatric disease [22,23]. More recent findings include loss-of-function variants in the ATP-binding cassette subfamily C member 8 (*ABCC8*) gene and in the endothelial cell-specific ligand BMP10 [24,25], whilst genome-wide gene burden approaches in predominantly adult cohorts identified the candidate risk genes ATPase 13A3 (*ATP13A3*), fibulin 2 (*FBLN2*) and platelet-derived growth factor D (*PDGFD*), among others [10,14]. These novel gene discoveries demonstrated the benefit of a granular approach to the genomic interrogation of discrete case cohorts.

While these data further developed the molecular architecture of PAH, the mutational landscape remains relatively poorly understood in paediatric cohorts, particularly for more recently reported genes [10,12]. Here, we sought to investigate a previously unsolved paediatric cohort, first, to better define the mutational background of cPAH, and second, to augment the existing body of data underlying this disease. Taken together, this study highlights the hitherto undefined role of *ATP13A3* in the pathogenesis of paediatric PAH and, importantly, provides independent validation of putative gene-disease associations in *BMP10* and *PDGFD*.

2. Materials and Methods

2.1. Patient Cohort and Clinical Assessment

Eighteen unrelated patients and their relatives were recruited from specialist PAH centres in the UK and the Netherlands. Ethical approval for the study was received from local ethics review boards and written informed consent was obtained from all participants or their legal guardian. The study was conducted in accordance with the Declaration of Helsinki, and the protocol was approved by the St Thomas' Hospital Research Ethics Committee (08/H0802/32).

Clinical diagnoses were established by PAH specialists in accordance with the classification of the World Symposium on Pulmonary Hypertension [26]. All patients previously underwent conventional genetic screening for *BMPR2*, *ACVRL1* and *ENG*, and were found to be mutation-negative [16,27]. Following pedigree analysis and clinical examination, patients were classified as having either idiopathic PAH (IPAH; *n* = 8), PAH associated with congenital heart disease (APAH-CHD; *n* = 7) or heritable PAH with one or more affected relatives (HPAH; *n* = 3).

2.2. Exome Sequencing

Genomic DNA was extracted from peripheral blood using standard methodologies. Whole exome sequencing for all probands was performed by Novogene Ltd. Briefly, exome libraries were captured using the SureSelect Human All Exon kit (Agilent Technologies, Santa Clara, CA, USA), and enriched fragments underwent paired-end sequencing on an Illumina platform. Quality control filtration of raw FASTQ files discarded read pairs containing adapter contamination and those with <80% of base calls achieving a Phred score of Q30. Sequence alignment to human reference genome assembly GRCh37 was undertaken using Burrows–Wheeler Aligner [28] and Picard MarkDuplicates (Genome Analysis Tool Kit (GATK), Broad Institute, Cambridge, MA, USA) was used to flag PCR duplicate reads for removal. GATK (v3.8) software was used for single nucleotide variant (SNV) and insertion and deletion (indel) detection [29]. Variant annotation was performed using ANNOtate VARiation (ANNOVAR) [30].

2.3. Candidate Gene Analysis

Annotated WES profiles for each patient were screened for rare, deleterious variants in 26 established and emerging PAH risk genes: *ABCC8*, *ACVRL1*, *AQP1*, *ATP13A3*, *BMP10*, *BMPR1B*, *BMPR2*, *CAV1*, *EIF2AK4*, *ENG*, *EVI5*, *FBLN2*, *GDF2*, *GGCX*, *KCNA5*, *KCNK3*, *KDR*, *KLF2*, *KLK1*, *NOTCH3*, *PDGFD*, *SMAD1*, *SMAD4*, *SMAD9*, *SOX17* and *TBX4*. Rare variants were defined as those with a minor allele frequency of ≤1 in 10,000 (MAF ≤ 0.0001), as reported in the Genome Aggregation Database control population (gnomAD v2.1.1; https://gnomad.broadinstitute.org). Splice region and exonic variants with likely functional consequences (nonsense, frameshift, nonframeshift indels and missense SNVs) were retained. Copy number variants were not considered in the analysis presented herein. The functional impact of missense variants was informed by in silico pathogenicity predictions. Specifically, variants with a Combined Annotation Dependent Depletion (CADD) score of ≥20 (https://cadd.gs.washington.edu) were considered potentially causal [31]. Additional evidence for pathogenicity was obtained from observing at least one deleterious score across Sorting Intolerant from Tolerant (SIFT; ≤0.05), PolyPhen-2 (≥0.909) and MutationTaster2 (annotation "disease causing") algorithms [32–34]. The location of each variant was inspected for correlation with conserved protein domains, providing further confirmation of likely pathogenicity.

Candidate variants were visualised and inspected for artefacts using Integrative Genomics Viewer (IGV) to exclude false positives [35,36]. All identified variants were validated in an independent sample by Sanger sequencing (Eurofins Genomics, Ebersberg, Germany) and tested for familial co-segregation where DNA samples were available.

3. Results

3.1. WES Analysis of the cPAH Cohort

Overall, 92% of base calls achieved a mean quality score of Q30. A mean depth of coverage of 65x (range: 54x–78x) was achieved for all samples, with an average of 91% (range: 86–94%) of bases achieving >20x coverage. Manual inspection of all PAH risk genes using IGV confirmed sufficient coverage of ≥10x across all exonic regions and intron–exon boundaries.

A targeted variant analysis of previously reported risk genes in our panel of childhood-onset IPAH, APAH and HPAH patients revealed putatively pathogenic missense variants in *ABCC8*, *ATP13A3*, *BMP10*, *PDGFD* and *SMAD9* (Table 1). All variants were validated by Sanger sequencing. Based on our exclusion criteria, no other likely causal variants in the reported PAH risk genes were identified in these patients.

Table 1. Variants Identified in Childhood-Onset Pulmonary Arterial Hypertension.

Patient	Sex	Age of Onset	Diagnosis	Variant Identified	Mutation Category	GnomAD MAF	Variant Identifier(s)	CADD Score	SIFT Score	PolyPhen2 (HumVar)	Mutation Taster
P1	M	9 y	APAH–CHD (ASD)	*ATP13A3* (exon 11) NM_024524.4 c.1148C>A (p.Thr383Lys)	Missense *	-	Novel	32	0.001, D	0.991, D	1.0, D
P2	F	3 y	IPAH	*BMP10* (exon 1) NM_014482.3 c.247G>A (p.Glu83Lys)	Missense	0.000009	dbSNP: rs1192957334; Not in ClinVar	28	0.047, D	0.999, D	1.0, D
P3	F	<5 y	IPAH	*PDGFD* (exon 4) NM_025208.5 c.550G>A (p.Glu184Lys)	Missense	-	dbSNP: rs769639743; Not in ClinVar	28.2	0.007, D	0.980, D	0.9998, D
P4	M	19 m	IPAH	*ABCC8* (exon 7) NM_000352.6 c.1069G>A (p.Val357Ile)	Missense	0.000018	dbSNP: rs771392416; Not in ClinVar	23	0.235, T	0.042, B	1.0, D
P5	F	4.5 y	HPAH (mother also affected)	*SMAD9* (exon 6) NM_001127217.3 c.1117G>A (p.Val373Ile)	Missense	0.00013	dbSNP: rs140504903; ClinVar: VCV000311894	27	0.001, D	0.936, D	1.0, D

APAH–CHD: PAH associated with congenital heart disease; ASD: atrial septal defect; B: benign; CADD: Combined Annotation Dependent Depletion; D: damaging (SIFT), probably damaging (PolyPhen2) or disease causing (Mutation Taster); IPAH: idiopathic pulmonary arterial hypertension; HPAH: heritable pulmonary arterial hypertension; MAF: minor allele frequency in gnomAD v2.1.1 (controls); m: months; SIFT: Sorting Independent from Tolerant; T: tolerated; y: years. * The ATP13A3 missense variant in Patient 1 is located within the consensus splice region.

3.2. Novel Association of ATP13A3 with Paediatric APAH-CHD

We identified a novel missense variant in *ATP13A3* (c.1148C>A, p.Thr383Lys) in a male child with APAH–CHD. To the best of our knowledge, this represents the first finding of a molecular defect in this risk gene in a paediatric case. The patient was diagnosed with PAH at nine years of age, associated with a secundum atrial septal defect (Figure 1A). All in silico predictions were consistent with a high probability of pathogenicity (Table 1). Parental samples were not available to conduct familial segregation analysis.

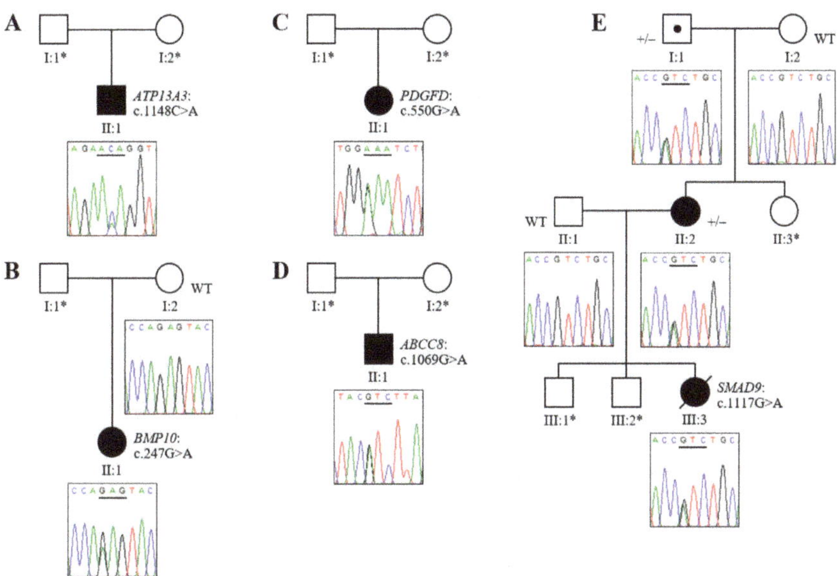

Figure 1. Sequence chromatograms and familial segregation of identified variants in childhood-onset pulmonary arterial hypertension (cPAH). (**A**) Patient 1 has a c.1148C>A missense mutation in *ATP13A3*; (**B**) Patient 2 has a heterozygous *BMP10* c.247G>A mutation; (**C**) Patient 3 is heterozygous for a c.550G>A mutation in *PDGFD*; (**D**) Patient 4 has a c.1069G>A missense variant in the *ABCC8* gene; (**E**) Patient 5 carries a heterozygous *SMAD9* c.1117G>A missense variant, which co-segregates with the disease (+/−) in his affected mother and maternal grandfather, who is an obligate carrier. Unaffected family members available for testing are all wild-type (WT). The horizontal line underlines the mutated codon. * DNA sample not available.

While annotated as a missense variant, further analysis of the *ATP13A3* c.1148C>A variant revealed that the mutation is located 3 bp from the intron–exon boundary and one nucleotide upstream of the highly conserved donor splicing motif of exon 12 (CAG|gtagga). An independent interrogation of in silico data using the NNSplice tool in MutationTaster revealed a potential disruption of the donor splice site, likely resulting in the formation of an aberrant gene transcript or premature stop codon leading to nonsense-mediated decay of the messenger RNA. However, further work is required on the cDNA level to experimentally validate this predicted deleterious effect.

3.3. Novel Mutations of BMP10, PDGFD and ABCC8 in Childhood IPAH

In the IPAH cohort we detected a *BMP10* variant (c.247G>A, p.Glu83Lys) in a female Dutch patient diagnosed with severe disease at 36 months. The patient was stably treated on oral dual therapy (bosentan and sildenafil) but died at 18 years of age. This mutation, which is observed in 1/109,376 control individuals in gnomAD and was not previously associated with disease in a

PAH clinical setting, provides the first independent confirmation of *BMP10* as a PAH risk gene [25]. All bioinformatic predictions were consistent with a high likelihood of pathogenicity (Table 1). Familial co-segregation analysis confirmed the unaffected mother was wild-type but, as no DNA was available from the father, it was not possible to determine whether this variant had arisen de novo in the proband (Figure 1B).

PDGFD is a newly identified risk gene in PAH [14]; this analysis offers the first independent validation of its causal association with PAH, as well as confirming a specific role in the development of paediatric disease through the identification of a rare missense variant (c.550G>A, p.Glu184Lys) in a female cPAH patient (Figure 1C). The variant, which is not present in population control databases, gave a CADD score of 28.2 and was predicted to be deleterious by all three missense prediction algorithms (Table 1).

A heterozygous missense variant in *ABCC8* (c.1069G>A, p.Val357Ile), novel in the context of PAH, was observed in a male patient diagnosed at 19 months of age (Figure 1D). While the CADD score and MutationTaster prediction were strongly indicative of a deleterious impact, the conclusions of pathogenicity based on other in silico markers were ambiguous, likely due to the interchangeability of valine and isoleucine residues at this position across related proteins in lower organisms (Table 1). Unavailability of parental DNA samples precluded familial segregation analysis of the *PDGFD* and *ABCC8* variants.

3.4. Analysis of HPAH Samples Identifies a Potential Molecular Defect of SMAD9

An examination of WES data from the three HPAH probands revealed a missense variant in *SMAD9* (c.1117G>A, p.Val373Ile) in a single family. Following a diagnosis of PAH, the proband died at five years of age. The affected amino acid residue displays high conservation across species, reflected in high deleterious scores across all in silico prediction algorithms employed (Table 1). Co-segregation was confirmed in two affected family members and an obligate carrier (Figure 1E). In addition, the candidacy of this variant as a causal defect in PAH was underpinned by the same finding in an HPAH patient reported in the ClinVar database (SCV000384082). However, as an observed allele frequency of >0.0001 in European controls exceeded our threshold for inclusion, this was resultantly considered to be a variant of unknown significance (VUS) in the absence of compelling functional data.

4. Discussion

This study applied a targeted 26 gene analysis to WES data in a European paediatric cohort, elucidating candidate molecular mechanisms in five patients to further illuminate the genetic architecture of early-onset PAH. It must be noted that all candidate variants presented in this analysis were categorised as variants of unknown significance in accordance with American College of Medical Genetics and Genomics (ACMG) guidelines [37], including those that met our criteria as being highly deleterious. Evidence of variant pathogenicity derived from population frequency, bioinformatic predictions and the biological relevance of these genes to PAH pathology is considered herein.

While *ATP13A3* was recently identified and functionally characterised as being causative in adult PAH [10], this study is the first to implicate heterozygous variation of *ATP13A3* in paediatric-onset cases. The gene encodes a P-type ATPase for which the substrate specificity and biological role remain poorly defined [38], but is a known transmembrane protein that localises to endosomal compartments and is likely involved in polyamine transport, key molecules for cell growth and proliferation [39,40]. Recent functional work demonstrated that *ATP13A3* mRNA is expressed in PASMCs, whilst loss of *ATP13A3* inhibits endothelial cell proliferation and increases apoptosis, consistent with disease initiation models in PAH [10]. The novel c.1148C>A (p.Thr383Lys) variant identified in this study alters a highly conserved amino acid residue located within the actuator domain, which functions to dephosphorylate the catalytic phosphorylation domain (Figure 2). The variant is predicted to disrupt splicing and result in nonsense-mediated decay. Although molecular defects in the *ATP13A3* gene clearly underlie susceptibility to PAH [10,12,20,41], the mechanism of pathogenicity remains to be defined. Further functional analyses are therefore warranted to elucidate whether ATP13A3 loss-of-function represents a molecular mechanism independent to the BMP signalling pathway, and thus a new therapeutic target.

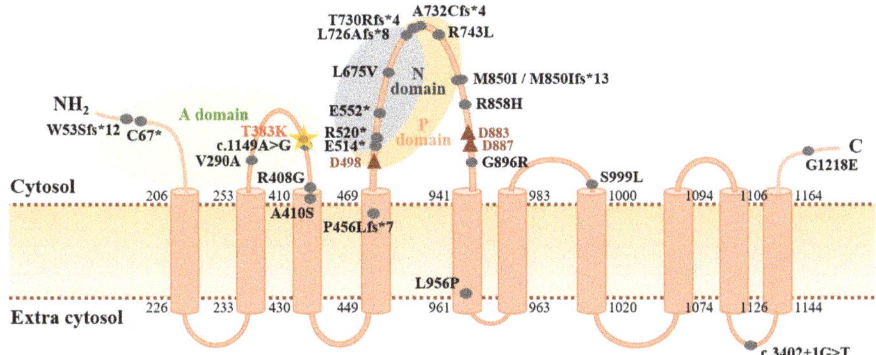

Figure 2. Protein structure of ATP13A3 highlighting mutations identified in PAH. Topological analysis of ATP13A3 according to UniProtKB protein component and site position data (ID: Q9H7F0) and published reports [38,40]. Likely pathogenic PAH mutations (CADD ≥ 15) reported in the literature are indicated by the filled grey circles [10,12,20,41]. The c.1148C>A (p.T383K) mutation identified in this study is highlighted by the gold star and is located adjacent to a previously reported splice-region variant [10]. Numbers indicate amino acid positions at each end of the 10 transmembrane domains. The red triangles denote essential asparagine residues (D498: active catalytic site; D883, D887: Mg^{2+} binding sites). A domain: actuator domain; C: carboxyl terminus; N domain: nucleotide binding domain; NH_2: amino terminus; P domain: phosphorylation domain.

The WES analysis workflow in this study identified a novel *BMP10* c.247G>A (p.Glu83Lys) missense variant with consistent in silico markers of pathogenicity. The residue is highly conserved across species and is located within the conserved TGF-beta propeptide domain, where a *BMP10* nonsense mutation was previously identified in cPAH [25]. This observation supports accumulating evidence implicating BMP10 loss-of-function in PAH pathogenesis [25,42]. *BMP10* encodes a high affinity ligand that activates the ALK1 and BMPR2 endothelial cell surface receptor within the BMP signalling pathway [43]. The gene product shares 65% amino acid sequence identity with its BMP9 paralogue, encoded by the *GDF2* gene. *GDF2* variant enrichment was recently reported in three large PAH cohorts, strongly implicating the BMP9/BMP10/ALK1/BMPR2 signalling pathway in PAH pathobiology [10,12,20]. Both in vitro and in vivo studies propose a model where BMP9 and BMP10 act interchangeably and with compensatory functionality via ALK1 to regulate endothelial cell migration and proliferation, maintaining endothelial quiescence [44,45]. Based on evidence of sequence homology, mature peptide structural similarities and reported functional redundancy of these gene products, it is proposed that the *BMP10* missense variant reported here may have a similar pathogenic mechanism to those observed in *GDF2*; further exploration on both the genetic and functional level is now warranted.

PDGFD missense variants were recently associated with PAH pathogenesis in a case-control analysis of a large, combined cohort [14]. The identification of a novel c.550G>A (p.Glu184Lys) variant of this gene in the cPAH panel analysed here serves to establish *PDGFD* as a contributing factor to childhood disease and provides independent validation of the discovery study. PDGF family signalling pathways play important roles in vascular pathologies, including mediation of remodelling processes that are known to occur in PAH [46]. The plausibility of this gene contributing to PAH risk is further enhanced by cell-type-specific gene expression studies using single cell RNA-seq data, demonstrating high expression in endothelial cells within heart and lung tissue, a key site of disease potentiation in PAH [14]. Of interest, *PDGFD* presents an exciting drug-targeting candidate, with studies of the tyrosine kinase inhibitor imatinib, a known pharmacological inhibitor of PDGF signalling, demonstrating 15-fold reductions in *PDGFD* expression in cardiac tissue [47] and suppression of medial thickening via receptor inhibition in patient-derived in vitro PASMC models [48].

The novel c.1069G>A (p.Val357Ile) missense variant identified in *ABCC8* describes an amino acid substitution in the highly conserved transmembrane domain 1 of the mature and functionally important SUR1 product. This finding supports previous investigations that indicate pathogenic variation in *ABCC8* underlies 2–5% of PAH cases, including 10 previous reports of paediatric-onset cases [8,12,24]. The gene encodes the SUR1 subunit of K_{ATP} potassium channels located in cardiac smooth muscle cells. Of note, the variant identified in this study was previously reported to cause hyperinsulinism in compound heterozygosity with a second *ABCC8* allele, although without any clinical signs of PAH [49]. Whole cell electrophysiology and radium efflux assays of reported missense variants have demonstrated loss of K_{ATP} channel activity [24]. Taken together, these data suggest that loss of SUR1-dependent channel function represents a pathogenic mechanism in PAH, although the primary physiological role remains unknown. These findings are of clinical pertinence, as *ABCC8* is considered a possible therapeutic target wherein impaired K_{ATP} activity could be pharmacologically rescued in vitro using diazoxide, a selective potassium-channel opening drug [24]. This may have further implications for the development of diazoxide-induced pulmonary hypertension in children with hyperinsulinaemic hypoglycaemia [50,51].

The *SMAD9* c.1117G>A (p.Val373Ile) variant detected in an HPAH case in this study is predicted to be deleterious by in silico analysis, further supported by familial co-segregation and its location with the C-terminal Mad Homology 2 (MH2) domain, a region of extensive evolutionary conservation. *SMAD1*, *SMAD4* and *SMAD9* encode signalling intermediaries that act as downstream mediators of the BMPR2 pathway. Variants in *SMAD9* were previously identified in three patients [17,52,53], and were recently validated in both adult-onset and paediatric PAH [12,20,22]. Functional analyses of the impact of *SMAD9* variants on SMAD-mediated signalling demonstrated impaired responses to ligand stimulation and reduced transcriptional activation of the downstream BMP target gene, *ID2* [17,52]. Although this variant was deemed a VUS based on our minor allele frequency threshold boundary, it is noteworthy that it was previously detected in an independent patient with PAH. This speaks to the challenges of setting an appropriate MAF cut-off when investigating rare diseases, especially in the context of notable reduced penetrance. Whereas most studies consider an approximation of disease allele frequency equivalent to population disease prevalence, inheritance-based calculations derived from Hardy–Weinberg equilibrium with adjustment for nonpenetrant alleles may be a more appropriate method to avoid false negative findings.

5. Conclusions

Taken together, these findings provide data supporting wider locus heterogeneity of paediatric PAH than previously reported and expand the allelic series of variation in known cPAH genetic risk factors. This study supports the inclusion of these genes in screening panels in future work for diagnostic surveillance of paediatric cases. In particular, we provide independent validation of the recently identified risk factors *BMP10* and *PDGFD*, and the first report of an *ATP13A3* variant in a paediatric case, warranting further analysis of understudied cellular pathways key to PAH pathogenesis. Notably, although novel gene detection was beyond the scope of this study in the context of extensive locus heterogeneity, this report emphasises the importance of the rigorous analysis of well-defined case cohorts to delineate the molecular basis of cPAH as a framework for future gene discovery initiatives.

Author Contributions: Conceptualization, R.D.M. and L.S.; formal analysis, S.M.G. and M.A.K.; funding acquisition, R.C.T., R.D.M. and L.S.; investigation, S.M.G., C.E.B., R.M.F.B., R.C.T., R.D.M. and L.S.; resources, R.M.F.B., R.C.T. and L.S.; visualization, S.M.G. and L.S.; writing—original draft preparation, S.M.G.; writing—review and editing, R.D.M. and L.S. All authors have read and agreed to the published version of the manuscript.

Funding: The study was funded by a British Heart Foundation programme grant (RG/08/006/25302) to R.C.T. and a pilot research grant from the Molecular and Clinical Sciences Research Institute at St George's, University of London (L.S. and R.D.M.). L.S. is supported by the Wellcome Trust Institutional Strategic Support Fund (204809/Z/16/Z) awarded to St. George's, University of London.

Conflicts of Interest: The University Medical Center Groningen contracts with Actelion, GSK and Lilly for steering committee and advisory board activities of Berger. All other authors declare no conflicts of interest. The funders had no role in the design of the study, in the collection, analyses, or interpretation of data, in the writing of the manuscript or in the decision to publish the results.

References

1. Machado, R.D.; Eickelberg, O.; Elliott, C.G.; Geraci, M.W.; Hanaoka, M.; Loyd, J.E.; Newman, J.H.; Phillips, J.A.; Soubrier, F.; Trembath, R.C.; et al. Genetics and genomics of pulmonary arterial hypertension. *J. Am. Coll. Cardiol.* **2009**, *54*, S32–S42. [CrossRef]

2. Machado, R.D.; Aldred, M.A.; James, V.; Harrison, R.E.; Patel, B.; Schwalbe, E.C.; Gruenig, E.; Janssen, B.; Koehler, R.; Seeger, W.; et al. Mutations of the TGF-beta type II receptor BMPR2 in pulmonary arterial hypertension. *Hum. Mutat.* **2006**, *27*, 121–132. [CrossRef]

3. O'Callaghan, D.S.; Humbert, M. A critical analysis of survival in pulmonary arterial hypertension. *Eur. Respir. Rev.* **2012**, *21*, 218–222. [CrossRef]

4. van Loon, R.L.E.; Roofthooft, M.T.R.; Hillege, H.L.; ten Harkel, A.D.J.; van Osch-Gevers, M.; Delhaas, T.; Kapusta, L.; Strengers, J.L.M.; Rammeloo, L.; Clur, S.-A.B.; et al. Pediatric pulmonary hypertension in the Netherlands: Epidemiology and characterization during the period 1991 to 2005. *Circulation* **2011**, *124*, 1755–1764. [CrossRef]

5. Li, L.; Jick, S.; Breitenstein, S.; Hernandez, G.; Michel, A.; Vizcaya, D. Pulmonary arterial hypertension in the USA: An epidemiological study in a large insured pediatric population. *Pulm. Circ.* **2017**, *7*, 126–136. [CrossRef]

6. Humbert, M.; Sitbon, O.; Chaouat, A.; Bertocchi, M.; Habib, G.; Gressin, V.; Yaici, A.; Weitzenblum, E.; Cordier, J.-F.; Chabot, F.; et al. Pulmonary arterial hypertension in France: Results from a national registry. *Am. J. Respir. Crit. Care Med.* **2006**, *173*, 1023–1030. [CrossRef]

7. Ivy, D. Pulmonary hypertension in children. *Cardiol. Clin.* **2016**, *34*, 451–472. [CrossRef]

8. Zhu, N.; Gonzaga-Jauregui, C.; Welch, C.L.; Ma, L.; Qi, H.; King, A.K.; Krishnan, U.; Rosenzweig, E.B.; Ivy, D.D.; Austin, E.D.; et al. Exome sequencing in children with pulmonary arterial hypertension demonstrates differences compared with adults. *Circ. Genom. Precis. Med.* **2018**, *11*, e001887. [CrossRef]

9. Chida, A.; Shintani, M.; Matsushita, Y.; Sato, H.; Eitoku, T.; Nakayama, T.; Furutani, Y.; Hayama, E.; Kawamura, Y.; Inai, K.; et al. Mutations of NOTCH3 in childhood pulmonary arterial hypertension. *Mol. Genet. Genom. Med.* **2014**, *2*, 229–239. [CrossRef]

10. Gräf, S.; Haimel, M.; Bleda, M.; Hadinnapola, C.; Southgate, L.; Li, W.; Hodgson, J.; Liu, B.; Salmon, R.M.; Southwood, M.; et al. Identification of rare sequence variation underlying heritable pulmonary arterial hypertension. *Nat. Commun.* **2018**, *9*, 1416. [CrossRef]

11. Morrell, N.W.; Aldred, M.A.; Chung, W.K.; Elliott, C.G.; Nichols, W.C.; Soubrier, F.; Trembath, R.C.; Loyd, J.E. Genetics and genomics of pulmonary arterial hypertension. *Eur. Respir. J.* **2019**, *53*, 1801899. [CrossRef] [PubMed]

12. Zhu, N.; Pauciulo, M.W.; Welch, C.L.; Lutz, K.A.; Coleman, A.W.; Gonzaga-Jauregui, C.; Wang, J.; Grimes, J.M.; Martin, L.J.; He, H.; et al. Novel risk genes and mechanisms implicated by exome sequencing of 2572 individuals with pulmonary arterial hypertension. *Genome Med.* **2019**, *11*, 69. [CrossRef] [PubMed]

13. Southgate, L.; Machado, R.D.; Gräf, S.; Morrell, N.W. Molecular genetic framework underlying pulmonary arterial hypertension. *Nat. Rev. Cardiol.* **2020**, *17*, 85–95. [CrossRef] [PubMed]

14. Zhu, N.; Swietlik, E.M.; Welch, C.L.; Pauciulo, M.W.; Hagen, J.J.; Zhou, X.; Guo, Y.; Karten, J.; Pandya, D.; Tilly, T.; et al. Rare variant analysis of 4241 pulmonary arterial hypertension cases from an international consortium implicate *FBLN2*, *PDGFD* and rare *de novo* variants in PAH. *bioRxiv* **2020**, 550327. [CrossRef]

15. Trembath, R.C.; Thomson, J.R.; Machado, R.D.; Morgan, N.V.; Atkinson, C.; Winship, I.; Simonneau, G.; Galie, N.; Loyd, J.E.; Humbert, M.; et al. Clinical and molecular genetic features of pulmonary hypertension in patients with hereditary hemorrhagic telangiectasia. *N. Engl. J. Med.* **2001**, *345*, 325–334. [CrossRef]

16. Harrison, R.E.; Berger, R.; Haworth, S.G.; Tulloh, R.; Mache, C.J.; Morrell, N.W.; Aldred, M.A.; Trembath, R.C. Transforming growth factor-beta receptor mutations and pulmonary arterial hypertension in childhood. *Circulation* **2005**, *111*, 435–441. [CrossRef]

17. Shintani, M.; Yagi, H.; Nakayama, T.; Saji, T.; Matsuoka, R. A new nonsense mutation of SMAD8 associated with pulmonary arterial hypertension. *J. Med. Genet.* **2009**, *46*, 331–337. [CrossRef]

18. Chida, A.; Shintani, M.; Nakayama, T.; Furutani, Y.; Hayama, E.; Inai, K.; Saji, T.; Nonoyama, S.; Nakanishi, T. Missense mutations of the BMPR1B (ALK6) gene in childhood idiopathic pulmonary arterial hypertension. *Circ. J.* **2012**, *76*, 1501–1508. [CrossRef]

19. Levy, M.; Eyries, M.; Szezepanski, I.; Ladouceur, M.; Nadaud, S.; Bonnet, D.; Soubrier, F. Genetic analyses in a cohort of children with pulmonary hypertension. *Eur. Respir. J.* **2016**, *48*, 1118–1126. [CrossRef]

20. Wang, X.-J.; Lian, T.-Y.; Jiang, X.; Liu, S.-F.; Li, S.-Q.; Jiang, R.; Wu, W.-H.; Ye, J.; Cheng, C.-Y.; Du, Y.; et al. Germline BMP9 mutation causes idiopathic pulmonary arterial hypertension. *Eur. Respir. J.* **2019**, *53*, 1801609. [CrossRef]

21. Kerstjens-Frederikse, W.S.; Bongers, E.M.H.F.; Roofthooft, M.T.R.; Leter, E.M.; Douwes, J.M.; Van Dijk, A.; Vonk-Noordegraaf, A.; Dijk-Bos, K.K.; Hoefsloot, L.H.; Hoendermis, E.S.; et al. TBX4 mutations (small patella syndrome) are associated with childhood-onset pulmonary arterial hypertension. *J. Med. Genet.* **2013**, *50*, 500–506. [CrossRef] [PubMed]

22. Zhu, N.; Welch, C.L.; Wang, J.; Allen, P.M.; Gonzaga-Jauregui, C.; Ma, L.; King, A.K.; Krishnan, U.; Rosenzweig, E.B.; Ivy, D.D.; et al. Rare variants in SOX17 are associated with pulmonary arterial hypertension with congenital heart disease. *Genome Med.* **2018**, *10*, 56. [CrossRef] [PubMed]

23. Hiraide, T.; Kataoka, M.; Suzuki, H.; Aimi, Y.; Chiba, T.; Kanekura, K.; Satoh, T.; Fukuda, K.; Gamou, S.; Kosaki, K. SOX17 mutations in Japanese patients with pulmonary arterial hypertension. *Am. J. Respir. Crit. Care Med.* **2018**, *198*, 1231–1233. [CrossRef] [PubMed]

24. Bohnen, M.S.; Ma, L.; Zhu, N.; Qi, H.; McClenaghan, C.; Gonzaga-Jauregui, C.; Dewey, F.E.; Overton, J.D.; Reid, J.G.; Shuldiner, A.R.; et al. Loss-of-function ABCC8 mutations in pulmonary arterial hypertension. *Circ. Genom. Precis. Med.* **2018**, *11*, e002087. [CrossRef]

25. Eyries, M.; Montani, D.; Nadaud, S.; Girerd, B.; Levy, M.; Bourdin, A.; Trésorier, R.; Chaouat, A.; Cottin, V.; Sanfiorenzo, C.; et al. Widening the landscape of heritable pulmonary hypertension mutations in paediatric and adult cases. *Eur. Respir. J.* **2019**, *53*, 1801371. [CrossRef]

26. Simonneau, G.; Montani, D.; Celermajer, D.S.; Denton, C.P.; Gatzoulis, M.A.; Krowka, M.; Williams, P.G.; Souza, R. Haemodynamic definitions and updated clinical classification of pulmonary hypertension. *Eur. Respir. J.* **2019**, *53*, 1801913. [CrossRef]

27. Haarman, M.G.; Kerstjens-Frederikse, W.S.; Vissia-Kazemier, T.R.; Breeman, K.T.N.; Timens, W.; Vos, Y.J.; Roofthooft, M.T.R.; Hillege, H.L.; Berger, R.M.F. The genetic epidemiology of pediatric pulmonary arterial hypertension. *J. Pediatr.* **2020**, *225*, 65–73.e5. [CrossRef]

28. Li, H.; Durbin, R. Fast and accurate short read alignment with Burrows-Wheeler transform. *Bioinformatics* **2009**, *25*, 1754–1760. [CrossRef]

29. DePristo, M.A.; Banks, E.; Poplin, R.; Garimella, K.V.; Maguire, J.R.; Hartl, C.; Philippakis, A.A.; del Angel, G.; Rivas, M.A.; Hanna, M.; et al. A framework for variation discovery and genotyping using next-generation DNA sequencing data. *Nat. Genet.* **2011**, *43*, 491–498. [CrossRef]

30. Wang, K.; Li, M.; Hakonarson, H. ANNOVAR: Functional annotation of genetic variants from high-throughput sequencing data. *Nucleic Acids Res.* **2010**, *38*, e164. [CrossRef]

31. Rentzsch, P.; Witten, D.; Cooper, G.M.; Shendure, J.; Kircher, M. CADD: Predicting the deleteriousness of variants throughout the human genome. *Nucleic Acids Res.* **2019**, *47*, D886–D894. [CrossRef] [PubMed]

32. Sim, N.-L.; Kumar, P.; Hu, J.; Henikoff, S.; Schneider, G.; Ng, P.C. SIFT web server: Predicting effects of amino acid substitutions on proteins. *Nucleic Acids Res.* **2012**, *40*, W452–W457. [CrossRef] [PubMed]

33. Adzhubei, I.; Jordan, D.M.; Sunyaev, S.R. Predicting functional effect of human missense mutations using PolyPhen-2. *Curr. Protoc. Hum. Genet.* **2013**, *76*, 7.20.1–7.20.41. [CrossRef] [PubMed]

34. Schwarz, J.M.; Cooper, D.N.; Schuelke, M.; Seelow, D. MutationTaster2: Mutation prediction for the deep-sequencing age. *Nat. Methods* **2014**, *11*, 361–362. [CrossRef]

35. Thorvaldsdóttir, H.; Robinson, J.T.; Mesirov, J.P. Integrative Genomics Viewer (IGV): High-performance genomics data visualization and exploration. *Brief. Bioinform.* **2013**, *14*, 178–192. [CrossRef]

36. Robinson, J.T.; Thorvaldsdóttir, H.; Wenger, A.M.; Zehir, A.; Mesirov, J.P. Variant Review with the Integrative Genomics Viewer. *Cancer Res.* **2017**, *77*, e31–e34. [CrossRef]

37. Richards, S.; Aziz, N.; Bale, S.; Bick, D.; Das, S.; Gastier-Foster, J.; Grody, W.W.; Hegde, M.; Lyon, E.; Spector, E.; et al. Standards and guidelines for the interpretation of sequence variants: A joint consensus recommendation of the American College of Medical Genetics and Genomics and the Association for Molecular Pathology. *Genet. Med.* **2015**, *17*, 405–424. [CrossRef]

38. Kühlbrandt, W. Biology, structure and mechanism of P-type ATPases. *Nat. Rev. Mol. Cell Biol.* **2004**, *5*, 282–295. [CrossRef]

39. Madan, M.; Patel, A.; Skruber, K.; Geerts, D.; Altomare, D.A.; Iv, O.P. ATP13A3 and caveolin-1 as potential biomarkers for difluoromethylornithine-based therapies in pancreatic cancers. *Am. J. Cancer. Res.* **2016**, *6*, 1231–1252.

40. Sørensen, D.M.; Holemans, T.; van Veen, S.; Martin, S.; Arslan, T.; Haagendahl, I.W.; Holen, H.W.; Hamouda, N.N.; Eggermont, J.; Palmgren, M.; et al. Parkinson disease related ATP13A2 evolved early in animal evolution. *PLoS ONE* **2018**, *13*, e0193228. [CrossRef]

41. Lerche, M.; Eichstaedt, C.A.; Hinderhofer, K.; Grünig, E.; Tausche, K.; Ziemssen, T.; Halank, M.; Wirtz, H.; Seyfarth, H.-J. Mutually reinforcing effects of genetic variants and interferon-β 1a therapy for pulmonary arterial hypertension development in multiple sclerosis patients. *Pulm. Circ.* **2019**, *9*, 2045894019872192. [CrossRef] [PubMed]

42. Hodgson, J.; Swietlik, E.M.; Salmon, R.M.; Hadinnapola, C.; Nikolic, I.; Wharton, J.; Guo, J.; Liley, J.; Haimel, M.; Bleda, M.; et al. Characterization of GDF2 Mutations and Levels of BMP9 and BMP10 in Pulmonary Arterial Hypertension. *Am. J. Respir. Crit. Care Med.* **2020**, *201*, 575–585. [CrossRef] [PubMed]

43. David, L.; Mallet, C.; Mazerbourg, S.; Feige, J.-J.; Bailly, S. Identification of BMP9 and BMP10 as functional activators of the orphan activin receptor-like kinase 1 (ALK1) in endothelial cells. *Blood* **2007**, *109*, 1953–1961. [CrossRef] [PubMed]

44. Ricard, N.; Ciais, D.; Levet, S.; Subileau, M.; Mallet, C.; Zimmers, T.A.; Lee, S.-J.; Bidart, M.; Feige, J.-J.; Bailly, S. BMP9 and BMP10 are critical for postnatal retinal vascular remodeling. *Blood* **2012**, *119*, 6162–6171. [CrossRef] [PubMed]

45. Tillet, E.; Bailly, S. Emerging roles of BMP9 and BMP10 in hereditary hemorrhagic telangiectasia. *Front. Genet.* **2014**, *5*, 456. [CrossRef] [PubMed]

46. Folestad, E.; Kunath, A.; Wågsäter, D. PDGF-C and PDGF-D signaling in vascular diseases and animal models. *Mol. Asp. Med.* **2018**, *62*, 1–11. [CrossRef] [PubMed]

47. Burke, M.J.; Walmsley, R.; Munsey, T.S.; Smith, A.J. Receptor tyrosine kinase inhibitors cause dysfunction in adult rat cardiac fibroblasts in vitro. *Toxicol. In Vitro* **2019**, *58*, 178–186. [CrossRef]

48. Morii, C.; Tanaka, H.Y.; Izushi, Y.; Nakao, N.; Yamamoto, M.; Matsubara, H.; Kano, M.R.; Ogawa, A. 3D in vitro model of vascular medial thickening in pulmonary arterial hypertension. *Front. Bioeng. Biotechnol.* **2020**, *8*, 482. [CrossRef]

49. Melikyan, M.; Kareva, M.; Petraykina, E.; Averyanova, J.; Christesen, H. Hypoglycaemia in children [abstract]. In Proceedings of the 51st Annual Meeting of the European Society for Paediatric Endocrinology (ESPE), Leipzig, Germany, 20–23 September 2012.

50. Timlin, M.R.; Black, A.B.; Delaney, H.M.; Matos, R.I.; Percival, C.S. Development of pulmonary hypertension during treatment with diazoxide: A case series and literature review. *Pediatr. Cardiol.* **2017**, *38*, 1247–1250. [CrossRef]

51. Chen, S.C.; Dastamani, A.; Pintus, D.; Yau, D.; Aftab, S.; Bath, L.; Swinburne, C.; Hunter, L.; Giardini, A.; Christov, G.; et al. Diazoxide-induced pulmonary hypertension in hyperinsulinaemic hypoglycaemia: Recommendations from a multicentre study in the United Kingdom. *Clin. Endocrinol.* **2019**, *91*, 770–775. [CrossRef]

52. Nasim, M.T.; Ogo, T.; Ahmed, M.; Randall, R.; Chowdhury, H.M.; Snape, K.M.; Bradshaw, T.Y.; Southgate, L.; Lee, G.J.; Jackson, I.; et al. Molecular genetic characterization of SMAD signaling molecules in pulmonary arterial hypertension. *Hum. Mutat.* **2011**, *32*, 1385–1389. [CrossRef] [PubMed]

53. Drake, K.M.; Zygmunt, D.; Mavrakis, L.; Harbor, P.; Wang, L.; Comhair, S.A.; Erzurum, S.C.; Aldred, M.A. Altered MicroRNA processing in heritable pulmonary arterial hypertension: An important role for Smad-8. *Am. J. Respir. Crit. Care Med.* **2011**, *184*, 1400–1408. [CrossRef] [PubMed]

Publisher's Note: MDPI stays neutral with regard to jurisdictional claims in published maps and institutional affiliations.

Article

Customized Massive Parallel Sequencing Panel for Diagnosis of Pulmonary Arterial Hypertension

Jair Antonio Tenorio Castaño [1,2,3], Ignacio Hernández-Gonzalez [4], Natalia Gallego [1,2,3], Carmen Pérez-Olivares [5], Nuria Ochoa Parra [5], Pedro Arias [1,2,3], Elena Granda [1,2,3], Gonzalo Gómez Acebo [1,2,3], Mauro Lago-Docampo [6,7,8], Julian Palomino-Doza [9,10], Manuel López Meseguer [11], María Jesús del Cerro [12], Spanish PAH Consortium [1,2,3,5], Diana Valverde [6,7,8], Pablo Lapunzina [1,2,3,*] and Pilar Escribano-Subías [5,9]

[1] Institute of Medical and Molecular Genetics (INGEMM)-IdiPAZ, Hospital Universitario La Paz-UAM Paseo de La Castellana, 261, 28046 Madrid, Spain; jairantonio.tenorio@gmail.com (J.A.T.C.); nataliagallegozazo@gmail.com (N.G.); palajara@gmail.com (P.A.); elena.granda@estudiante.uam.es (E.G.); gonzalo.gomezacebo@gmail.com (G.G.A.); spanishpahconsortium@gmail.com (S.P.C.)

[2] CIBERER, Centro de Investigación Biomédica en Red de Enfermedades Raras, ISCIII, Melchor Fernández Almagro, 3, 28029 Madrid, Spain

[3] ITHACA, European Reference Network on Rare Congenital Malformations and Rare Intellectual Disability, 1000 Brussels, Belgium

[4] Department of Cardiology, Hospital Universitario Río Hortega, 47012 Valladolid, Spain; hgnacho@hotmail.com

[5] Unidad Multidisciplinar de Hipertensión Pulmonar, Servicio de Cardiología, Hospital Universitario 12 de Octubre, 28034 Madrid, Spain; carmenperezolivaresd@gmail.com (C.P.-O.); nuriaochoaparra@hotmail.com (N.O.P.); pilar.escribano.subias@gmail.com (P.E.-S.)

[6] CINBIO, Universidade de Vigo, 36310 Vigo, Spain; maurolagodocampo@gmail.com (M.L.-D.); dianaval@uvigo.es (D.V.)

[7] Instituto de Investigación Sanitaria Galicia Sur, Hospital Álvaro Cunqueiro, 36213 Vigo, Spain

[8] Centro de Investigaciones Biomédicas (CINBIO), 36310 Vigo, Spain

[9] J. CIBERCV, Centro de Investigación Biomédica en Red de Enfermedades Cardiovasculares, ISCIII, 28029 Madrid, Spain; julian.palomino@salud.madrid.org

[10] Unidad de Miocardiopatías Familiares, Servicio de Cardiología, Hospital Universitario 12 de Octubre, 28041 Madrid, Spain

[11] Servicio de Neumología, Hospital Universitario Vall d'Hebron, 08035 Barcelona, Spain; manuelopez@vhebron.net

[12] Paediatric cardiology and Grown up Congenital Heart Disease Unit, Hospital Ramón y Cajal, 28034 Madrid, Spain; majecerro@yahoo.es

* Correspondence: pablo.lapunzina@salud.madrid.org; Tel.: +34-91-727-72-17; Fax: +34-91-207-10-40

Received: 7 September 2020; Accepted: 28 September 2020; Published: 30 September 2020

Abstract: Pulmonary arterial hypertension is a very infrequent disease, with a variable etiology and clinical expressivity, making sometimes the clinical diagnosis a challenge. Current classification based on clinical features does not reflect the underlying molecular profiling of these groups. The advance in massive parallel sequencing in PAH has allowed for the describing of several new causative and susceptibility genes related to PAH, improving overall patient diagnosis. In order to address the molecular diagnosis of patients with PAH we designed, validated, and routinely applied a custom panel including 21 genes. Three hundred patients from the National Spanish PAH Registry (REHAP) were included in the analysis. A custom script was developed to annotate and filter the variants. Variant classification was performed according to the ACMG guidelines. Pathogenic and likely pathogenic variants have been found in 15% of the patients with 12% of variants of unknown significance (VUS). We have found variants in patients with connective tissue disease (CTD) and congenital heart disease (CHD). In addition, in a small proportion of patients (1.75%), we observed a possible digenic mode of inheritance. These results stand out the importance of the genetic testing of patients with associated forms of PAH (i.e., CHD and CTD) additionally to the classical IPAH and

HPAH forms. Molecular confirmation of the clinical presumptive diagnosis is required in cases with a high clinical overlapping to carry out proper management and follow up of the individuals with the disease.

Keywords: pulmonary arterial hypertension; massive parallel sequencing; NGS; digenic inheritance; and genetics

---

1. Introduction

Pulmonary arterial hypertension (PAH) is an uncommon disease, characterized by the increase in blood pressure in the lung arteries, which causes right ventricle failure and potentially can lead to death if not treated. According to the Spanish PAH Registry (REHAP), the prevalence of the disease in individuals older than 14 years is approximately 15 to 50 cases per million of the population [1].

Pulmonary hypertension (PH) classification has been recently proposed and encompasses five groups [2]. Classically, PAH included in group one diagnosis was based on hemodynamic parameters assessed by right heart catheterization: mean pulmonary arterial pressure (mPAP) ≥ 25 mmHg at rest, pulmonary artery wedge pressure (PAWP) ≤ 15 mmHg, and pulmonary vascular resistance (PVR) > 3 UW, in the absence of other causes of precapillary PH such as PH due to lung diseases, chronic thromboembolic disease or other rare diseases [3].

Idiopathic (IPAH) and heritable PAH (HPAH), included within group one, have been classically associated with *BMPR2* mutations [4], with a penetrance of 20%, which means that approximately 80% of individuals with *BMPR2* will not develop clinically detectable PAH [5]. Mutations in this gene only explain 20% of idiopathic cases and about 60% of heritable PAH individuals [5,6], suggesting that additional genes are responsible for PAH. Thus, several genes were reported in the past few years associated with PAH.

Genes described in PAH are associated with a variety of molecular pathways [6–12]. Currently, 12 genes have been clearly related with PAH with a higher level of evidence: *BMPR2*, *EIF2AK4*, *TBX4*, *ATP13A3*, *GDF2*, *SOX17*, *AQP1*, *ACVRL1*, *SMAD9*, *ENG*, *KCNK3*, and *CAV1*, and five with a lower level of evidence: *SMAD4*, *SMAD1*, *KLF2*, *BMPR1B*, and *KCNA5* [13]. Also, there is increasing evidence for the involvement of other genes in PAH with less frequency and in different pathways. *KLK1* and *GGCX* have been recently related to IPAH [14].

Pulmonary venooclusive disease (PVOD) is the most lethal subtype of PAH, with different clinical, histological, and genetic features. It is characterized by very low diffusion capacity (DLCO), resting hypoxemia, severe desaturation on exercise, and a characteristic radiological pattern. Heritable forms are caused by bi-allelic mutations in *EIF2AK4*, a member of the kinases that phosphorylate the alpha subunit of eukaryotic translation initiation factor-2. We showed that consanguineous families with PVOD from Gipsy origin have a founder effect mutation in *EIF2AK4*, which causes a missense variant and haploinsufficiency of the gene [15]. Nowadays PVOD and pulmonary capillary disease are considered similar entities with overlapping clinical features, with the same underlying genetic mechanism [16]. Differential diagnosis between PAH and PVOD remains a clinical challenge and misdiagnosis is a common problem with important clinical implications.

More recently, a "second hit" hypothesis came up due to the identification of two pathogenic variants in two genes related to PAH [17]. Eichstaedt et al. identified a family with two variants, one in *EIF2AK4* and one in *BMPR2*, in a family with HPAH [18]. Additionally, the group of Yuxin Fan described a severe early-onset PAH individual that had two variants probably contributing to the disease, one in *BMPR2* and one in *KCNA5* [19]. This was also reported in two families from Lebanon in a combination of variants in *BMPR2-GDF2* and *BMPR2-TBX4*, respectively [20]. Taking this into account and based on the low penetrance of *BMPR2*-associated PAH, the co-occurrence of more than one variant associated with the disease might explain why some individuals with only *BMPR2* mutations do not develop PAH. This can explain in part why the presence of *BMPR2* pathogenic variants alone is not always enough to develop the disease.

Thus, the main objective of this study was to study a cohort of three hundred patients with idiopathic, heritable, PVOD and APAH from the Spanish registry of patients with PAH (REHAP), through a custom, in-house NGS panel of 21 genes. Secondly, phenotype–genotype correlation and survival rate were also assessed.

2. Material and Methods

2.1. Cohort Description

All patients from group one and idiopathic, heritable, and PVOD and patients with associated PAH to congenital heart disease and connective tissue disease included in the national Spanish Registry of Pulmonary Arterial Hypertension (REHAP) were eligible for this study (Supplementary Figure S1). REHAP is an observational national registry with the aim to serve as a database for research purposes. It includes patients with any form of PAH [21,22]. Clinical parameters for PAH diagnosis were obtained from the 2015 ERS/ESC guidelines [3].

In Spain, genetic testing of patients included in the REHAP has been offered since 2011, with a focus on IPAH, HPAH, and PVOD at the beginning but later expanded to associated forms as well. This study followed the ethical principles of the European board of medical genetics. All patients signed a consent form and genetic counseling pre and post-test was offered.

The study was approved by the ethical committee for scientific research of each participant center and also by the ethical committee of the La Paz University Hospital (CEIC-HULP PI-1210).

Before genetic testing, family history information was collected, and DNA samples from the probands were extracted. Segregation analysis in patients carrying a variant of unknown significance was performed when possible. In cases of unaffected carriers, a complete diagnostic and specific follow up was performed to rule out PAH.

2.2. Statistical Analysis

Categorical variables are presented as absolute frequency and proportions, and continuous variables as the mean ± standard deviation or median (interquartile range). In order to validate the normality of the statistical distribution, the Kolmogorov–Smirnov test was applied, and to compare clinical features and continuous variables among patients, t-student was calculated. For variables that did not follow a normal distribution, the Wilcoxon signed-rank test was used. The rate of percentages was compared by Chi-square test, Fisher exact test, or Wilcoxon test, as appropriate.

Survival analysis was performed using the Kaplan-Meier analysis, matching the date of diagnosis with the date of the first diagnostic right heart catheterization (RHC). All-cause mortality or bilateral lung transplantation were defined as the endpoint and the log-rank test was used for comparison between groups. Multivariate Cox regression models identified significant predictors of mortality. A comparison of the survival rate among different groups was also performed (Supplementary Figure S2).

A two-sided *p*-value of less than 0.05 was considered statistically significant. Statistical analyses were done using Stata (v12.1 for Mac; StataCorp, College Station, Brazos County, TX, USA).

2.3. Genetic Analysis

A customized panel of 21 genes (HAP v1.2) including *ABCC8, ACVRL1, BMPR1B, BMPR2, CAV1, CBLN2, CPS1, EIF2AK4, ENG, GDF2, KCNA5, KCNK3, MMACHC, NOTCH3, SARS2, SMAD1, SMAD4, SMAD5, SMAD9, TBX4,* and *TOPBP1*. This custom panel was designed with NimbleDesign (Roche, Indianapolis, IN, USA). Fragmentation and library preparation were performed with SeqCap EZ Choice Enrichment Kit (Roche, Indianapolis, IN, USA). Sequencing was performed with the Illumina MiSeq platform (Illumina, San Diego, CA, USA) following the manufacturer's instructions. Genes were chosen because of their probed association with PAH or based on previous research data. An in-house pipeline for bioinformatic analysis was developed to perform quality control and variant annotation.

Our Institute (INGEMM) has developed a suite of quality control (QC) scripts that simplify data quality assessment. QC included an assessment of total reads, library complexity, capture efficiency, coverage distribution (95% at ≥20×), capture uniformity, raw error rates, Ti/Tv ratio in coding regions (typically 3.2 for known sites and 2.9 for novel sites), and distribution of known and novel variants relative to dbSNP and zygosity.

Annotation: an automated pipeline for the annotation of variants derived from targeted sequences in the capture genes was developed. INGEMM script application returned several variant annotations including dbSNPrs identification, gene names with accession numbers, predicted functional effect, protein positions and, for amino-acid changes (dbNSFP, CADD), conservation scores and several population frequency databases, and known clinical associations along with a vast array of annotations for non-coding sequences derived from ENCODE. Databases for pathogenic variants such as ClinVar, Human Gene Mutation Database (HGMD), Mastermind, LOVD, Alamut, and Varsome have also been reviewed.

After that, the variant prioritization was performed according to the filtering strategy described in Figure 1. Finally, variants were classified according to the ACMG guidelines [23]. A custom script developed in-house called "LACONv" has also been developed by INGEMM in order to analyze large genomic DNA gain and losses or copy number variation (CNVs) which can be found in the following repository (https://github.com/kibanez/LACONv) [24].

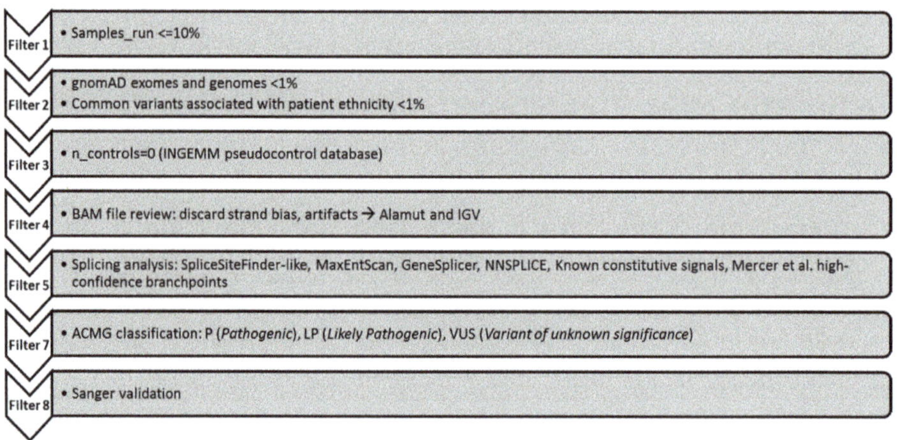

Figure 1. Filtering pipeline for NGS panel HAP v1.2. A set of filters applied for variant prioritization. Only variants does not fulfil the criteria for variant quality were validated with Sanger sequencing [25].

3. Results

After the quality control analysis, 33 out of the 300 sequenced samples were discarded and 267 were subsequently filtered. We have identified 86 variants in 81 patients (32.2%). Out of them, 34 (39.5%) have been classified as pathogenic variants, 14 (16.3%) as likely pathogenic, and 38 (44.2%) as variants of unknown significance (VUS). In 186 samples non pathogenic, likely pathogenic, or VUS were detected in the analyzed genes. *BMPR2* was the predominantly mutated gene with 25 pathogenic or likely pathogenic variants (29%, 25/86) followed by *EIF2AK4*, *TBX4*, and *ACVRL1*.

Regarding the VUS variants detected, *NOTCH3* and *ABCC8* had the highest frequency of these variants (Figure 2). Based on the etiology, 41 variants were detected in 38 IPAH individuals, 15 in 15 individuals with HPAH, and 10 in 8 individuals with PVOD. In associated -PAH, we have found 15 variants (Supplementary Tables S1 and S2). Finally, in all three patients with suspected heritable hemorrhagic telangiectasia (HHT), an *AVCRL1* pathogenic variant was detected, confirming the clinical suspicion.

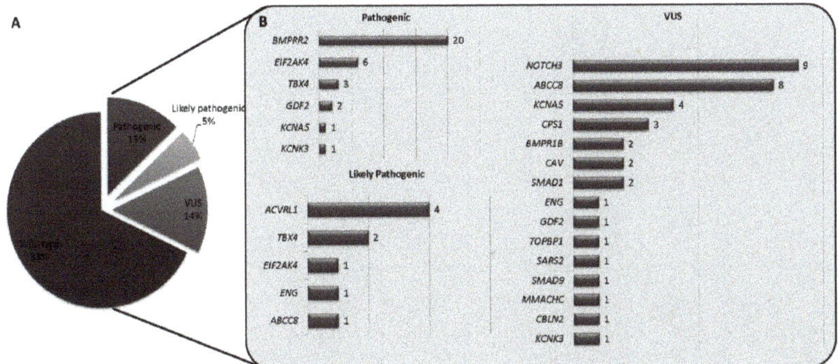

Figure 2. Frequency of gene mutation detected after the analysis of 21 genes. (**A**) A total of 14% of pathogenic and likely pathogenic variants were identified, with a 15% of variants of unknown significance (VUS) and in approximately 65% of individuals no candidate gene variants were identified. After quality check, 6% of the analyzed samples were discarded due to low quality. (**B**) The gene with the highest pathogenic variants detected was BMPR2, followed by EIF2AK4, because many patients with PVOD were also included in the analysis.

Strikingly, we have found five unrelated families in whom there was more than one candidate causative variant (Table 1, Figure 3). In two families, one of the variants was found in *BMPR2* and the co-occurrence was with *NOTCH3* (x2). In the other three families, the variants were located in *ABCC8-NOTCH3*, *ABCC8-SARS2*, and *TBX4-SMAD1*. Three of these families were classified as IPAH and the other two as PAH-CHD.

In all consanguineous individuals with a clinical diagnosis of PVOD, we found the known Spanish founder missense pathogenic variant in *EIF2AK4*:NM_001013703.3:c.3344C>T(p.Pro1115Leu) in a homozygous state [15]. Additionally, in two siblings from a non-consanguineous couple, we found compound heterozygous pathogenic variants in *EIF2AK4* ((*EIF2AK4*:NM_001013703.3(EIF2AK4_v001):c.3766C>T:p.(Arg1256*) and *EIF2AK4*:NM_001013703.3(EIF2AK4_v001):c.4392dup:p.(Lys1465*)). Segregation analysis demonstrated that parents were carriers of each variant. One of the siblings is a Caucasian male with a clinical presumptive diagnosis of PVOD at 30 years of age. He underwent bilateral lung transplantation 3.5 years after diagnosis. His sister was also diagnosed with PVOD at 38 years of age leading to lung transplantation one year after diagnosis.

We have found five patients from five unrelated families carrying a pathogenic or likely pathogenic variant in *TBX4* [24]. All patients had a variable clinical expressivity with a moderate to severe reduction in diffusion capacity (DLCO).

Table 1. Individuals in whom more than one candidate gene was detected. Variant prioritization showed that in five unrelated families more than one candidate gene was identified.

Patient ID	PAH Etiology	Variant 1	ACMG Classification	Variant 2	ACMG Classification
HTP070	IPAH	*BMPR2*:NM_001204.6:c.77-5_77-2delTTTA	P	*NOTCH3*:NM_000435.2:c.181C>T:p.(Arg61Trp)	VUS
HTP081	IPAH	*BMPR2*:Exon2 duplication	N/A	*NOTCH3*:NM_000435.2:c.1964A>G:p.(Asn655Ser)	VUS
HTP466	IPAH	*ABCC8*:NM_000352.4:c.3976G>A:p.(Glu1326Lys)	LP	*NOTCH3*:NM_000435.2:c.6532C>T:p.(Pro2178Ser)	VUS
HTP474	CHD	*ABCC8*:NM_000352.4(ABCC8):c.3394G>A:p.(Asp1132Asn)	VUS	*SARS2*:NM_017827.3:c.136G>A:p.(Glu46Lys)	VUS
HTP611	CHD	*TBX4*:NM_018488.3:c.1018C>T:p.(Arg340 *)	P	*SMAD1*:NM_005900.2:c.738G>C:p.(Met246Ile)	VUS

* Stop codon.

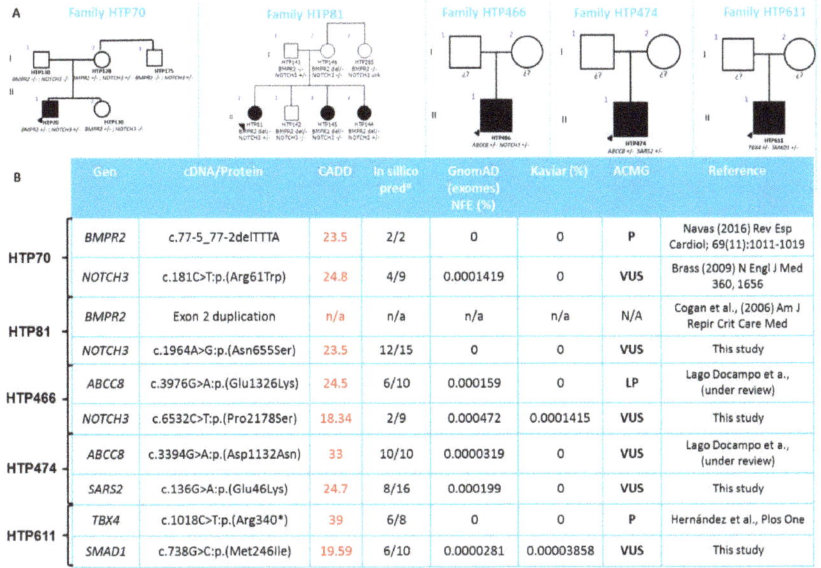

A

B

	Gen	cDNA/Protein	CADD	In sillico pred[α]	GnomAD (exomes) NFE (%)	Kaviar (%)	ACMG	Reference
HTP70	BMPR2	c.77-5_77-2delTTTA	23.5	2/2	0	0	P	Navas (2016) Rev Esp Cardiol; 69(11):1011-1019
	NOTCH3	c.181C>T:p.(Arg61Trp)	24.8	4/9	0.0001419	0	VUS	Brass (2009) N Engl J Med 360, 1656
HTP81	BMPR2	Exon 2 duplication	n/a	n/a	n/a	n/a	N/A	Cogan et al., (2006) Am J Repir Crit Care Med
	NOTCH3	c.1964A>G:p.(Asn655Ser)	23.5	12/15	0	0	VUS	This study
HTP466	ABCC8	c.3976G>A:p.(Glu1326Lys)	24.5	6/10	0.000159	0	LP	Lago Docampo et a., (under review)
	NOTCH3	c.6532C>T:p.(Pro2178Ser)	18.34	2/9	0.000472	0.0001415	VUS	This study
HTP474	ABCC8	c.3394G>A:p.(Asp1132Asn)	33	10/10	0.0000319	0	VUS	Lago Docampo et a., (under review)
	SARS2	c.136G>A:p.(Glu46Lys)	24.7	8/16	0.000199	0	VUS	This study
HTP611	TBX4	c.1018C>T:p.(Arg340*)	39	6/8	0	0	P	Hernández et al., Plos One
	SMAD1	c.738G>C:p.(Met246Ile)	19.59	6/10	0.0000281	0.00003858	VUS	This study

Figure 3. Pedigree and in silico analysis of the families with more than one variant detected. (**A**) Genealogy of the families in which segregation analysis was performed (when available). (**B**) Table of in sillico analysis of the candidate variants, including the ACMG classification as well as the reference in which the variant was described. n/a: not applicable, P: pathogenic, VUS: Variant of unknown significance. α: In silico analysis was performed by dbNSFP counting the ones that predict a deleterious effect over the total available.

Finally, in *CBLN2*, we found one candidate variant in a patient with IPAH classified as VUS (*CBLN2*: NM_182511.3:c.263C>T:p.(Ser88Phe)).

4. Discussion

PAH is a multifactorial disease, in which many different phenotypes and etiologies can explain the prognosis and severity of the disease. During the last few years, mainly due to the advances in genomics technologies, many genes and genomic associations were described in PAH. Therefore, nowadays the diagnostic approach for PAH should include genetic analysis because it is crucial for follow up, genetic counseling, to grade the severity, or even for future personalized therapies [26,27]. Since 2014, we have been evaluating patients not only with IPAH, HPAH, and PVOD but also secondary forms from which there is not so much evidence for the relation with genetic mutations.

Thus, the HAP-panel v1.2 included 21 genes based on the literature, previous experience, and well-established genes in PAH, which influence the pathophysiology of the disease. We included patients with PAH, no matter the type of PAH (idiopathic, heritable, or associated forms). We found that 30.34% (81/267) of individuals with pathogenic, likely pathogenic, and/or VUS variants (Figure 2). A total of 53.41% (47/88) of these variants were classified as pathogenic or likely pathogenic, and the remaining 31.82% (38/88) VUS, meaning that we have a relatively high number of variants without clear effect. The majority of the VUS have been found in *NOTCH3*, a gene from which mutations have been recently associated with IPAH [9,28,29]. It has been suggested that *NOTCH3* mutations are involved in proliferation and cell viability and impairs the NOTCH3-HES5 signaling pathway, a crucial pathway in the development of PAH [30].

Regarding *NOTCH3*, it has been suggested that the majority of the mutations in *NOTCH3* cause a missense change in the protein. Out of the 38 VUS variants reported herein, 23.7% (9/38) have been found in *NOTCH3*, which means that functional studies and segregation analysis in available cases are strongly recommended to confirm or discard the possible effect for these variants in the protein function and pathways in which *NOTCH3* is involved.

This means that, if all the *NOTCH3* variants were classified as pathogenic or likely pathogenic, the frequency of mutations in this gene in PAH would be much more common than initially reported [9]. Strikingly, among the five unrelated families with a suggested digenic inheritance, two have had pathogenic variants in *BMPR2* and *NOTCH3*, which also segregates with the disease (Table 1). One of the index patients was classified as IPAH and clinical features of the proband showed one adult diagnosed at 39 years, with severe hemodynamic parameters (very low cardiac output). Oral dual therapy (riociguat and ambrisentan) was initiated with clinical and hemodynamic improvement. Eight years after diagnosis, he presents a low risk of death under dual oral therapy. It is important to remark that, as far as we know, there is no previous evidence of segregation of variants in *BMPR2* and *NOTCH3*, and this fact, might explain in part, why penetrance of *BMPR2* in idiopathic cases of PAH is only about 20% (14% in males and 42% in females) [6,13].

Furthermore, three additional families have been found to have two variants in two genes (Table 1, Figure 3); adding evidence that digenic inheritance may be a mechanism associated with PAH as a previous report suggested [18]. Two out of these three families were diagnosed with APAH-CDH and one with IPAH.

Two siblings from an unrelated couple were found to have a compound heterozygous mutation in *EIF2AK4*, each one inherited from healthy parents. Clinical features of the patients correlated with the molecular findings and with the presumptive diagnosis of PVOD.

In *ABCC8*, a gene initially included with minimal evidence of relation with PAH, we found nine variants (Table 2, Figure 4), seven in IPAH individuals, one APAH, and one patient with CHD. All detected variants are distributed along SUR1, the protein encoded by *ABCC8*, so there are no hotspots in the gene, in concordance with previous reports [31] where the mechanism suggested was the loss of the ATP-sensitive potassium channel function. Functional assays by minigenes (data not shown), confirmed that the variant p.Asp1132Asn affects the splicing by exon skipping of exon 27 [32] and the variants p.(Glu100Lys), p.(Val477Met), and p.(Glu1326Lys) do not alter the splicing and maybe the pathogenic effect is not by splicing defects. In fact, variant p.(Glu1326Lys) has been found together with another variant in *NOTCH3* (NM_000435.2:c.6532C>T:p.(Pro2178Ser)), which means that further analysis must be performed to confirm the role of these changes.

Table 2. Variants identified in *ABCC8*. ID and etiology of patients in which a variant in *ABCC8* was identified. Transcript for ABCC8 variant annotation was NM_000352.4.

Patient ID	PAH Etiology	Variant
HTP114	IPAH	*ABCC8*:NM_000352.4: c.1643C>T:p.(Thr548Met)
HTP88	IPAH	*ABCC8*:NM_000352:exon26: c.3288_3289del:p.(His1097Profs*16)
HTP151	IPAH	*ABCC8*:NM_000352.4: c.3238G>A:p.(Val1080Ile)
HTP159	IPAH	*ABCC8*:NM_000352.4: c.2422C>A:p.(Gln808Lys)
HTP162	IPAH	*ABCC8*:NM_000352.4: c.1429G>A:p.(Val477Met)
HTP466	IPAH	*ABCC8*:NM_000352.4: c.3976G>A:p.Glu1326Lys
HTP37	IPAH	*ABCC8*:NM_000352.3: c.579+5G>A
HTP78	CREST-PAH	*ABCC8*:NM_000352.3: c.2694+1G>A
HTP474	CHD	*ABCC8*:NM_000352.4: c.3394G>A:p.Asp1132Asn

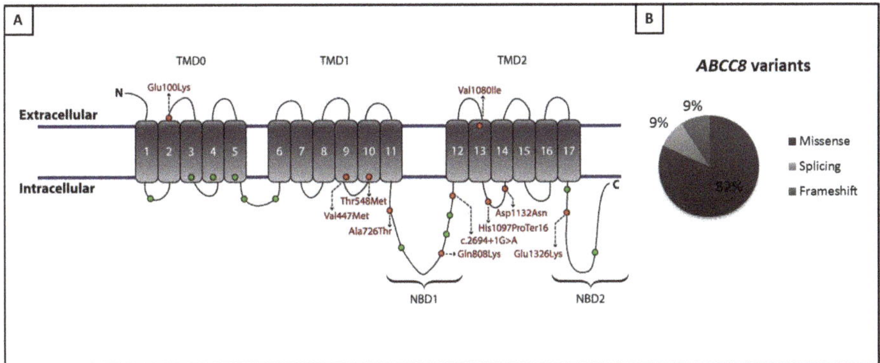

Figure 4. Variants detected in *ABCC8*. (**A**) Schematic representation of SUR1 (encoded by *ABCC8*) protein with its 17 transmembrane domains, and the variants detected in our cohort (red) and in Bohnen et al series (green). (**B**) Distribution of variants detected in *ABCC8* in our cohort based on the type of variant.

Only one patient has been found to have a VUS in *CBLN2*, but further analysis needs to be done to elucidate the possible role of this gene in PAH.

In patients with PAH associated with CHD, we found ten variants in eight samples, three classified as pathogenic (two in *BMPR2* in two siblings and one in *TBX4*) and six VUS in *CPS1*, *ABCC8*, *SMAD5*, *SARS2*, *SMAD1*, and *NOTCH3* (Table 3).

The two *BMPR2* mutation carriers came from unrelated families, inherited the variant from their healthy mother, and no additional candidate variants were detected in the NGS panel in both siblings, suggesting that there may be other factors influencing the PAH-phenotype in the siblings, such as second hits in other genes not included in the panel. One of these siblings is a male diagnosed with PH at 19 years of age. In the first work-up at diagnosis, an interventricular septal defect with a right to left shunt was observed. Right heart catheterization (RHC) confirmed suprasystemic PH. Considering these findings, ventricular septal defect closure was contraindicated, and PAH-targeted treatment was initiated. Clinical follow-up included a goal-oriented PAH therapy and the risk profile was assessed periodically according to the guidelines. Finally, he underwent bilateral lung transplantation four years after diagnosis.

His sister was also diagnosed with PAH at 26 years of age. At diagnosis, an interatrial septal defect with a left-to-right shunt was observed. In the RHC, PH with elevated pulmonary vascular resistance was confirmed. Dual oral PAH treatment was initiated, and she currently presents a low risk of death under oral therapy three years after diagnosis according to the hemodynamic parameters.

Previous studies suggested that pathogenic variants in *BMPR2* are present in up to 7.5% of patients with PAH associated with CHD [33,34]. Furthermore, Liu et al. observed that *BMPR2* pathogenic variants are more frequent in patients with pulmonary vascular disease (PVD) associated with CHD in comparison with those patients without PVD. To date, there is no effective method to predict the development of PH after the correction of a CHD. According to current guidelines, correction is contraindicated when PVD is established [35]. However, approximately 10% of patients with repaired atrial septal defect or ventricular septal defect develop PAH and therefore genetic background can contribute to explaining these differences. The role of germline mutations in these patients is unknown. These findings suggest that PAH mutations might contribute to the development of PAH in CHD patients. Genetic testing may allow us to identify high-risk patients. However, more studies are needed to assess the importance of genetic abnormalities in CHD.

Table 3. Variants detected in associated PAH forms (APAH). Description of variants found in patients with APAH, specifically with congenital heart disease (CHD) and connective tissue disease (CTD). ID: patient identifier, GT: genotype, ACMG: American College of Medical Genetics. Hom: homozygous; Het: heterozygous; P: pathogenic; VUS: variant of unknown significance.

ID	PAH Etiology	cDNA and Protein Position	GT	ACMG Classification
HTP501	CHD	*CPS1*:NM_001122633.2(CPS1): c.3047C>T:p.(Thr1016Met)	Hom	VUS
HTP536	CHD	*BMPR2*:NM_001204.6: c.2674delG: p.(Glu892Asnfs*4)	Het	P
HTP541	CHD	*BMPR2*:NM_001204.6: c.2674delG: p.(Glu892Asnfs*4)	Het	P
HTP474	CHD	*ABCC8*:NM_000352.4(ABCC8): c.3394G>A:p.(Asp1132Asn) *SARS2*:NM_017827.3: c.136G>A:p.(Glu46Lys)	Het Het	VUS
HTP558	CHD	*SMAD5*:NM_001001420.2: c.763A>G:p.(Ile255Val)	Het	VUS
HTP262	CHD	*NOTCH3*:NM_000435.3: c.6097C>G:p.(Pro2033Ala)	Het	VUS
HTP472	CHD	*CPS1*:NM_001122633.2: c.1036G>A:p.(Ala346Thr)	Het	VUS
HTP611	CHD	*TBX4*:NM_018488.3: c.1018C>T:p.(Arg340*) *SMAD1*:NM_005900.2: c.738G>C:p.(Met246Ile)	Het Het	P VUS
HTP551	CTD	*GDF2*:NM_016204: c.642G>A:p.(Trp214*)	Het	VUS
HTP355	CTD	*CPS1*:NM_001875.4: c.4252C>T:p.(Pro1418Ser)	Het	VUS
HTP452	CTD	*NOTCH3*:NM_000435: c.5203G>A:p.(Glu1735Lys)	Het	VUS
HTP564	CTD	*TBX4*:NM_018488.2: c.1112dupC:p.(Pro372Serfs*14)	Het	P
HTP78	CTD	*ABCC8*:NM_000352.3: c.2694+1G>A	Het	VUS

In all patients where a pathogenic variant in *TBX4* was detected, the clinical expressivity was highly variable, including an initial suspicion of PVOD, interstitial lung disease, pulmonary vascular abnormalities, and CHD. In one individual, a variant in *SMAD1* has been also observed in addition to the *TBX4* variant. This variant in *SMAD1* was classified as VUS. He suffered with PAH associated with a non-repaired atrial septal defect. The variant observed in *TBX4* is a very infrequent variant with an extremely low frequency in the analyzed control populations (gnomAD exomes: 0.0000281, gnomAD genomes, 1000G, Kaviar, Beacon, ESP, and Bravo) although the majority of the in silico bioinformatic tools did not suggest a deleterious effect of the variant. Thus, the relationship of the variant detected in *SMAD1* and the phenotype of the patients remains unclear. The nonsense *TBX4* variant detected in this patient (*TBX4*(NM_018488.3):c.1018C>T:(p.Arg340*)) was reported last year by Galambos et al. [36]. In their study, the two siblings in whom the same variant was detected presented a clinically specific phenotype with transient patent ductus arteriosus, a patent foramen oval, and interstitial lung disease (ILD). Strikingly, one individual did not manifest pulmonary hypertension; reflecting the possible contribution of other genetic, epigenetic, or external factors that can contribute to the development of the disease. The presence of left-to-right shunt causes an overflow in pulmonary circulation that can lead to PH. However, this does not fully explain these complex cases. It is necessary to further investigate the possible involvement of *TBX4* in complex cardiac diseases such as CHD and ILD, and how it can contribute to increasing the risk of development of PH in these individuals.

Additionally, we found three variants in *CPS1*, one in a homozygous state, a gene for which neonatal susceptibility to PAH has been suggested with an increase of pulmonary arterial pressure after a surgical repair of congenital heart defects [37]. Furthermore, polymorphisms in *CPS1* have also been related to persistent pulmonary hypertension of the newborn (PPHN) [38]. All variants were classified as VUS and unfortunately, no parental samples were available for segregation analysis.

Between 15 and 30% of PAH cases are secondary to a CTD [39] such as systemic sclerosis or systemic lupus erythematosus. Five-year survival in PAH-CTD is around 44% in the US population [40]. In Spain, one-year survival in PAH-CTD is 81% [1,41]. There is increasing evidence of the contribution of genetic mutation in PAH-CTD, as previous reports suggested [17]. In our CTD-PAH cohort, we have found four variants (two classified as VUS in *ABCC8* and *NOTCH3*, and two as pathogenic in *TBX4* and *GDF2*) (Table 3), adding further evidence of the potential relation between variants in well-known genes for PAH and CTD-PAH. Furthermore, as far as we know, this is the first time a variant in *TBX4* is associated with CTD-PAH. All variants were found in different genes and de novo events were confirmed in segregation studies when available.

We have compared the clinical features of all patients with a confirmed genetic defect (Supplementary Tables S3–S5). We observed that those patients who carried a pathogenic or likely pathogenic variant had a higher mean pulmonary artery pressure (PAPm) and pulmonary vascular resistance (PVR), with a lower cardiac index. However, they were younger and less likely to respond to acute vasodilator testing. Regarding functional class, no statistical differences were observed in the clinical severity assessed by the WHO functional class. However, the six-minute walking test (6MWT) was statistically higher in the group of patients with pathogenic variants compared with the non-mutated group.

In terms of the overall survival rate, we have found no significant difference between both groups (Figure 5). Additional analysis of survival by Kaplan–Meier does not show any statistical differences between patients with different gene variants except for *EIF2AK4*, as we expected due to the severity of the associated PAH form. Also, we have not found statistical differences in survival between idiopathic, hereditary, and associated PAH forms, nor between carriers and non-carriers with APAH and PVOD (Supplementary Figure S2).

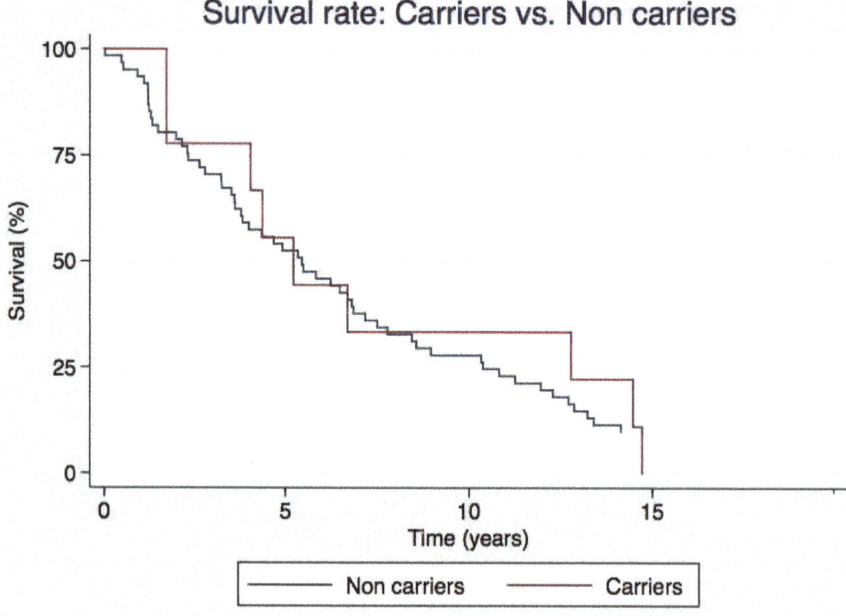

Figure 5. Survival analysis between carriers' vs. non-carriers. Kaplan-Meier analysis did not reflect statistical differences between individuals with PAH and pathogenic or likely pathogenic variants and individuals without any pathogenic, likely pathogenic variant detected.

5. Conclusions

To summarize, we applied a custom personalized panel of genes for PAH diagnosis, which demonstrated to be very useful to confirm a presumptive diagnosis in cases with overlapping clinical features of PAH. The panel allows the identification of variants in the PAH related genes in a reasonable turnaround time for clinical purposes. We also detected variants in genes that have been recently described in PAH, and that could potentially be new therapeutic targets for personalized medicine. Lastly, in secondary PAH, we found variants that may influence the phenotype of the disease. Further studies to confirm the role of these variants are needed.

Supplementary Materials: The following are available online at http://www.mdpi.com/2073-4425/11/10/1158/s1, Table S1: List of variants detected after prioritization. Table S2: List of variants detected after prioritization. Table S3: PAH Spanish Biobank cohort demographic and hemodynamic data. Table S4: PAH Spanish Biobank cohort demographic and hemodynamic data. Table S5: Clinical comparison between carriers' vs. non carriers. Figure S1. Analyzed cohort of PAH. Figure S2: Survival analysis in PAH.

Author Contributions: Conceptualization, J.A.T.C., P.L. and P.E.-S.; methodology, J.A.T.C., N.G., P.L., C.P.-O., M.L.M., M.J.d.C.; software, J.A.T.C., N.G., P.A.; validation, P.A., N.G., M.L.-D., D.V.; formal analysis, J.A.T.C., N.G., I.H.-G.; investigation, P.A., E.G., G.G.A., N.O.P., J.P.-D., M.L.M., M.J.d.C.; resources, N.O.P.; data curation, S.P.C., J.A.T.C., N.G., I.H.-G.; writing—original draft preparation, J.A.T.C., I.H.-G.; writing—review and editing, J.A.T.C., I.H.-G., N.G.; supervision, P.E.-S., P.L.; project administration, N.O.P.; funding acquisition, P.E.-S., P.L., J.A.T.C., D.V. All authors have read and agreed to the published version of the manuscript.

Funding: This research was funded by Instituto de Salud Carlos III, FISPI18/01233, Janssen unrestricted grant and Annual grant supported by the Foundation against PAH (FCHP, https://www.fchp.es/). MLD is supported by a Xunta de Galicia predoctoral fellowship (ED481A-2018/304).

Acknowledgments: We would like to thank all patients and families for their collaboration and participation in this study and all the center's participants in the PAH Consortium, including the REHAP registry.

Conflicts of Interest: The authors declare that there is no conflict of interest regarding the publication of this article. IHG and NG contributed equally to this work.

References

1. Escribano-Subias, P.; Blanco, I.; Lopez-Meseguer, M.; Lopez-Guarch, C.J.; Roman, A.; Morales, P.; Castillo-Palma, M.J.; Segovia, J.; Gomez-Sanchez, M.A.; Barbera, J.A.; et al. Survival in pulmonary hypertension in Spain: Insights from the Spanish registry. *Eur. Respir. J.* **2012**, *40*, 596–603. [CrossRef] [PubMed]
2. Simonneau, G.; Montani, D.; Celermajer, D.S.; Denton, C.P.; Gatzoulis, M.A.; Krowka, M.; Williams, P.G.; Souza, R. Haemodynamic definitions and updated clinical classification of pulmonary hypertension. *Eur. Respir. J.* **2019**, *53*. [CrossRef] [PubMed]
3. Galie, N.; Humbert, M.; Vachiery, J.L.; Gibbs, S.; Lang, I.; Torbicki, A.; Simonneau, G.; Peacock, A.; Vonk Noordegraaf, A.; Beghetti, M.; et al. 2015 ESC/ERS guidelines for the diagnosis and treatment of pulmonary hypertension. *Rev. Esp. Cardiol.* **2016**, *69*, 177. [CrossRef] [PubMed]
4. Thomson, J.R.; Machado, R.D.; Pauciulo, M.W.; Morgan, N.V.; Humbert, M.; Elliott, G.C.; Ward, K.; Yacoub, M.; Mikhail, G.; Rogers, P.; et al. Sporadic primary pulmonary hypertension is associated with germline mutations of the gene encoding BMPR-II, a receptor member of the TGF-beta family. *J. Med. Genet.* **2000**, *37*, 741–745. [CrossRef]
5. Navas, P.; Tenorio, J.; Quezada, C.A.; Barrios, E.; Gordo, G.; Arias, P.; Lopez Meseguer, M.; Santos-Lozano, A.; Palomino Doza, J.; Lapunzina, P.; et al. Molecular analysis of BMPR2, TBX4, and KCNK3 and genotype-phenotype correlations in Spanish patients and families with idiopathic and hereditary pulmonary arterial hypertension. *Rev. Esp. Cardiol.* **2016**, *69*, 1011–1019. [CrossRef]
6. Machado, R.D.; Eickelberg, O.; Elliott, C.G.; Geraci, M.W.; Hanaoka, M.; Loyd, J.E.; Newman, J.H.; Phillips, J.A., 3rd; Soubrier, F.; Trembath, R.C.; et al. Genetics and genomics of pulmonary arterial hypertension. *J. Am. Coll. Cardiol.* **2009**, *54* (Suppl. S1), S32–S42. [CrossRef]
7. Antigny, F.; Hautefort, A.; Meloche, J.; Belacel-Ouari, M.; Manoury, B.; Rucker-Martin, C.; Pechoux, C.; Potus, F.; Nadeau, V.; Tremblay, E.; et al. Potassium channel subfamily K member 3 (KCNK3) contributes to the development of pulmonary arterial hypertension. *Circulation* **2016**, *133*, 1371–1385. [CrossRef]
8. Austin, E.D.; Ma, L.; LeDuc, C.; Berman Rosenzweig, E.; Borczuk, A.; Phillips, J.A., 3rd; Palomero, T.; Sumazin, P.; Kim, H.R.; Talati, M.H.; et al. Whole exome sequencing to identify a novel gene (caveolin-1) associated with human pulmonary arterial hypertension. *Circ. Cardiovasc. Genet.* **2012**, *5*, 336–343. [CrossRef]
9. Chida, A.; Shintani, M.; Matsushita, Y.; Sato, H.; Eitoku, T.; Nakayama, T.; Furutani, Y.; Hayama, E.; Kawamura, Y.; Inai, K.; et al. Mutations of NOTCH3 in childhood pulmonary arterial hypertension. *Mol. Genet. Genom. Med.* **2014**, *2*, 229–239. [CrossRef]
10. De Jesus Perez, V.A.; Yuan, K.; Lyuksyutova, M.A.; Dewey, F.; Orcholski, M.E.; Shuffle, E.M.; Mathur, M.; Yancy, L., Jr.; Rojas, V.; Li, C.G.; et al. Whole-exome sequencing reveals TopBP1 as a novel gene in idiopathic pulmonary arterial hypertension. *Am. J. Respir. Crit. Care Med.* **2014**, *189*, 1260–1272. [CrossRef]
11. Eyries, M.; Montani, D.; Girerd, B.; Perret, C.; Leroy, A.; Lonjou, C.; Chelghoum, N.; Coulet, F.; Bonnet, D.; Dorfmuller, P.; et al. EIF2AK4 mutations cause pulmonary veno-occlusive disease, a recessive form of pulmonary hypertension. *Nat. Genet.* **2014**, *46*, 65–69. [CrossRef] [PubMed]
12. Kerstjens-Frederikse, W.S.; Bongers, E.M.; Roofthooft, M.T.; Leter, E.M.; Douwes, J.M.; Van Dijk, A.; Vonk-Noordegraaf, A.; Dijk-Bos, K.K.; Hoefsloot, L.H.; Hoendermis, E.S.; et al. TBX4 mutations (small patella syndrome) are associated with childhood-onset pulmonary arterial hypertension. *J. Med. Genet.* **2013**, *50*, 500–506. [CrossRef] [PubMed]
13. Morrell, N.W.; Aldred, M.A.; Chung, W.K.; Elliott, C.G.; Nichols, W.C.; Soubrier, F.; Trembath, R.C.; Loyd, J.E. Genetics and genomics of pulmonary arterial hypertension. *Eur. Respir. J.* **2018**, 1801899. [CrossRef] [PubMed]
14. Zhu, N.; Pauciulo, M.W.; Welch, C.L.; Lutz, K.A.; Coleman, A.W.; Gonzaga-Jauregui, C.; Wang, J.; Grimes, J.M.; Martin, L.J.; He, H.; et al. Novel risk genes and mechanisms implicated by exome sequencing of 2572 individuals with pulmonary arterial hypertension. *Genome Med.* **2019**, *11*, 69. [CrossRef] [PubMed]

15. Tenorio, J.; Navas, P.; Barrios, E.; Fernandez, L.; Nevado, J.; Quezada, C.A.; Lopez-Meseguer, M.; Arias, P.; Mena, R.; Lobo, J.L.; et al. A founder EIF2AK4 mutation causes an aggressive form of pulmonary arterial hypertension in Iberian Gypsies. *Clin. Genet.* **2015**, *88*, 579–583. [CrossRef] [PubMed]

16. Best, D.H.; Sumner, K.L.; Austin, E.D.; Chung, W.K.; Brown, L.M.; Borczuk, A.C.; Rosenzweig, E.B.; Bayrak-Toydemir, P.; Mao, R.; Cahill, B.C.; et al. EIF2AK4 mutations in pulmonary capillary hemangiomatosis. *Chest* **2014**, *145*, 231–236. [CrossRef]

17. Pousada, G.; Baloira, A.; Valverde, D. Complex inheritance in pulmonary arterial hypertension patients with several mutations. *Sci. Rep.* **2016**, *6*, 33570. [CrossRef]

18. Eichstaedt, C.A.; Song, J.; Benjamin, N.; Harutyunova, S.; Fischer, C.; Grunig, E.; Hinderhofer, K. EIF2AK4 mutation as "second hit" in hereditary pulmonary arterial hypertension. *Respir. Res.* **2016**, *17*, 141. [CrossRef]

19. Wang, G.; Knight, L.; Ji, R.; Lawrence, P.; Kanaan, U.; Li, L.; Das, A.; Cui, B.; Zou, W.; Penny, D.J.; et al. Early onset severe pulmonary arterial hypertension with 'two-hit' digenic mutations in both BMPR2 and KCNA5 genes. *Int. J. Cardiolo.* **2014**, *177*, e167. [CrossRef]

20. Abou Hassan, O.K.; Haidar, W.; Nemer, G.; Skouri, H.; Haddad, F.; BouAkl, I. Clinical and genetic characteristics of pulmonary arterial hypertension in Lebanon. *BMC Med. Genet.* **2018**, *19*, 89. [CrossRef]

21. Lazaro Salvador, M.; Quezada Loaiza, C.A.; Rodriguez Padial, L.; Barbera, J.A.; Lopez-Meseguer, M.; Lopez-Reyes, R.; Sala Llinas, E.; Alcolea, S.; Blanco, I.; Escribano Subias, P.; et al. Portopulmonary hypertension: Prognosis and management in the current treatment era. Results from the REHAP Registry. *Intern. Med. J.* **2020**. [CrossRef] [PubMed]

22. Alonso-Gonzalez, R.; Lopez-Guarch, C.J.; Subirana-Domenech, M.T.; Ruiz, J.M.; Gonzalez, I.O.; Cubero, J.S.; del Cerro, M.J.; Salvador, M.L.; Dos Subira, L.; Gallego, P.; et al. Pulmonary hypertension and congenital heart disease: An insight from the REHAP national registry. *Int. J. Cardiol.* **2015**, *184*, 717–723. [CrossRef] [PubMed]

23. Richards, S.; Aziz, N.; Bale, S.; Bick, D.; Das, S.; Gastier-Foster, J.; Grody, W.W.; Hegde, M.; Lyon, E.; Spector, E.; et al. Standards and guidelines for the interpretation of sequence variants: A joint consensus recommendation of the American college of medical genetics and genomics and the association for molecular pathology. *Genet. Med. Off. J. Am. Coll. Med. Genet.* **2015**, *17*, 405–424. [CrossRef]

24. Hernandez-Gonzalez, I.; Tenorio, J.; Palomino-Doza, J.; Martinez Menaca, A.; Morales Ruiz, R.; Lago-Docampo, M.; Valverde Gomez, M.; Gomez Roman, J.; Enguita Valls, A.B.; Perez-Olivares, C.; et al. Clinical heterogeneity of pulmonary arterial hypertension associated with variants in TBX4. *PLoS ONE* **2020**, *15*, e0232216. [CrossRef]

25. Mercer, T.R.; Clark, M.B.; Andersen, S.B.; Brunck, M.E.; Haerty, W.; Crawford, J.; Taft, R.J.; Nielsen, L.K.; Dinger, M.E.; Mattick, J.S. Genome-wide discovery of human splicing branchpoints. *Genome Res.* **2015**, *25*, 290–303. [CrossRef]

26. Song, J.; Eichstaedt, C.A.; Viales, R.R.; Benjamin, N.; Harutyunova, S.; Fischer, C.; Grunig, E.; Hinderhofer, K. Identification of genetic defects in pulmonary arterial hypertension by a new gene panel diagnostic tool. *Clin. Sci.* **2016**, *130*, 2043–2052. [CrossRef]

27. Barozzi, C.; Galletti, M.; Tomasi, L.; De Fanti, S.; Palazzini, M.; Manes, A.; Sazzini, M.; Galie, N. A combined targeted and whole exome sequencing approach identified novel candidate genes involved in heritable pulmonary arterial hypertension. *Sci. Rep.* **2019**, *9*, 753. [CrossRef]

28. Zhang, H.S.; Liu, Q.; Piao, C.M.; Zhu, Y.; Li, Q.Q.; Du, J.; Gu, H. Genotypes and phenotypes of Chinese pediatric patients with idiopathic and heritable pulmonary arterial hypertension—A single-center study. *Can. J. Cardiol.* **2019**, *35*, 1851–1856. [CrossRef]

29. Hansmann, G.; Koestenberger, M.; Alastalo, T.P.; Apitz, C.; Austin, E.D.; Bonnet, D.; Budts, W.; D'Alto, M.; Gatzoulis, M.A.; Hasan, B.S.; et al. 2019 updated consensus statement on the diagnosis and treatment of pediatric pulmonary hypertension: The European pediatric pulmonary vascular disease network (EPPVDN), endorsed by AEPC, ESPR and ISHLT. *J. Heart Lung Transplant. Off. Publ. Int. Soc. Heart Transplant.* **2019**, *38*, 879–901. [CrossRef]

30. Li, X.; Zhang, X.; Leathers, R.; Makino, A.; Huang, C.; Parsa, P.; Macias, J.; Yuan, J.X.; Jamieson, S.W.; Thistlethwaite, P.A. Notch3 signaling promotes the development of pulmonary arterial hypertension. *Nat. Med.* **2009**, *15*, 1289–1297. [CrossRef]

31. Bohnen, M.S.; Ma, L.; Zhu, N.; Qi, H.; McClenaghan, C.; Gonzaga-Jauregui, C.; Dewey, F.E.; Overton, J.D.; Reid, J.G.; Shuldiner, A.R.; et al. Loss-of-function ABCC8 mutations in pulmonary arterial hypertension. *Circ. Genom. Precis. Med.* **2018**, *11*, e002087. [CrossRef] [PubMed]

32. Lago-Docampo, M.; Tenorio, J.; Hernandez-Gonzalez, I.; Perez-Olivares, C.; Escribano-Subias, P.; Pousada, G.; Baloira, A.; Arenas, M.; Lapunzina, P.; Valverde, D. Characterization of rare ABCC8 variants identified in Spanish pulmonary arterial hypertension patients. *Sci. Rep.* **2020**, *10*, 15135. [CrossRef] [PubMed]

33. Roberts, K.E.; McElroy, J.J.; Wong, W.P.; Yen, E.; Widlitz, A.; Barst, R.J.; Knowles, J.A.; Morse, J.H. BMPR2 mutations in pulmonary arterial hypertension with congenital heart disease. *Eur. Respir. J.* **2004**, *24*, 371–374. [CrossRef] [PubMed]

34. Liu, D.; Liu, Q.Q.; Guan, L.H.; Jiang, X.; Zhou, D.X.; Beghetti, M.; Qu, J.M.; Jing, Z.C. BMPR2 mutation is a potential predisposing genetic risk factor for congenital heart disease associated pulmonary vascular disease. *Int. J. Cardiol.* **2016**, *211*, 132–136. [CrossRef] [PubMed]

35. Stout, K.K.; Daniels, C.J.; Aboulhosn, J.A.; Bozkurt, B.; Broberg, C.S.; Colman, J.M.; Crumb, S.R.; Dearani, J.A.; Fuller, S.; Gurvitz, M.; et al. 2018 AHA/ACC guideline for the management of adults with congenital heart disease: Executive summary: A Report of the American college of cardiology/American heart association task force on clinical practice guidelines. *Circulation* **2019**, *139*, e637–e697. [CrossRef]

36. Galambos, C.; Mullen, M.P.; Shieh, J.T.; Schwerk, N.; Kielt, M.J.; Ullmann, N.; Boldrini, R.; Stucin-Gantar, I.; Haass, C.; Bansal, M.; et al. Phenotype characterisation of TBX4 mutation and deletion carriers with neonatal and paediatric pulmonary hypertension. *Eur. Respir. J.* **2019**, *54*. [CrossRef] [PubMed]

37. Canter, J.A.; Summar, M.L.; Smith, H.B.; Rice, G.D.; Hall, L.D.; Ritchie, M.D.; Motsinger, A.A.; Christian, K.G.; Drinkwater, D.C., Jr.; Scholl, F.G.; et al. Genetic variation in the mitochondrial enzyme carbamyl-phosphate synthetase I predisposes children to increased pulmonary artery pressure following surgical repair of congenital heart defects: A validated genetic association study. *Mitochondrion* **2007**, *7*, 204–210. [CrossRef]

38. Kaluarachchi, D.C.; Smith, C.J.; Klein, J.M.; Murray, J.C.; Dagle, J.M.; Ryckman, K.K. Polymorphisms in urea cycle enzyme genes are associated with persistent pulmonary hypertension of the newborn. *Pediatr. Res.* **2018**, *83*, 142–147. [CrossRef]

39. McGoon, M.D.; Benza, R.L.; Escribano-Subias, P.; Jiang, X.; Miller, D.P.; Peacock, A.J.; Pepke-Zaba, J.; Pulido, T.; Rich, S.; Rosenkranz, S.; et al. Pulmonary arterial hypertension: Epidemiology and registries. *Turk. Kardiyol. Dern. Ars.* **2014**, *42* (Suppl. S1), 67–77. [CrossRef]

40. Benza, R.L.; Miller, D.P.; Barst, R.J.; Badesch, D.B.; Frost, A.E.; McGoon, M.D. An evaluation of long-term survival from time of diagnosis in pulmonary arterial hypertension from the REVEAL Registry. *Chest* **2012**, *142*, 448–456. [CrossRef]

41. Rosenkranz, S.; Dumitrescu, D. Pulmonary hypertension-back to the future. *Rev. Esp. Cardiol.* **2017**, *70*, 901–904. [CrossRef] [PubMed]

Article

Genetic Evaluation in a Cohort of 126 Dutch Pulmonary Arterial Hypertension Patients

Lieke M. van den Heuvel [1,2,3,†], Samara M. A. Jansen [4,†], Suzanne I. M. Alsters [1], Marco C. Post [5,6], Jasper J. van der Smagt [3], Frances S. Handoko-De Man [4], J. Peter van Tintelen [1,3], Hans Gille [1], Imke Christiaans [7], Anton Vonk Noordegraaf [4], HarmJan Bogaard [4] and Arjan C. Houweling [1,*]

[1] Department of Clinical Genetics, Amsterdam UMC (location VUmc), 1081HV Amsterdam, The Netherlands; l.m.vandenheuvel@amsterdamumc.nl (L.M.v.d.H.); s.alsters@amsterdamumc.nl (S.I.M.A.); J.P.vanTintelen-3@umcutrecht.nl (J.P.v.T.); jjp.gille@amsterdamumc.nl (H.G.)

[2] Netherlands Heart Institute, 3511EP Utrecht, The Netherlands

[3] Department of Genetics, University Medical Centre Utrecht, Utrecht University, 3584CX Utrecht, The Netherlands; j.j.vandersmagt@umcutrecht.nl

[4] Department of Lung Disease, Amsterdam UMC (location VUmc), 1081HV Amsterdam, The Netherlands; s.jansen1@amsterdamumc.nl (S.M.A.J.); fs.deman@amsterdamumc.nl (F.S.H.-D.M.); a.vonk@amsterdamumc.nl (A.V.N.); hj.bogaard@amsterdamumc.nl (H.B.)

[5] Department of Cardiology, St. Antonius hospital, 3435CM Nieuwegein, The Netherlands; m.post@antoniusziekenhuis.nl

[6] Department of Cardiology, University Medical Centre Utrecht, Utrecht University, 3584CX Utrecht, The Netherlands

[7] Department of Clinical Genetics, University Medical Centre Groningen, 9713GZ Groningen, The Netherlands; i.christiaans@umcg.nl

* Correspondence: a.houweling@amsterdamumc.nl; Tel.: +31-20-444-0150

† These authors contributed equally to this work.

Received: 9 September 2020; Accepted: 10 October 2020; Published: 13 October 2020

Abstract: Pulmonary arterial hypertension (PAH) is a severe, life-threatening disease, and in some cases is caused by genetic defects. This study sought to assess the diagnostic yield of genetic testing in a Dutch cohort of 126 PAH patients. Historically, genetic testing in the Netherlands consisted of the analysis of BMPR2 and SMAD9. These genes were analyzed in 70 of the 126 patients. A (likely) pathogenic (LP/P) variant was detected in 22 (31%) of them. After the identification of additional PAH associated genes, a next generation sequencing (NGS) panel consisting of 19 genes was developed in 2018. Additional genetic testing was offered to the 48 *BMPR2* and *SMAD9* negative patients, out of which 28 opted for NGS analysis. In addition, this gene panel was analyzed in 56 newly identified idiopathic (IPAH) or pulmonary veno occlusive disease (PVOD) patients. In these 84 patients, NGS panel testing revealed LP/P variants in BMPR2 (N = 4), GDF2 (N = 2), EIF2AK4 (N = 1), and TBX4 (N = 3). Furthermore, 134 relatives of 32 probands with a LP/P variant were tested, yielding 41 carriers. NGS panel screening offered to IPAH/PVOD patients led to the identification of LP/P variants in GDF2, EIF2AK4, and TBX4 in six additional patients. The identification of LP/P variants in patients allows for screening of at-risk relatives, enabling the early identification of PAH.

Keywords: pulmonary arterial hypertension; genetic analysis; NGS gene panel; BMPR2; TBX4; GDF2; EIF2AK4

1. Introduction

Pulmonary arterial hypertension (PAH) is a rare and life-threatening disease with an estimated prevalence of 5.9–60 cases per a million adults [1–3]. Due to variable and non-specific symptoms, patients are often diagnosed well after the development of right heart failure, indicating a considerable delay in diagnosis.

PAH can be classified into different subgroups, including idiopathic PAH (IPAH), heritable PAH (HPAH), pulmonary veno-occlusive disease (PVOD), drug and toxin induced PAH, and PAH associated with other conditions, including connective tissue disease, human immunodeficiency virus (HIV) infection, portal hypertension, or congenital heart disease [4]. IPAH and PVOD have similar clinical and hemodynamic characteristics but different pathophysiological and histological characteristics and progression rates [5]. IPAH is diagnosed when any potential cause or risk factor has been excluded [4]. HPAH can be diagnosed after the identification of a (likely) pathogenic (LP/P) genetic variant in a PAH associated gene [5]. HPAH generally follows an autosomal dominant inheritance pattern with incomplete penetrance [6]. Both IPAH and HPAH can occur sporadically or familial, when two or more members of a family are affected with PAH.

Pathogenic variants in the *BMPR2* gene (MIM# 131195) were identified as the main genetic cause of HPAH, explaining 75–90% of familial cases and 10–20% of sporadic cases [7–10]. The *BMPR2* gene belongs to the transforming growth factor β (TGF-β) superfamily and is involved in the regulation of cell growth and apoptosis. Evans et al. showed that compared to *BMPR2*-negative PAH patients, carriers of a pathogenic variant in *BMPR2* were, on average, diagnosed at a younger age and had a more severe progression of the disease [7].

Until 2018, DNA diagnostics in IPAH patients in the Netherlands consisted of sequencing *BMPR2* and *SMAD9* (OMIM # 615342), encoding an intracellular signal tran Material S2ucer of the TGF-B pathway. The additional use of next-generation sequencing (NGS) techniques has resulted in the identification of several additional genes potentially associated with PAH [9]. However, these recently identified gene–disease relationships remain to be established beyond case-control studies and with more certainty over time. In a minority of IPAH cases, (potentially) associated variants have been reported in several genes including *KCNK3* (encoding a PH sensitive potassium channel), *CAV1* (encoding an integral membrane protein), *TBX4* (playing a critical role in the development of respiratory system), *GDF2* (also called BMP9 and a member of the highly conserved transforming growth factor-β superfamily), *AQP1* (encoding aquaporin-1), *SOX17* (encoding a transcription factor involved in Wnt/β-catenin and Notch signaling during development), and *BMPR1B* (encoding a type I BMP receptor) [5,9,11]. *EIF2AK4* variants have been reported as an autosomal recessive cause of PVOD whereas variants in the *BMPR2* gene have also been reported to cause PVOD [12–14].

The identification of a pathogenic variant in a patient diagnosed with IPAH/PVOD allows for the early detection of disease in genotype positive relatives. This enables timely treatment, which is important as a recent study indicated that early treatment of PAH delays disease progression and improves transplant-free survival [15]. Furthermore, predictive DNA testing enables reproductive choices, including the possibility of preimplantation genetic diagnosis [8].

Previous research has shown that genetic testing may reveal a genetic predisposition to the disease also in sporadic cases. In two studies, disease-causing variants were identified in 13–17% of sporadic IPAH/PVOD cases [8,16]. This can likely be explained by the relatively low penetrance in HPAH and the de novo occurrence of pathogenic variants [17–19]. While newly diagnosed IPAH patients can benefit from genetic counselling and DNA testing, only a limited number of patients undergo genetic testing and this is generally limited to *BMPR2/SMAD9*. The recent identification of additional genes involved in the disease, in combination with increasing evidence on health benefits of early detection of the disease in unaffected relatives, prompted us to offer all IPAH/PVOD patients genetic counselling and (extended) genetic testing using an NGS panel of 19 PAH-associated genes if necessary. Here, we describe the diagnostic yield of genetic testing in a Dutch cohort, in terms of identified (likely) pathogenic variants and the number of identified genotype positive relatives.

2. Materials and Methods

2.1. Subjects

Genetic testing has been offered to IPAH/PVOD patients referred to the PAH center of expertise in the Amsterdam UMC since 2002. Prior to the introduction of NGS technologies, genetic testing was not routinely offered to all patients but often only in familial cases. Patients with IPAH or PVOD were included in this study from January 2018 until April 2020. Patients in whom prior analysis of BMRP2 and SMAD9 did not result in the detection of a LP/P variant and patients who were not previously tested were offered the option of (extended) genetic testing using NGS via a letter of their treating physician. Interested patients received genetic counselling and could opt for genetic testing with an NGS panel consisting of 19 genes. All included patients were unrelated. The clinical geneticist or genetic counsellor obtained informed consent for genetic testing from all patients. All patients provided consent for the use of their data for research. The study was conducted in accordance with the Declaration of Helsinki. The Medical Ethical Committee of the Amsterdam UMC (location VUmc) assessed the study protocol and confirmed that the study was exempt from ethics review according to the Dutch Medical Research Involving Human Subjects Act (2017.541).

All patients in whom a LP/P variant as a cause of their PAH was detected were encouraged to inform their first-degree relatives about the option of predictive DNA testing. For assistance, we provided personalized family letters for patients' relatives with information about the disease and access to genetic counselling and—if desired—predictive DNA testing. Patients in whom a causal BMPR2 mutation had been detected via genetic testing (i.e., HPAH patients) prior to January 2018 were informed about current preventive and treatment options for themselves and their at-risk relatives by letter, including the information on preimplantation genetic diagnosis (PGD). They were also asked to inform their relatives, supported by an updated family letter if desired. Relatives could opt for genetic testing when interested.

2.2. Data Collection

Sociodemographic data (age at diagnosis and sex), clinical data (NYHA functional class, mean pulmonary artery pressure (mPAP), mean right atrial pressure (mRAP), pulmonary vascular resistance (PVR) and right ventricle end-diastolic volume index (RVEDVI), right ventricle end-systolic volume index (RVESVI), and right ventricular ejection fraction (RVEF) at diagnosis), as well as family history were collected. DNA diagnostics was performed at the DNA laboratory of Amsterdam UMC, location VUmc. Up until 2017, DNA testing of BMPR2 and SMAD9 were performed using Sanger sequencing. From 2018 onwards, a WES-based virtual panel was analyzed, which included 19 PAH-associated genes (*ABCA3, ACVRL1, BMPR1B, BMPR2, CAV1, EIF2AK4, ENG, FOXF1, GDF2, KCNA5, KCNK3, NOTCH1, NOTCH3, RASA1, SMAD1, SMAD4, SMAD9, TBX4*, and *TOPBP1*). AQP1 and SOX17 were not included in this NGS panel. Using the NGS test point, mutations and small insertions and deletions can be detected. Additionally, multiplex ligation-dependent probe amplification (MLPA) was used to detect large deletions or duplications of BMPR2, because our WES-based panel test does not allow the detection of exon deletions and duplications.

All variants detected were classified using the ACGS/VKGL guidelines [20]. For variant classification, a 5-class variant classification system was used: class 1 (benign), class 2 (likely benign), class 3 (variant of uncertain significance), class 4 (likely pathogenic), and class 5 (pathogenic) [21]. The classification of variants was based on the occurrence of the variant in control populations (gnomAD database), in silico predictions of the impact of an amino acid change on the function of the protein (PolyPhen2, SIFT, AlignGVGD), and in silico prediction of the potential impact of the nucleotide change on splicing. Variants causing frameshifts or premature stop codons were considered likely pathogenic (class 4) or pathogenic (class 5). Variants occurring in a control population at a frequency > 1% were considered polymorphisms (class 1). Detailed information on the analysis methods is given in Supplementary Materials S1.

2.3. Statistical Analysis

Statistical analysis was performed with SPSS (Version 25.0) and R Studio (Version 4.0.2, 2020-06-22). Data were visualized using the R ggplot2 package. Data were described as mean and SD, or median and IQR, as appropriate. Chi-square tests, log-rank tests, t-tests, and Wilcoxon signed rank tests were used to assess the differences between clinical observations in patients with and without a (likely) pathogenic variant, and between patients with a positive and negative family history, where appropriate. A *p*-value of <0.05 was considered statistically significant.

3. Results

3.1. Genetic Analyses

Figure 1 shows a flow-chart of the number of patients in whom genetic testing was performed. In total, 126 patients were included in our study. Until 2018, genetic testing at the Amsterdam UMC consisted of analysis of BMPR2 and SMAD9. These genes were analyzed in 70 of the 126 patients. In 22 of these 70 patients (31%), a LP/P variant of BMPR2 was identified. Additional genetic testing (NGS panel) was offered to the 48 BMPR2 and SMAD9 negative patients, including 2 patients with a VUS in the BMPR2 gene. Of them, 28 opted for NGS analysis. Fifty-six patients diagnosed with PAH after 2018 were directly tested with this NGS panel. In total, 84 patients were genetically tested using NGS. The NGS panel yielded 10 additional patients with a LP/P variant, 3 LP/P variants in the PAH patients previously tested negative for LP/P variants in BMPR2 and SMAD9 and 7 LP/P variants in the PAH group diagnosed after 2018. NGS panel testing revealed LP/P variants in BMPR2 ($N = 4$), GDF2 ($N = 2$), EIF2AK4 ($N = 1$), and TBX4 ($N = 3$). In 11 out of 84 patients (9%), NGS panel revealed a VUS (class 3 variants). In total, a LP/P variant was identified in 32 out of 126 patients (25%).

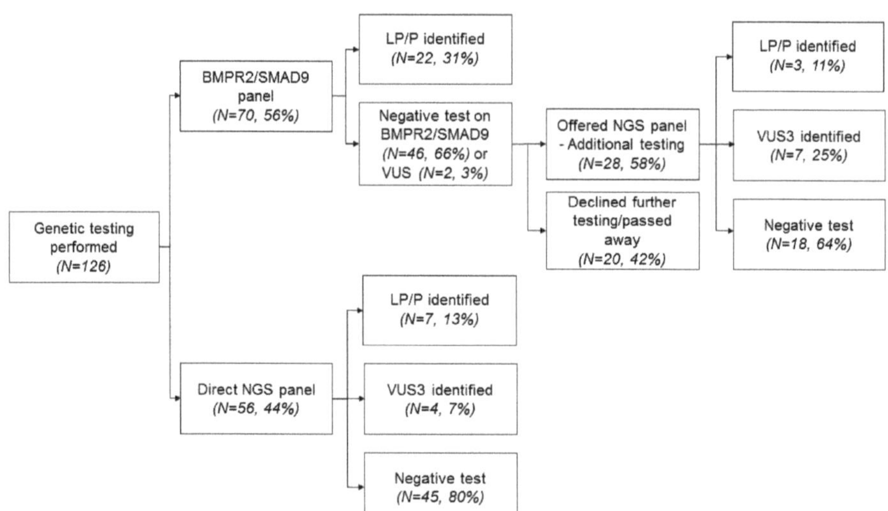

Figure 1. Flow-chart of genetic analyses in idiopathic/pulmonary veno occlusive disease (IPAH/PVOD) patients.

Table 1 shows the characteristics of the study population. The mean age at diagnosis was 49 years ($SD = 16$) in all patients, with a mean age of 52 years in patients without and 37 years in patients with a LP/P variant ($p = 0.001$). Seventy-one percent of patients were female. The majority of patients ($N = 114$, 90%) were diagnosed with IPAH, whereas only 12 patients were diagnosed with PVOD. Seventeen patients in our total cohort had a positive family history for PAH. Of the patients with a

LP/P variant (n = 32), nine had a positive family history. In eight patients with a positive family history no LP/P variant could be detected. A negative family history was reported in 13 patients (41%) with a LP/P variant compared to 55 patients (59%) without a LP/P variant. During follow-up (median 3 years, range 0–7), seventeen patients died, and four patients received a lung transplant.

Table 1. Sociodemographic and clinical characteristics of patients eligible for genetic testing at diagnosis.

Characteristic	All Patients N = 126 N (%)	No LP/P Variant N = 94 N (%) [a]	LP/P Variant N = 32 N (%) [a]	p Value
Sex				
Female	89 (71)	66 (70)	23 (72)	0.922
Male	37 (29)	28 (30)	9 (28)	
Age at diagnosis, years	49 ± 16	52 ± 17	37 ± 13	0.001
Clinical diagnosis [b]				
IPAH	114 (90)	84 (89)	30 (94)	0.031
PVOD	12 (10)	10 (11)	2 (6)	
NYHA functional class				
NYHA I-II	36 (29)	25 (29)	11 (34)	0.587
NYHA III-IV	78 (62)	60 (71)	18 (56)	
BMI, kg/m^2	26.9 ± 5.2	27.0 ± 5.3	26.6 ± 5.0	0.757
Hemodynamics				
mPAP, mmHg	52 ± 17	50 ± 17	59 ± 14	0.008
PVR, WU	9.9 (5.7–12.7)	8.8 (4.8–12.2)	11.9 (9.2–15.2)	0.022
mRAP, mmHg	9 (6–11)	9 (6–11)	8 (6–11)	0.214
PCWP, mmHg	10 ± 3	10 ± 3	9 ± 3	0.039
CI, L/min/m^2	2.5 ± 0.8	2.6 ± 0.8	2.2 ± 0.6	0.061
RVEDVI, mL/m^2	84 ± 27	82 ± 27	90 ± 27	0.243
RVESVI, mL/m^2	53 ± 26	50 ± 25	64 ± 28	0.038
RV EF, %	38 ± 13	41 ± 12	30 ± 12	<0.001
Family history				
No	68 (54)	55 (59)	13 (41)	0.025
Yes	17 (13)	8 (9)	9 (28)	
Unclear	31 (25)	22 (23)	9 (28)	
Unknown	10 (8)	9 (10)	1 (3)	
Death	17 (14)	11 (12)	6 (20)	0.324
Lung transplant	4 (3)	4 (4)	0 (0)	NA [c]
Median FU, years	3 (0–7)	2 (0–5)	6 (2–11)	0.001

Data are given as mean (SD), median (range) or number (percentage). IPAH = Idiopathic pulmonary arterial hypertension; PVOD = pulmonary veno-occlusive disease; NYHA = New York Association functional class; BMI = body mass index; mPAP = mean pulmonary arterial pressure; PVR = pulmonary vascular resistance; mRAP = mean right atrial pressure; RVEF = right ventricular ejection fraction; RDEVI = right ventricular end-diastolic volume index; RVESVI = right ventricular end-systolic volume index. [a] Not all numbers add up to the total number of patients due to missing values. [b] Concerns p value of chi-square test performed on difference 'idiopathic PAH' versus 'PVOD', due to >20% of cells having an expected count less than 5. [c] Significance testing not possible due to small numbers.

The variants in the group of patients with a LP/P (N = 32) were identified in BMPR2 (N = 26), TBX4 (N = 3), GDF2 (N = 2), and EIF2AK4 (N = 1), see Table 2. One identical pathogenic variant in BMPR2 was identified in three unrelated probands (c.1471C > T, p.(Arg491Trp)), indicating a potential founder effect of this variant. VUS were found in FOXF1 (N = 3), NOTCH3 (N = 4), BMPR2 (N = 3), and TBX4 (N = 2), of whom two VUS were found in one patient (see Supplementary Table S1).

Table 2. Overview of the detected class 4 and 5 variants in 32 unrelated pulmonary arterial hypertension (PAH) patients.

Gene	Nucleotide Change	Amino Acid Change	Class	Novel	Reference	Remark Pathogenicity
TBX4 *	c.40_49del	p. (Phe14Argfs*28)	Class 5 [a]	No	[9]	NA
TBX4 *	c.916G > T	p.(Glu306 *)	Class 4	Yes	NA	Premature stopcodon; not present in gnomAD and ClinVar
TBX4 *	c.1112del	p.(Pro371Leufs*8)	Class 5 [a]	No	[10]	NA
BMPR2	c.(?_-1)_(*1_?)del (entire gene)	p.0	Class 5	No	[22]	NA
BMPR2	c.(?_-1)_(76 + 1_77-1)del (exon 1)	p.? [e]	Class 5	No	[13] [a]	
BMPR2	c.76 + 2T > G	p.? [e]	Class 5	No	[9] [a]	NA
BMPR2	c.246A > G	p.(Glu48_Gly83del) (splice defect)	Class 4	Yes	NA	Defective splice donor site exon 2; use of a cryptic splice donor site in exon 2 (mRNA analysis performed in our lab)
BMPR2	c.348C > G	p.Cys116Trp	Class 4	Yes	NA	Variant not present in controls (gnomAD). Highly conserved region; AlignGVGD: class C65; SIFT: deleterious; PolyPhen2: probably damaging, score 1.000. Variant not present in ClinVar
BMPR2	c.350G > C	p.(Cys117Ser)	Class 5	No	[13]	NA
BMPR2	c.399del	p.(Pro134Leufs*18)	Class 5	No	[23] [a]	NA
BMPR2	c.619dup	p.(Glu207Glyfs* 13)	Class 5	No	[23] [a]	NA
BMPR2	c.690del	p.(Val231Cysfs*21)	Class 5	Yes	[23] [a]	NA
BMPR2	c.852_852 + 1insA	p.(Gly285Argfs*13)	Class 4	No	[9]	NA
BMPR2	c.994C > T	p.(Arg332 *)	Class 5	No	[24]	NA
BMPR2	c.1133G > T	p.(Gly378Val)	Class 5 [c]	No	[9]	NA
BMPR2	c.1217T > G	p.(Met406Arg)	Class 4	No	[9]	NA
BMPR2	c.1454A > G	p.(Asp485Gly)	Class 5	No	[23] [a]	NA
BMPR2	c.1459G > T	p.(Asp487Tyr)	Class 4 [c]	Yes	NA	Not present in controls (gnomAD); Highly conserved region; AlignGVGD: class C65; SIFT: deleterious; PolyPhen2: probably damaging, score 1.000. ClinVar: 1 entry, likely pathogenic (VCV000212812.2)

Table 2. *Cont.*

Gene	Nucleotide Change	Amino Acid Change	Class	Novel	Reference	Remark Pathogenicity
BMPR2 (N = 3)	c.1471C > T	p.(Arg491Trp)	Class 5	No	[23] [a]	NA
BMPR2	c.1487G > A	p.(Cys496Tyr)	Class 5	No	[13]	NA
BMPR2	c.1525G > T	p.(Glu509 *)	Class 5	No	[25] [a]	
BMPR2	c.1978G > T	p.(Glu660 *)	Class 5	No	[25] [a]	
BMPR2	c.2161C > T	p.(Gln721 *)	Class 5	Yes	NA	Premature stopcodon. Not present in gnomAD or ClinVar
BMPR2	c.2752C > T	p.(Gln918 *)	Class 5	No	[25] [a]	
BMPR2	c.(418 + 1_419-1)_(2866 + 1_2867-1)del (exon 4-12)	p.? [e]	Class 5	No	[23] [a]	NA
BMPR2	c.(529 + 1_530-1)_(967 + 1_968-1)del (exon 5-7)	p.? [e]	Class 5	No	[13]	NA
BMPR2	c.(967 + 1_968-1)_(1128 + 1_1129-1)dup (exon 8)	p.(Val377Ilefs*12)	Class 5	No	[23] [a]	A tandem exon duplication was confirmed by Sanger sequencing on cDNA
EIF2AK4	c.1739dup	p.(Arg581Glufs*9)	Class 5 [d]	No	[26]	NA
GDF2	c.328C > T	p.(Arg110Trp)	Class 4	No	[9] [b]	NA
GDF2	c.451C > T	p.(Arg151 *)	Class 4	No	[16]	NA

NA not applicable. Reference sequences: BMPR2 NM_001204.6; EIF2AK4 NM_001013703.3; GDF2 NM_016204.3 and TBX4 NM_018488.2. [a] These patients only tested on BMPR2 and SMAD9 were reported previously in Van der Bruggen et al. [23] and/or Girerd et al. [25]. [b] This patient was previously reported by Gräf et al. [9].[c] Both parents tested negative for this variant. [d] Homozygotic. [e] An effect on the protein level is expected, but it is not possible to give a reliable prediction of the consequences. * Was previously tested on BMPR2/SMAD9.

3.2. Genotype-Phenotype Correlation

The main clinical characteristics of the patients with and without a LP/P variant identified are presented in Table 1. A significant difference in age at diagnosis in patients with and without a LP/P variant was observed, those with a LP/P variant had a significantly younger age at diagnosis ($p = 0.001$), as shown in Figure 2. In addition, patients with a LP/P variant had significantly higher mPAP and PVR at diagnosis compared to those without LP/P variant ($p = 0.008$ and $p = 0.022$, respectively). Moreover, a significantly higher RVESVI and significantly lower RVEF was observed in patients with a LP/P variant, ($p = 0.038$ and $p < 0.001$ respectively). Furthermore, a LP/P variant was identified more often in patients with a positive family history compared to those with a negative family history ($p = 0.025$).

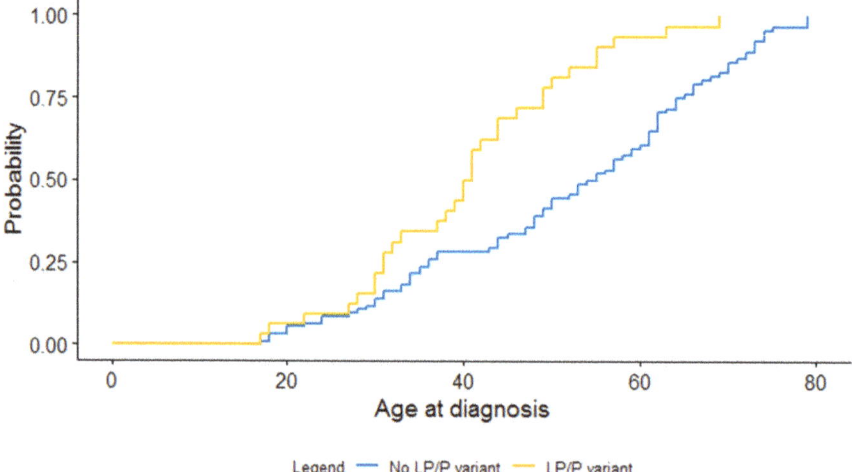

Figure 2. Kaplan–Meier curve for difference in age at diagnosis between patients with and without a LP/P variant.

3.3. Relatives

Following the initial identification of LP/P variants in 32 probands, 134 relatives (range per family: 0–42) have been tested to date. Of the relatives pursuing predictive DNA testing, 41 were shown to be a carrier, of whom 36 relatives carried a variant in BMPR2 (88%), 3 in TBX4 (7%), and 2 in GDF2 (5%). These relatives were offered annual check-ups at the out-patient clinic to detect early signs of PAH and receive treatment accordingly. In one asymptomatic carrier, mild PAH was diagnosed at the first visit and treatment was subsequently started. Further cascade genetic testing and clinical screening is currently ongoing. One relative, who had opted for predictive DNA testing and turned out to be a carrier, successfully pursued PGD.

4. Discussion

In this study, we describe the results of genetic testing and the characteristics in a Dutch cohort of 126 adult probands diagnosed with non-associated PAH or PVOD. A LP/P variant was identified in 25% of patients in this cohort. The vast majority of these patients had a LP/P variant in *BMPR2* (81%). LP/P variants were detected in *BMPR2* ($N = 26$), *GDF2* ($N = 2$), *EIF2AK4* ($N = 1$), and *TBX4* ($N = 3$). Expanding genetic testing in 28 patients previously tested negative for LP/P variants in *BMPR2* and *SMAD9* resulted in the identification of three disease-causing variants (11%). Patients with a LP/P variant had worse hemodynamics and a younger age at diagnosis.

In eight families with familial PAH, we were unable to identify a disease-causing variant with our NGS panel, pointing to the possibility of other genetic causes [27]. In these gene-elusive families, first-degree relatives of PAH patients were offered clinical screening. Furthermore, these patients and other PAH patients were asked to participate in international efforts aimed at the identification of novel PAH associated genes [9,28,29]. With these international efforts, an *AQP1* variant (c.583C > T; p.Arg195Trp) was identified in one PAH patient who tested negative on our NGS panel [9]. Novel gene–disease relations are established at a rapid pace for PAH and other genetic diseases. Especially when the identification of a genetic cause can result in health benefits in relatives, it is important to establish patient databases and to obtain informed consent by which mutation negative patients can be re-contacted for future additional genetic testing.

In our cohort, three patients had a disease-causing variant in *TBX4*. Variants in *TBX4* have previously been recognized as a cause of neonatal and paediatric pulmonary hypertension [30,31]. However, recent studies also reported pathogenic *TBX4* variants in adult-onset pulmonary hypertension [9,16]. We previously described the clinical characteristics of our *TBX4* patients indicating a female predominance, bronchial diverticulosis, distinct skeletal anomalies, and a history of asthma in all [32]. PAH associated with variants in *TBX4* is clinically highly variable [33]. In addition to *TBX4*, we identified LP/P variants in *GDF2*, resulting in loss of *BMP9* function [34]. Causal variants in *GDF2* (*BMP9*) were first described in adult-onset PAH patients by Gräf et al. [9], and subsequently reported in two studies describing *GDF2* variants in respectively 6.7% and 1.1% of sporadic PAH patients [31]. Interestingly, no LP/P variants were found in the other 15 PAH-associated genes, including *SMAD9*. Although *SMAD9* has been repeatedly shown to cause PAH when disrupted, it is considered to be a very rare cause of PAH. This is confirmed by the absence of LP/P variants in this gene in our cohort.

In this study, only *BMPR2* was screened for larger deletions or duplications using MLPA. Therefore, larger deletions or duplications in the other genes on our panel may have been missed. Larger Copy Number Variants (CNVs) including the *TBX4* gene have been reported to cause PAH [4]. As these CNVs often include additional genes located in proximity to the TBX4 gene, they often cause additional features, such as intellectual disability and congenital defects. It is therefore unlikely that large CNVs are common in the cohort reported here, as such features were not present. Small deletions/CNVs, however, may have been missed in our cohort. In a previous study in patients with hereditary thoracic aortic aneurysms small CNVs were identified as a genetic cause in 6/66 (9%) patients with a LP/P variant using the exome hidden Markov model (XHMM; an algorithm to identify CNVs in targeted NGS data) [35]. Subsequent cohort studies are required to further establish gene–disease relations with certainty and to elucidate the role of small CNVs in PAH genes.

In conclusion, genetic testing in a Dutch cohort of 126 non-associated PAH/PVOD patients revealed a LP/P variant in 32 patients (25%). BMPR2 was the main cause (88%) of the LP/P variants. NGS identified a genetic cause in an additional six patients. *TBX4* and *GDF2* variants were found in three and two patients with PAH, respectively. In addition, a homozygous variant in *EIF2AK4* was identified in one PVOD patient. A genetic cause was identified in 21% of sporadic cases, underscoring the importance of genetic testing in PAH/PVOD. The identification of LP/P variants in patients allows for screening of at-risk relatives; in this study, 41 out of 134 unaffected tested relatives (31%) were shown to be a carrier. Predictive DNA testing allows for clinical screening of at-risk relatives, supporting the early identification of PAH and the possibility of PGD.

Supplementary Materials: The following are available online at http://www.mdpi.com/2073-4425/11/10/1191/s1, Supplementary material S1: Detailed information on the genetic analysis methods; Table S1: Overview of the detected class 3 variants in unrelated PAH patients.

Author Contributions: Conceptualization, J.P.v.T., I.C., A.V.N., H.B. and A.C.H.; Methodology, H.G., H.B., and A.C.H.; Software, H.G.; Validation, H.G. and A.C.H.; Formal Analysis, L.M.v.d.H. and S.M.A.J.; Investigation, L.M.v.d.H. and S.M.A.J.; Resources, L.M.v.d.H. and S.M.A.J.; Data Curation, L.M.v.d.H. and S.M.A.J.; Writing—Original Draft Preparation, L.M.v.d.H. and S.M.A.J.; Writing—Review & Editing, S.I.M.A., M.C.P., J.J.v.d.S., J.P.v.T., H.G., I.C., A.V.N., F.S.H.-D.M., H.B. and A.C.H.; Visualization, L.M.v.d.H.and S.M.A.J.; Supervision, J.P.v.T., I.C., F.S.H.-D.M., H.B. and A.C.H.; Project Administration, J.P.v.T. and H.B.; Funding Acquisition, J.P.v.T., I.C., A.V.N., F.S.H.-D.M., H.B. and A.C.H. All authors have read and agreed to the published version of the manuscript.

Funding: This work was supported the Netherlands Cardiovascular Research Initiative: An initiative with the support of the Dutch Heart Foundation (CVON2017-4 DOLPHIN-GENESIS).

Conflicts of Interest: The authors declare no conflict of interest.

References

1. Humbert, M.; Sitbonn, O.; Chaouat, A.; Bertocchi, M.; Habib, G.; Gressin, V.; Yaici, A.; Weitzenblum, E.; Cordier, J.-F.; Chabot, F.; et al. Pulmonary arterial hypertension in France: Results from a national registry. *Am. J. Respir. Crit. Care Med.* **2006**, *173*, 1023–1030. [CrossRef] [PubMed]

2. McGoon, M.D.; Benza, R.L.; Escribano-Subias, P.; Jiang, X.; Miller, D.P.; Peacock, A.J.; Pepke-Zaba, J.; Pulido, T.; Rich, S.; Rosenkranz, S.; et al. Pulmonary arterial hypertension: Epidemiology and registries. *J. Am. Coll. Cardiol.* **2013**, *62*, D51–D59. [CrossRef] [PubMed]

3. Peacock, A.J.; Murphy, N.F.; McMurray, J.J.; Caballero, L.; Stewart, S. An epidemiological study of pulmonary arterial hypertension. *Eur. Respir. J.* **2007**, *30*, 104–109. [CrossRef] [PubMed]

4. Galie, N.; Humbert, M.; Vachiery, J.L.; Gibbs, S.; Lang, I.; Torbicki, A.; Simonneau, G.; Peacock, A.; Vonk Noordegraaf, A.; Beghetti, M.; et al. 2015 ESC/ERS Guidelines for the diagnosis and treatment of pulmonary hypertension: The Joint Task Force for the Diagnosis and Treatment of Pulmonary Hypertension of the European Society of Cardiology (ESC) and the European Respiratory Society (ERS): Endorsed by: Association for European Paediatric and Congenital Cardiology (AEPC), International Society for Heart and Lung Transplantation (ISHLT). *Eur. Heart J.* **2016**, *37*, 67–119. [PubMed]

5. Yang, H.; Zeng, Q.; Ma, Y.; Liu, B.; Chen, Q.; Li, W.; Xiong, C.; Zhou, Z. Genetic analyses in a cohort of 191 pulmonary arterial hypertension patients. *Respir. Res.* **2018**, *19*, 87. [CrossRef]

6. Ma, L.; Chung, W.K. The role of genetics in pulmonary arterial hypertension. *J. Pathol.* **2017**, *241*, 273–280. [CrossRef]

7. Evans, J.D.; Girerd, B.; Montani, D.; Wang, X.J.; Galie, N.; Austin, E.D.; Elliott, G.; Asano, K.; Grunig, E.; Yan, Y.; et al. BMPR2 mutations and survival in pulmonary arterial hypertension: An individual participant data meta-analysis. *Lancet Respir. Med.* **2016**, *4*, 129–137. [CrossRef]

8. Girerd, B.; Montani, D.; Jais, X.; Eyries, M.; Yaici, A.; Sztrymf, B.; Savale, L.; Parent, F.; Coulet, F.; Godinas, L.; et al. Genetic counselling in a national referral centre for pulmonary hypertension. *Eur. Respir. J.* **2016**, *47*, 541–552. [CrossRef]

9. Graf, S.; Haimel, M.; Bleda, M.; Hadinnapola, C.; Southgate, L.; Li, W.; Hodgson, J.; Liu, B.; Salmon, R.M.; Southwood, M.; et al. Identification of rare sequence variation underlying heritable pulmonary arterial hypertension. *Nat. Commun.* **2018**, *9*, 1416. [CrossRef]

10. Zhu, N.; Pauciulo, M.W.; Welch, C.L.; Lutz, K.A.; Coleman, A.W.; Gonzaga-Jauregui, C.; Wang, J.; Grimes, J.M.; Martin, L.J.; He, H.; et al. Novel risk genes and mechanisms implicated by exome sequencing of 2572 individuals with pulmonary arterial hypertension. *Genome Med.* **2019**, *11*, 69. [CrossRef]

11. Arora, R.; Metzger, R.J.; Papaioannou, V.E. Multiple roles and interactions of TBX4 and TBX5 in development of the respiratory system. *PLoS Genet.* **2012**, *8*, e1002866. [CrossRef] [PubMed]

12. Best, D.H.; Sumner, K.L.; Smith, B.P.; Damjanovich-Colmenares, K.; Nakayama, I.; Brown, L.M.; Ha, Y.; Paul, E.; Morris, A.; Jama, M.A.; et al. EIF2AK4 Mutations in Patients Diagnosed With Pulmonary Arterial Hypertension. *Chest* **2017**, *151*, 821–828. [CrossRef] [PubMed]

13. Machado, R.D.; Aldred, M.A.; James, V.; Harrison, R.E.; Patel, B.; Schwalbe, E.C.; Gruenig, E.; Janssen, B.; Koehler, R.; Seeger, W.; et al. Mutations of the TGF-beta type II receptor BMPR2 in pulmonary arterial hypertension. *Hum. Mutat.* **2006**, *27*, 121–132. [CrossRef] [PubMed]

14. Montani, D.; Achouh, L.; Dorfmuller, P.; Le Pavec, J.; Sztrymf, B.; Tcherakian, C.; Rabiller, A.; Haque, R.; Sitbon, O.; Jais, X.; et al. Pulmonary veno-occlusive disease: Clinical, functional, radiologic, and hemodynamic characteristics and outcome of 24 cases confirmed by histology. *Medicine* **2008**, *87*, 220–233. [CrossRef] [PubMed]

15. Lau, E.M.; Humbert, M.; Celermajer, D.S. Early detection of pulmonary arterial hypertension. *Nat. Rev. Cardiol.* **2015**, *12*, 143–155. [CrossRef]

16. Eyries, M.; Montani, D.; Nadaud, S.; Girerd, B.; Levy, M.; Bourdin, A.; Tresorier, R.; Chaouat, A.; Cottin, V.; Sanfiorenzo, C.; et al. Widening the landscape of heritable pulmonary hypertension mutations in paediatric and adult cases. *Eur. Respir. J.* **2019**, *53*, 1801371. [CrossRef]

17. Momose, Y.; Aimi, Y.; Hirayama, T.; Kataoka, M.; Ono, M.; Yoshino, H.; Satoh, T.; Gamou, S. De novo mutations in the BMPR2 gene in patients with heritable pulmonary arterial hypertension. *Ann. Hum. Genet.* **2015**, *79*, 85–91. [CrossRef]

18. Newman, J.H.; Phillips, J.A.; Loyd, J.E. Narrative review: The enigma of pulmonary arterial hypertension: New insights from genetic studies. *Ann. Intern. Med.* **2008**, *148*, 278–283. [CrossRef]

19. Newman, J.H.; Wheeler, L.; Lane, K.B.; Loyd, E.; Gaddipati, R.; Phillips, J.A.; Loyd, J.E. Mutation in the gene for bone morphogenetic protein receptor II as a cause of primary pulmonary hypertension in a large kindred. *N. Engl. J. Med.* **2001**, *345*, 319–324. [CrossRef]

20. Wallis, Y.; Payne, S.; McAnulty, C.; Bodmer, D.; Sistermans, E.; Robertson, K.; Moore, D.; Abbs, S.; Deans, Z.; Devereau, A. *Practice Guidelines for the Evaluation of Pathogenicity and the Reporting of Sequence Variants in Clinical Molecular Genetics*; Association for Clinical Genetic Science: London, UK; The Dutch Society of Clinical Genetic Laboratory Specialists: Arnhem, The Netherlands, 2013.

21. Richards, S.; Aziz, N.; Bale, S.; Bick, D.; Das, S.; Gastier-Foster, J.; Grody, W.W.; Hegde, M.; Lyon, E.; Spector, E.; et al. Standards and guidelines for the interpretation of sequence variants: A joint consensus recommendation of the American College of Medical Genetics and Genomics and the Association for Molecular Pathology. *Genet. Med.* **2015**, *17*, 405–424. [CrossRef]

22. Aldred, M.A.; Vijayakrishnan, J.; James, V.; Soubrier, F.; Gomez-Sanchez, M.A.; Martensson, G.; Galie, N.; Manes, A.; Corris, P.; Simonneau, G.; et al. BMPR2 gene rearrangements account for a significant proportion of mutations in familial and idiopathic pulmonary arterial hypertension. *Hum. Mutat.* **2006**, *27*, 212–213. [CrossRef] [PubMed]

23. Van der Bruggen, C.E.; Happe, C.M.; Dorfmuller, P.; Trip, P.; Spruijt, O.A.; Rol, N.; Hoevenaars, F.P.; Houweling, A.C.; Girerd, B.; Marcus, J.T.; et al. Bone Morphogenetic Protein Receptor Type 2 mutation in pulmonary arterial hypertension: A view on the right ventricle. *Circulation* **2016**, *133*, 1747–1760. [CrossRef]

24. Thomson, J.R.; Machado, R.D.; Pauciulo, M.W.; Morgan, N.V.; Humbert, M.; Elliot, G.C.; Ward, M.; Yacoub, M.; Mikhail, G.; Rogers, P.; et al. Sporadic primary pulmonary hypertension is associated with germline mutations of the gene encoding BMPR-II, a receptor member of the TGF-beta family. *J. Med. Genet.* **2000**, *37*, 741–745. [CrossRef] [PubMed]

25. Girerd, B.; Coulet, F.; Jais, X.; Eyries, M.; Van Der Bruggen, C.; De Man, F.; Houweling, A.C.; Dorfmuller, P.; Savale, L.; Sitbon, O.; et al. Characteristics of pulmonary arterial hypertension in affected carriers of a mutation located in the cytoplasmic tail of bone morphogenetic protein receptor type 2. *Chest* **2015**, *147*, 1385–1394. [CrossRef] [PubMed]

26. Hadinnapola, C.; Bleda, M.; Haimel, M.; Screaton, N.; Swift, A.; Dorfmuller, P.; Preston, S.D.; Southwood, M.; Hernandez-Sanchez, J.; Martin, J.; et al. Phenotypic characterization of EIF2AK4 mutation carriers in a large cohort of patients diagnosed clinically with pulmonary arterial hypertension. *Circulation* **2017**, *136*, 2022–2033. [CrossRef] [PubMed]

27. Barozzi, C.; Galletti, M.; Tomasi, L.; De Fanti, S.; Palazzini, M.; Manes, A.; Sazzini, M.; Galie, N. A Combined targeted and whole exome sequencing approach identified novel candidate genes involved in heritable pulmonary arterial hypertension. *Sci. Rep.* **2019**, *9*, 753. [CrossRef] [PubMed]

28. Bohnen, M.S.; Ma, L.; Zhu, N.; Qi, H.; McClenaghan, C.; Gonzaga-Jauregui, C.; Dewey, F.E.; Overton, J.D.; Reid, J.G.; Shuldiner, A.R.; et al. Loss-of-function ABCC8 mutations in pulmonary arterial hypertension. *Circ. Genom. Precis. Med.* **2018**, *11*. [CrossRef] [PubMed]

29. Turro, E.; Astle, W.J.; Megy, K.; Graf, S.; Greene, D.; Shamardina, O.; Allen, H.L.; Sanchis-Juan, A.; Frontini, M.; Thys, C.; et al. Whole-genome sequencing of patients with rare diseases in a national health system. *Nature* **2020**, *583*, 96–102. [CrossRef]

30. Galambos, C.; Mullen, M.P.; Shieh, J.T.; Schwerk, N.; Kielt, M.J.; Ullmann, N.; Boldrini, R.; Stucin-Gantar, I.; Haass, C.; Bansal, M.; et al. Phenotype characterisation of TBX4 mutation and deletion carriers with neonatal and paediatric pulmonary hypertension. *Eur. Respir. J.* **2019**, *54*, 1801965. [CrossRef]

31. Levy, M.; Eyries, M.; Szezepanski, I.; Ladouceur, M.; Nadaud, S.; Bonnet, D.; Soubrier, F. Genetic analyses in a cohort of children with pulmonary hypertension. *Eur. Respir. J.* **2016**, *48*, 1118–1126. [CrossRef]

32. Jansen, S.M.A.; Van den Heuvel, L.; Meijboom, L.J.; Alsters, S.I.M.; Vonk Noordegraaf, A.; Houweling, A.; Bogaard, H.J. Correspondence regarding "T-box protein 4 mutation causing pulmonary arterial hypertension and lung disease": A single-centre case series. *Eur. Respir. J.* **2020**, *55*, 1902272. [CrossRef] [PubMed]

33. Hernandez-Gonzalez, I.; Tenorio, J.; Palomino-Doza, J.; Martinez Menaca, A.; Morales Ruiz, R.; Lago-Docampo, M.; Valverde Gomez, M.; Gomez Roman, J.; Enguita Valls, A.B.; Perz-Olivares, C.; et al. Clinical heterogeneity of pulmonary arterial hypertension associated with variants in TBX4. *PLoS ONE* **2020**, *15*, e0232216. [CrossRef] [PubMed]
34. Hodgson, J.; Swietlik, E.M.; Salmon, R.M.; Hadinnapola, C.; Nikolic, I.; Wharton, J.; Guo, J.; Liley, J.; Haimel, M.; Bleda, M.; et al. Characterization of GDF2 mutations and levels of BMP9 and BMP10 in pulmonary arterial hypertension. *Am. J. Respir. Crit. Care Med.* **2020**, *20*, 575–585. [CrossRef] [PubMed]
35. Overwater, E.; Marsili, I.; Baars, M.J.H.; Baas, A.F.; Van de Beek, I.; Dulfer, E.; Van Hagen, J.M.; Hilhorst-Hofstee, Y.; Kempers, M.; Krapels, I.P.; et al. Results of next-generation sequencing gene panel diagnostics including copy-number variation analysis in 810 patients suspected of heritable thoracic aortic disorders. *Hum. Mutat.* **2018**, *39*, 1173–1192. [CrossRef]

 *genes*

Review

'There and Back Again'—Forward Genetics and Reverse Phenotyping in Pulmonary Arterial Hypertension

Emilia M. Swietlik [1,2,3], Matina Prapa [1,3], Jennifer M. Martin [1], Divya Pandya [1], Kathryn Auckland [1], Nicholas W. Morrell [1,2,3,4] and Stefan Gräf [1,4,5,*]

1   Department of Medicine, University of Cambridge, Cambridge Biomedical Campus, Cambridge CB2 0QQ, UK; es740@cam.ac.uk (E.M.S.); mp931@medschl.cam.ac.uk (M.P.); jm2055@medschl.cam.ac.uk (J.M.M.); dp623@medschl.cam.ac.uk (D.P.); ka328@cam.ac.uk (K.A.); nwm23@cam.ac.uk (N.W.M.)
2   Royal Papworth Hospital NHS Foundation Trust, Cambridge CB2 0AY, UK
3   Addenbrooke's Hospital NHS Foundation Trust, Cambridge CB2 0QQ, UK
4   NIHR BioResource for Translational Research, University of Cambridge, Cambridge Biomedical Campus, Cambridge CB2 0QQ, UK
5   Department of Haematology, University of Cambridge, Cambridge Biomedical Campus, Cambridge CB2 0PT, UK
*   Correspondence: sg550@cam.ac.uk

Received: 26 October 2020; Accepted: 23 November 2020; Published: 26 November 2020

Abstract: Although the invention of right heart catheterisation in the 1950s enabled accurate clinical diagnosis of pulmonary arterial hypertension (PAH), it was not until 2000 when the landmark discovery of the causative role of bone morphogenetic protein receptor type II (*BMPR2*) mutations shed new light on the pathogenesis of PAH. Since then several genes have been discovered, which now account for around 25% of cases with the clinical diagnosis of idiopathic PAH. Despite the ongoing efforts, in the majority of patients the cause of the disease remains elusive, a phenomenon often referred to as "missing heritability". In this review, we discuss research approaches to uncover the genetic architecture of PAH starting with forward phenotyping, which in a research setting should focus on stable intermediate phenotypes, forward and reverse genetics, and finally reverse phenotyping. We then discuss potential sources of "missing heritability" and how functional genomics and multi-omics methods are employed to tackle this problem.

Keywords: forward phenotyping; forward genetics; reverse genetics; reverse phenotyping; pulmonary arterial hypertension; intermediate phenotypes; whole-genome sequencing; epigenetic inheritance; genetic heterogeneity; phenotypic heterogeneity

---

1. Introduction

Rare diseases, such as pulmonary arterial hypertension (PAH), are enriched with underlying genetic causes and are defined as life-threatening or chronically debilitating disorders with a prevalence of less than 1 in 2000 [1]. Although individually characterised by low prevalence, in total, rare diseases pose a significant burden to health care systems and a diagnostic challenge. Over 6000 rare diseases have been reported to date (ORPHANET [2]) and new genotype-phenotype associations are discovered every month [3]. Despite the several-fold increase in genetic diagnoses in the area of rare diseases [4], the cause of the disease remains elusive in a significant proportion of cases.

Since the 2008 World Symposium on Pulmonary Hypertension (WSPH), the term "heritable PAH" (HPAH) has been used to describe both familial PAH and sporadic PAH with an identified underlying pathogenic variant [5] (Table 1). Family history is a vital disease component directly linked to the proportion of variation attributable to genetic factors, known as heritability [6,7]. Familial studies have been used historically as a tool for gene mapping, with the classical example of twin studies commonly used to disentangle the relative contribution of genes and environment to complex human traits [2].

In statistical terms, heritability is defined as a proportion of the phenotypic variance that can be attributed to the variance of genotypic values:

$$H^2 \ (broad\ sense) \ = \ Var \ (G) \ \div \ Var \ (p) \tag{1}$$

H^2 estimates are specific to the population, disease and circumstances on which they are estimated [8]. In theory, total genotypic variance ($Var(G)$) can be divided into multiple components: $Var(A)$total additive variance (breeding values); $Var(D)$ —dominance variance (interactions between alleles at the same locus); $Var(I)$ —epistatic variance (interactions between alleles at different loci); $Var(G \times E)$—variation arising from interactions between genes and the environment.

$$Var \ (G) \ = \ Var \ (A) \ + \ Var \ (D) \ + \ Var \ (I) \ + \ Var \ (G \times E) \tag{2}$$

In practice, total genotypic variation is difficult to measure, although estimates can be made, and the components of genotypic variation are nearly unattainable. Hence genetic studies usually refer to heritability in its narrow sense, which is the proportion of the phenotypic variance that can be attributed to favourable or unfavourable alleles.

$$h^2 \ (narrow\ sense) \ = \ Var \ (A) \ \div \ Var \ (p) \tag{3}$$

Recently, genome-wide association studies (GWAS) have enabled the estimation of additive heritability attributed to common genetic variation (single nucleotide polymorphisms—SNPs), albeit with a typically small effect size [9]. This has led to the issue of "missing heritability" [10], whereby SNP-based estimates were not sufficient to explain prior heritability predictions arising from twin or familial recurrence studies. To that effect, several hypotheses have been proposed (Figures 1 and 2) including the complex interplay between genes and environment and the often overlooked potential contribution of structural variation [11].

To date, about 75% of patients with a clinical diagnosis of idiopathic pulmonary arterial hypertension (IPAH) have no defined genetic cause of the disease. This review outlines the role of forward and reverse genomic and phenomic approaches as well as other omic technologies in the search for missing heritability of PAH (Figure 1). Understanding the genetic architecture of PAH and its dynamic interplay with the environment is a prerequisite to predict personalized patient risk, decipher interaction pathways underlying the disease and develop strategies for therapy and prevention.

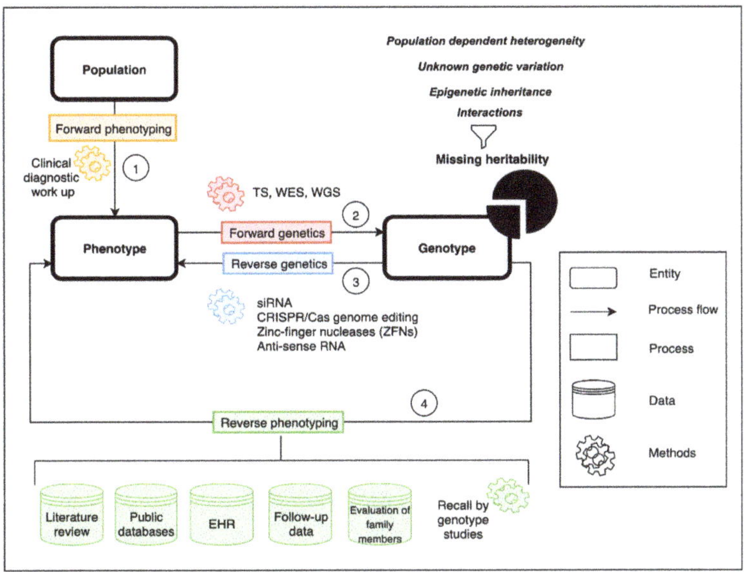

Figure 1. Graphical abstract. Schematic representation of the concepts of forward phenotyping (1) and genetics (2) and reverse genetics (3) and phenotyping (4). Public databases include, but are not limited to, Online Mendelian Inheritance in Man (OMIM), Human Phenotype Ontology (HPO), DatabasE of genomiC varIation and Phenotype in Humans using Ensembl Resources (DECIPHER), ClinVar databases. Forward genetics—'genotype to phenotype' approach; reverse genetics—analysis of the impact of induced variation within a specific gene on gene function; reverse phenotyping—clinical assessment directed by genetic results. Abbreviations: TS—targeted sequencing; WES—whole-exome sequencing; WGS—whole-genome sequencing; siRNA—small interfering RNA; EHR—Electronic Healthcare Records.

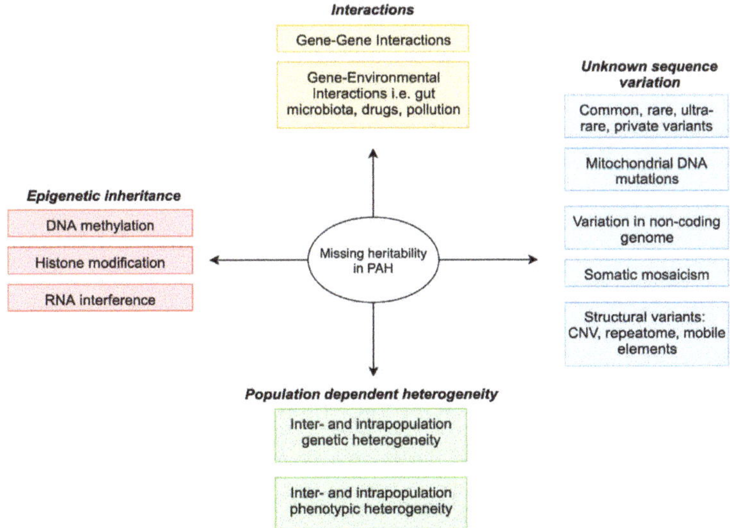

Figure 2. Potential factors contributing to missing heritability in PAH.

Table 1. Definitions and changes to the classification of Group 1 PAH. Abbreviations: WSPH—World Symposium on Pulmonary Hypertension; ERS/ESC—European Respiratory Society and European Society of Cardiology; mPAP—mean pulmonary artery pressure; (R)—resting; (E)—exercise; PAWP—pulmonary artery wedge pressure; PVR—pulmonary vascular resistance; LV—left ventricle; PPH—primary pulmonary hypertension; *BMPR2*—Bone morphogenetic protein receptor, type II; *ACVRL1*—Activin A receptor like type 1; *SMAD9*—SMAD Family Member 9; *CAV1*—Caveolin 1; *KCNK3*—Potassium two pore domain channel subfamily K member 3; PVOD/PCH—pulmonary veno-occlusive disease/pulmonary capillary haemangiomatosis; PAH—pulmonary arterial hypertension; PH—pulmonary hypertension; PPHN—persistent pulmonary hypertension of the newborn; RHC—right heart catheterization.

WSPH Proceedings and ERS/ESC Guidelines	Definition of Group 1	Comments	Changes to the Classification
1st WSPH, Geneva, 1973 [12]	No haemodynamic definition mentioned		
2nd WSPH, Evian, 1998 [13]	No haemodynamic definition mentioned, but RHC recommended for diagnosis		Introduction of the terms primary (PPH) and secondary (related to other conditions) pulmonary hypertension, recognition of familial forms of PH
3rd WSPH, Venice, 2003 [14]	mPAP(R) > 25 mmHg; mPAP(E) > 30 mmHg; PAWP < 15 mmHg; PVR > 3 WU		Abandonment of the term primary pulmonary hypertension, the introduction of terms idiopathic and familial PAH as well as associated PAH, *BMPR2* and *ACVRL1* implicated in the pathogenesis of PAH
4th WSPH, Dana Point, 2008 [5]	mPAP(R) ≥ 25 mmHg; PAWP ≤ 15 mmHg	Exercise-induced PH removed from the definition as although (R) mPAP has been shown to be stable across age groups, (E) mPAP increases with age hence based on the available data it was not possible to define a cutoff	Introduction of the terms idiopathic (no family history, no precipitating risk factor) and hereditary (encompassing familial cases with or without identified germline mutations and PAH). Inclusion of PH associated with Schistosomiasis and PH associated with chronic hemolytic anaemia to Group 1
ERS/ESC Guidelines, 2009 [15]	mPAP(R) ≥ 25 mmHg; PAWP ≤ 15 mmHg; CO normal or reduced	No definition of PH on exercise	
5th WSPH, Nice, 2013 [16]	mPAP(R) ≥ 25 mmHg; PAWP ≤ 15 mmHg; PVR > 3 WU	Introduction of PVR to the definition, a recommendation to report PVR in WU; fluid challenge may be helpful to unmask occult LV diastolic dysfunction	*SMAD9*, *CAV1* and *KCNK3* included as risk genes for HPAH
ERS/ESC Guidelines, 2015 [17]	mPAP(R) ≥ 25 mmHg; PAWP ≤ 15 mmHg	The clinical significance of a mPAP between 21 and 24 mmHg is unclear	Group 1′ PVOD/PCH has been expanded and includes idiopathic, heritable, drug-, toxin- and radiation-induced and associated forms; PPHN includes a heterogeneous group of conditions that may differ from classical PAH. As a consequence, PPHN has been sub-categorised as group I″.
6th WSPH, Nice, 2018 [18]	mPAP(R) ≥ 20 mmHg; PAWP ≤ 15 mmHg; PVR ≥ 3 WU	PVR ≥ 3 WU should be used as a diagnostic criterion for all forms of PH	PAH long-term responses to calcium channel blockers established as a subtype of Group 1; PAH with overt features of venous/capillaries (PVOD/PCH) involvement established as a subtype of Group 1

2. Genetic and Phenotypic Heterogeneity

Discoveries in the field of rare diseases have been hampered by genetic and phenotypic heterogeneity of these entities. Genetic heterogeneity is a situation in which sequence variation in two or more genes results in the same or very similar phenotype. The degree of genetic heterogeneity varies between different diseases (Figure 3A). For instance, sickle cell anaemia has only been associated with mutations in one gene [19], Haemoglobin subunit β (*HBB)* and tuberous sclerosis [20] in two genes, retinitis pigmentosa [21] with over 60 and intellectual disability with over 800 [22]. PAH also shows high genetic heterogeneity (Figure 3B), with so far around ~20 risk genes reported [23] and more expected to be found.

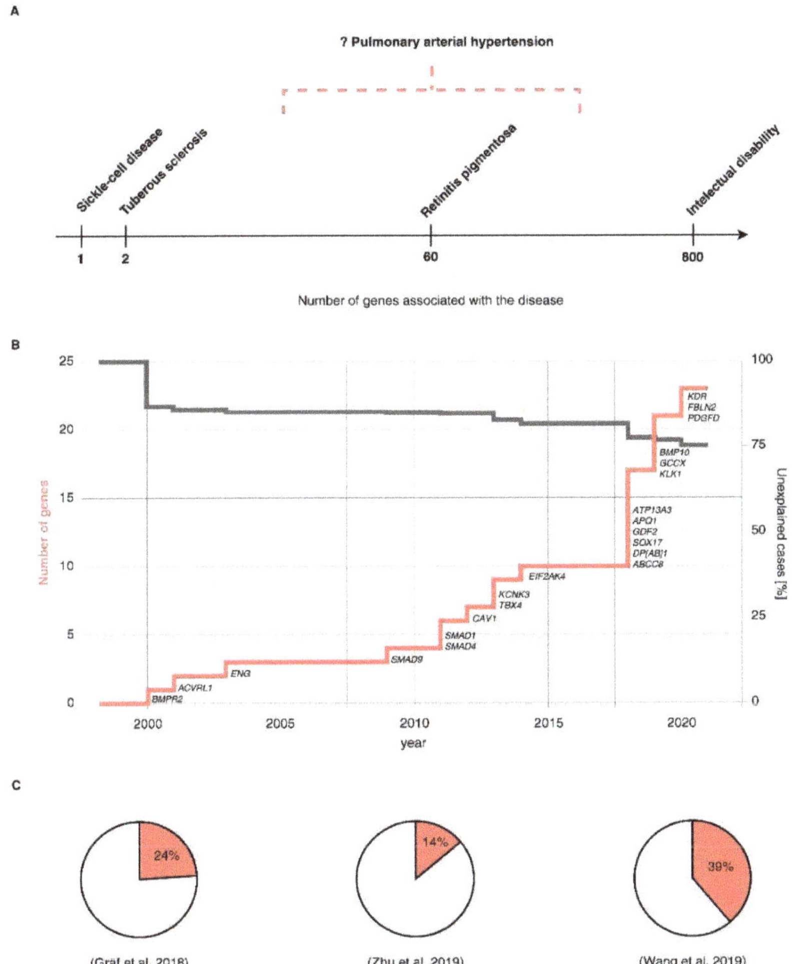

Figure 3. (**A**). Genetic heterogeneity in various genetic disorders, (**B**). Genetic discoveries in PAH, (**C**). Proportion of explained cases by cohort (Gräf et al., 2018—I/HPAH [24]; Zhu et al., 2019—Group 1 PAH [25]; Wang et al., 2019—IPAH [26]). '?' denotes uncertainty around number of genes involved in the pathogenesis of PAH.

Genotypic heterogeneity is further complicated by phenotypic heterogeneity. Pulmonary hypertension (PH) is a highly heterogeneous condition, defined as an elevation of mean pulmonary artery pressure (mPAP) equal to or greater than 25 mmHg measured by right heart catheterisation (RHC) in the supine position at rest [17]. This somewhat arbitrary threshold for defining PH was proposed at the 1st WSPH in Geneva in 1973 and has now been challenged by a new body of evidence showing that normal mPAP is 14 ± 3.3 mmHg, which suggests an upper limit of normal at 20 mmHg (14 mmHg + 2SD). This new threshold has been endorsed by the 6th WSPH along with the inclusion of pulmonary vascular resistance (PVR) and pulmonary artery wedge pressure (PAWP) cut-offs into the new haemodynamic definition. This new definition categorises PH into three groups based on haemodynamic criteria: pre-capillary PH, isolated post-capillary PH (IpcPH) and combined pre- and post-capillary PH (CpcPH) [18] (Table 1).

Haemodynamic definitions of PH encompass multiple cardiopulmonary entities that have been classified into five groups: PAH, PH secondary to left heart disease, PH due to lung disease and/or hypoxia, chronic thromboembolic pulmonary hypertension (CTEPH) or other pulmonary artery obstructions, and PH with unclear and/or multifactorial mechanisms.

The clinical classification aims to categorise patients into groups according to pathophysiological mechanisms, haemodynamics and therapeutic management. Despite this, there is persistent heterogeneity within groups and subgroups, and not all patients fit easily into a single category, which can be a reflection of genetic pleiotropy (Figure 4).

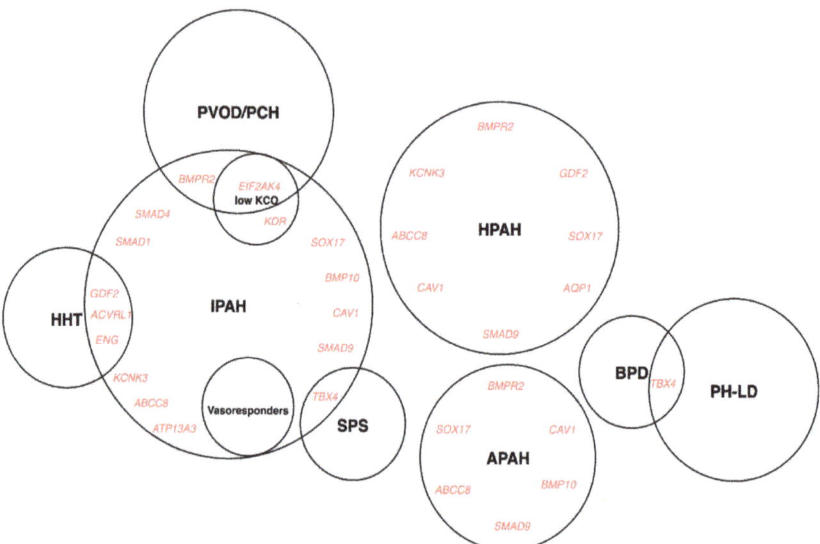

Figure 4. Genotype-phenotype associations in PAH. Abbreviations: I/H/APAH—idiopathic/hereditary/associated pulmonary arterial hypertension; PVOD/PCH—pulmonary veno-occlusive disease/pulmonary capillary haemangiomatosis; SPS—small patella syndrome; BPD—bronchopulmonary dysplasia; HHT—hereditary hemorrhagic telangiectasia.

Based on the recent advances in mechanistic understanding of the disease, the 6th WSPH proposed new changes to the classification of PH, including two previously recognised phenotypes as the subgroups of Group 1, namely "PAH long-term responders to calcium channel blockers" (Group 1.5) and "PAH with overt features of venous/capillaries (Pulmonary veno-occlusive disease(PVOD)/Pulmonary capillary haemangiomatosis (PCH)) involvement" (Group 1.6) [18]. Other phenotypes previously reported in the literature and summarised in An Official American Society Statement: Pulmonary Hypertension Phenotypes [27] such as "severe" PH in respiratory disease, maladaptive right ventricular

(RV) hypertrophy, PH in elderly individuals, PAH in children and PAH with metabolic syndrome [28] are awaiting clinical validation and confirmation of utility both in clinical practice and research settings.

3. Forward Phenotyping for Genetic Studies

A precise definition of the phenotype of interest is a cornerstone of any genetic study (Figure 1, process 1). As described above, clinical diagnosis relies on clustering patients based on observable and measurable traits, signs and symptoms, which are the product of genetic, epigenetic and environmental factors. As a consequence, clinical phenotypes can be dynamic and reactive, which is useful and desirable in the clinical setting but unsuitable for genetic studies. A distinction must be made between clinical and research diagnosis, particularly diagnosis for genetic analysis. The former is usually spread over time and acquired in several stages: history taking, physical examination, differential diagnosis and confirmation, the latter usually needs to be ascertained during a single encounter. To make this feasible and reliable, standardised checklist and operating procedures need to be in place, diagnostic criteria should follow simple inclusion/exclusion rules and phenotypes need to be described using controlled vocabulary to avoid ambiguity. Additionally, the data must be in a format amenable to computational analysis. Finally, the validity of phenotypes is confirmed in test cohorts, through functional studies and ultimately via reverse phenotyping (see below). The accuracy and precision of phenotype measurements are of paramount importance. In genetic studies, the diagnosis misclassification or admixture of phenocopies can significantly affect power to detect an association [29]. Equally, categorising biologically continuous phenotypes (i.e., age, mPAP, diffusing capacity of the lungs for carbon monoxide (DLCO)) is prone to errors due to flaws in quantification methods and arbitrary thresholds.

Phenotype optimisation for genetic studies aims at finding homogenous groups of patients that likely share the same genetic architecture. This can be approached through various strategies. For example, an extreme phenotype strategy aims at identifying rare variants with large effect sizes through recruitment of patients with traits at either end of the phenotypic spectrum. These phenotypes can be based on family history, age of onset, outcome, severity scores, biomarker levels, disease trajectory or response to treatment [30–32]. Such stratification was proven to increase the power to detect novel disease risk genes [30,33,34] and to be cost-effective [35]. Other strategies include covariate-based methods which jointly estimate the effect of multiple variables and data reduction techniques. Alternatively, intermediate phenotypes can be used. Intermediate phenotypes are features closer to underlying biology that are at least as heritable as the phenotype itself, stable over time, and are associated with the disease of interest [36].

Although clinical phenotyping remains the most widely used method of patient stratification both in clinical practice and research, it requires substantial domain knowledge and is time-consuming. Computational phenotyping based on clinical and/or "omics" datasets using machine learning might be an alternative due to unparalleled diagnostic precision, accuracy and speed. Two recent publications exemplify the power of computational tools in identifying disease phenotypes. Based on blood cytokine profiles, a prospective observational study of Group 1 PAH discovered and validated four immune phenotypes; importantly, these phenotypes differed in clinical outcomes despite the fact that demographics, PAH aetiologies, comorbidities, and treatments were similar across clusters [37]. Likewise, clinical data mining using the Comparative Prospective Registry of Newly Initiated Therapies for Pulmonary Hypertension (COMPERA) again revealed four clusters with differing survival and response to therapy [38]. A viable alternative approach for clinical data mining is utilising phenotype ontologies, such as Human Phenotype Ontology (HPO), which allow standardised, highly granular and precise phenotyping across different disease domains [39]. Use of ontologies to define phenotypes has already proven useful in identifying novel candidate genes for rare disorders [40]. Ontology-based analysis of phenotypes has been further facilitated by the implementation of methods for manipulation, visualisation and computation of semantic similarity between ontological terms and sets of terms [41].

4. Forward Genetics

4.1. Concepts

Forward genetics is a classic molecular genetics approach used to elucidate the genetic underpinnings of a mutant phenotype of interest [42]. Forward genetics is typically considered a 'phenotype to genotype' approach as mutant phenotypes are first observed before their corresponding genes are identified (Figure 1, process 2). In humans, forward genetics approaches most commonly include family-based linkage studies and/or genome-wide association studies (GWAS) and, more recently, rare variant association studies (RVAS).

4.2. Methodology

4.2.1. Study Design

The two main approaches for studying the underlying genetics of PAH are family-based studies and case-control studies. The former is based on studying inheritance patterns of genetic polymorphisms, the second involves comparing genotype frequencies between cases and controls. Family-based studies are effective when parental samples along with phenotype information are available and the disease in question has a high penetrance; they are particularly useful for studying dichotomous traits and are robust to population stratification [43]. Case-control studies are a viable alternative if the above criteria are not met, although they have their own challenges which need to be addressed. To name the most important:

1. Selection of cases (recognition of selection bias, incident vs. prevalent cases recruitment)
2. Case definition (precise definition of the phenotype that can be ascertained in a research setting)
3. Selection of controls (healthy vs. disease controls, matched in respect to age, sex and ethnicity, having a comparable evaluation of presence or absence of the phenotype in question)

Power calculations in genetic studies are an absolute necessity as ignoring this basic step can lead to both underpowered (risking false rejection of null hypothesis and characterised by wide sampling distributions for sample estimates) and overpowered (wasteful and often unethical) experiments. Factors limiting power to detect new genotype-phenotype associations which need to be accounted for are phenotypic variance, phenocopies, the effect size of risk alleles and minor allele frequency (MAF), with the last two factors driving the difference of sample sizes between GWAS and RVAS.

4.2.2. Statistical Methods

Prior to the widespread use of GWAS, the most important tool in genetics were linkage studies in families, these were particularly useful in single-gene disorders in which implicated genes have large effect sizes. GWAS on the other hand compares the frequency of common SNPs between unrelated cases and controls. The associated SNPs are then considered markers of relevant regions that influence the risk of the phenotype. In fact, power calculations provided evidence that GWAS are better than linkage studies at detecting variation with small effect sizes [44]. Multiple statistical methods can be applied in GWAS, for example, Pearson X^2 test, normal approximation to Fisher's Exact Test, logistic regression, categorical model tests, Cochran–Armitage Trend test, and allele tests. The best method depends on the mode of inheritance and trait frequency. Importantly, the assumptions used in various tests may differ; these assumptions directly impact the results, as tests that assume the same mode of inheritance should yield the same results (i.e., Cochran–Armitage Trend Additive test, Logistic Regression Additive test and Allele test). Due to a large number of comparisons, adjustment for multiple testing is necessary; therefore, the p-value threshold is Bonferroni corrected (which encourages high type II error). Additionally, genotype and phenotype misclassification errors impact on power in GWAS. Epistatic scenarios and modelling gene–environment interactions require yet another set of methods and are computationally challenging although feasible [45].

GWAS is unsuitable for single-variant testing, due to the potentially low prevalence of mutation carriers and small effect size, which would both require unfeasibly large sample sizes. Instead, gene and region-based aggregation approaches have been developed which compare mutation frequencies between cases and controls within the boundaries of the gene. These techniques are appropriate when different variants exhibit an equal risk of disease and thus have the same phenotypic impact. For instance, several variants may result in LoF (e.g., nonsense, frameshift, essential splice site), and thus analysis would determine the association by counting the presence of LoF variants between cases and controls. Prior to association testing, quality control and filtering methods are utilised, namely sequencing quality scores, MAF filters [46] (usually MAF of 1:10,000 for autosomal dominant disorders and MAF of 1:1000 for autosomal recessive disorders) and in silico predictions. Predictions include deleteriousness scores for missense variants such as PolyPhen-2 [47], Sorting Intolerant From Tolerant (SIFT) [47,48], and rare exome variant ensemble learner (REVEL) [49], conservation scores such as Genomic Evolutionary Rate Profiling (GERP) [50], PhyloP [51] or PhastCons [52], or the Combined Annotation Dependent Depletion (CADD) score [53], which combines several metrics in one score. Analysis of the protein-coding region, consisting of ~20,000 genes, requires adjustment for multiple-testing. This can be done using the Bonferroni correction, where $\alpha = (0.05/20,000) \approx 2.5 \times 10^{-6}$). Where several models are applied, this adjustment must be made more stringent by dividing by the number of models tested. Region-based collapsing approaches hinge on the notion that different regions within genes may vary in their tolerance to missense variation. An alternative approach, particularly useful in smaller studies, is collapsing variants that belong to the same gene set (i.e., genes that belong to the same pathway). Candidate gene testing is a powerful approach to avoid overcorrection and, therefore, false-negative results. This proved useful when investigating members of the transforming growth factor-β (TGF-β) pathway, such as *SMAD9* [54], which did not reach statistical significance in the exome-wide analysis [24]. More recently, the same approach revealed an association between *TET2* and PAH, which was further supported by experimental evidence [55] but did not reach exome-wide significance [25].

Complex genetic models such as recessive inheritance pose additional challenges. In the recessive mode of inheritance, the MAF threshold must be more lenient as heterozygotes are unaffected (higher MAF in reference populations); also, variants in cis configuration (affecting the same allele) might be wrongly counted (as a pose to trans variants, which are those present on opposing alleles). Similarly, testing for digenic inheritance is particularly problematic due to the large number of possible combinations requiring testing and adjusting for [56].

A number of statistical methods have been developed to test for rare variant associations. Burden tests [57–59] aggregate the information found within a predefined genetic region into a summary dose variable. In weighted burden tests [60], variants are weighted according to their frequency or functional significance. Adaptive burden tests [61] aim to account for bidirectional effects by selecting appropriate weights. Variance component (kernel) tests such as (Sequence) Kernel Association Test (SKAT) [62] allow to test risk and protective variants simultaneously but are underpowered when most variants are causal, and effects are unidirectional. Omnibus tests such as SKAT-O [63], which combines burden tests with the variance-component test, might be particularly useful when there is little knowledge of the underlying disease architecture. In addition to frequentist approaches, a Bayesian statistical framework offers a robust alternative. Bayesian model comparison methods such as BeviMed [64] allow for the testing of associations between rare Mendelian disease and a genomic locus by comparing support for a model where disease risks depend on genotypes at rare variant sites in the locus and a genotype-independent "null" model. The prior probability in such models can vary across variants (reflective of external biological information, i.e., depending on MAF, conservation scores, gene ontologies, expression in the tissue of interest) or be constant for all genes/variants reflecting the prior belief of the overall proportion of variants that are associated with a given phenotype.

Last but not least, an essential step in rare variant discovery is to ascertain the pathogenicity of a given variant and its causative role in the disease. Not all damaging variants are pathogenic and in silico approaches alone are not enough to predict if the variant is disease-causing [65]. Viability and phenotyping inferred from knockout mice screens, as well as essentiality screens on human cell lines, may further help predict variant impact [66]. To aid both research and clinical decision making, the American College of Medical Genetics and the Association for Molecular Pathology (ACMG) issued recommendations that combine and weigh the computational, functional, population and clinical evidence to determine pathogenicity [67]. Other initiatives such as ClinGen and ClinVar aim to define the clinical relevance of genes and variants reported in the literature for use in precision medicine and research [68].

4.2.3. Molecular Genetic Techniques

Molecular genetics techniques used for genetic diagnosis, including the detection of specific gene mutations and copy number variants, have been recently summarized [69]. Traditional methods used to identify candidate genes involved in the pathogenesis of PAH include linkage analysis, but more recently next-generation sequencing (NGS) has taken centre stage. The advent of NGS technologies has opened a plethora of opportunities both for clinical diagnostics and research. Such technologies that fall under the umbrella of NGS include targeted panel sequencing, whole-exome sequencing (WES) and whole-genome sequencing (WGS). Nevertheless, the need for more conventional methods such as Sanger sequencing and multiplex ligand-dependent probe amplification (MLPA) remains. Disease-specific panels include a set of genes or regions of genes that are known to be causative of a specific phenotype. This is particularly beneficial in the clinical context when assessing highly heterogeneous traits, such as intellectual disability. Although these panels are not always consistent across laboratories, efforts are being made to produce guidance around their design and development [70]. Targeted panel testing has been introduced in PAH including known and candidate disease genes [71].

WES includes < 2% of the genome i.e., the coding regions only. This method is clinically useful given that 85% of all described disease-causing sequence variants are in this region [72]. For diseases that are more genetically heterogeneous, WES has proven to be a fruitful method, especially when incorporating segregation analysis, which increases the diagnostic yield from 23.6% in probands to 31% in child–parent trios [73]. WES has been used for both the identification and discovery of candidate genes in PAH and has been applied to family-based [74] and case-control studies [75]. Limitations of WES include poor coverage of some exons such as GC rich regions and low confidence to identify structural variation [76].

WGS is a high-throughput sequencing technology predominantly used in the research setting [77]. WGS is massively parallel, DNA fragments are aligned to form a contiguous sequence. The cost of this technology is halving approximately every two years [78]; however, the $1000 genome often referred to is still some way off unless sequencing is performed at scale [79]; as an example, WGS is currently being introduced to the UK National Health Service in collaboration with Genomics England.

European Respiratory Society and European Society of Cardiology (ERS/ESC) guidelines published in 2015 recommended sequential testing starting with bone morphogenetic protein receptor type-2 (*BMPR2*) sequencing and MLPA in patients with sporadic/familial PAH and *EIF2AK4* sequencing in sporadic/familial PVOD/PCH [17], and this approach has been successfully used in many clinical and research settings across the world [80]. With the decreasing cost of WES/WGS, a common practice is now that of virtual panel testing whereby a selected number of genes are chosen for bioinformatics analysis based on the individual's phenotype. This also allows for data reanalysis when a novel disease-gene is identified and/or another condition is suspected (emerging phenotype over time). In the UK, this is coordinated at a national level via the Genomics England PanelApp tool; virtual disease gene panels applied to WGS data are continuously curated and include a PAH panel [81]. As more patients with PAH are being tested via NGS methods, a diagnostic benefit is starting to emerge (Figure 3C).

Despite the benefits of NGS technologies, there are some challenges that require attention and systematic solutions, among these, storage and handling of big data remains a significant consideration, also management of incidental findings as well as the reporting of variants of unknown significance (VUS).

Along with the growing number of research projects using NGS to uncover the genetic basis of various diseases, there has been an ongoing effort to aggregate and harmonise WES and WGS data from large-scale disease and population projects and to make them publicly available as a reference variome. This started in 2012 with a funder project called Exome Aggregation Consortium (ExAC) which harvested WES data from over 60,000 individuals; this was followed by The Genome Aggregation Database (gnomAD), of which three versions have been released so far, covering 71,702 genomes from unrelated individuals aligned against GRCh38 (v3). Also, in 2012, came the announcement of the 100,000 genomes project by the UK Government. The project sequenced the genomes of 100,000 NHS patients with particular focus on those with rare disease(s) and cancer. Another useful resource is the Trans-Omics for Precision Medicine (TopMed) program, which aims to sequence over 120,000 well-phenotyped individuals as well as collect other omics datasets. The assertion of ethnic diversity is an important consideration, and several initiatives such as KoVariome [82], Genes and Health [83] and BioBank Japan [84] are addressing this issue. Furthermore, new disease cohort genomic databases are being established. Examples include: NIHR BioResource Rare Disease Study (NBR) [77], Inflammatory Bowel Disease BioResource [85], Genetic Links to Anxiety and Depression (GLAD) [86] and Eating Disorders Genetics Initiative (EDGI) [87]. The advantages of these datasets are numerous; first, they provide allele frequencies for diverse populations, second, they help to address the overestimation of disease penetrance arising from the historical focus on multiplex pedigrees [88], and third, through acknowledging variable penetrance, they help to identify genetic and environmental disease modifiers [89].

4.2.4. Reference Genome

The Human Genome Project was completed in 2003 [90,91] and since then, successive iterations of the human reference genome have been published, updated, and refined by the Genome Reference Consortium (GRC). Recent versions include GRCh37 (hg19) and GCRh38 (hg38) released in 2009 and 2013, respectively. These are both composite genomes, i.e., derived from the sequence of several anonymous donors; the make-up of these two assemblies is largely similar, with approximately 93% of the primary assembly composed of sequences from 11 genomic clone libraries.

To date, most large-scale PAH studies have aligned their data to the GRCh37 reference genome (Table 2), with only a couple of recent studies aligning their data to GRCh38 [25,55]. Even though GRCh38 was released seven years ago, the transition from GRCh37 to GRCh38 has been a long process and recent analyses have sought to compare the two reference panels. Guo et al. [92] demonstrated that GRCh38 provides a more accurate analysis of human sequencing data due to the improved annotation of the exome and the additional reads aligned to GRCh38, findings which indicate better structural and sequence representation. In addition, Pan et al. [93] noted that GRCh38 had better genome coverage, with a 5% increase in the number of SNVs identified. In comparison to GRCh37, GRCh38 altered 8000 nucleotides, corrected several misassembled regions, filled in gaps, and increased the number of genes and protein-coding transcripts [92]; additionally, GRCh38 is the first human reference genome to contain sequence-based representations for the centromeres [94]. Whilst GRCh37 is a single representation of multiple genomes, with only three regions containing alternative sequences (UDP-glucuronosyltransferases 2B subfamily (*UGT2B*) on chromosome 4, the major histocompatibility complex (MHC) region on chromosome 6, and the *MAPT* gene on chromosome 17) [95], GRCh38 includes 261 alternate loci across 178 genomic regions, providing a more robust representation of human population variation [94]. The increased level of alternative sequence representation requires new analysis methods to support their inclusion yet at present, most tools and pipelines do not make use of these [95]. Despite the advantages, GRCh38 still contains gaps and errors at repetitive and structurally diverse regions [96]. Additionally, as it is a mosaic haploid representation of the human genome [94], in which poor alignment can affect the detection of alleles in regions of high variation, such as the MHC locus and KIR, it is unlikely to truly represent human diversity [96]. It does, however, provide the starting point for a more inclusive population-based reference genome, or pan-genome [97] and as such, will play an evolving role in the generation of individual diploid genome assemblies and graph-based representations of genome-wide population variation [98–100], thereby providing unique opportunities for data analysis.

Table 2. Landmark forward genetics studies in Group 1 PAH. Abbreviations: *BMPR2*—Bone morphogenic protein receptor type 2; *ENG*—Endoglin; *ACVRL1*—Activin A Receptor Like Type 1; *SMAD*—SMAD Family Member; *CAV1*—Caveolin 1; *KCNK3*—Potassium Two Pore Domain Channel Subfamily K Member 3; *TBX4*—T-Box Transcription Factor 4; *EIF2AK4*—Eukaryotic Translation Initiation Factor 2 α Kinase 4; *GDF2*—Growth Differentiation Factor 2; *SOX17*—SRY-Box Transcription Factor 17; *ATP13A3*—ATPase 13A3; *AQP1*—Aquaporin 1; *ABCC8*—ATP Binding Cassette Subfamily C Member 8; *BMP10*—Bone Morphogenetic Protein 10; *KLK1*—Kallikrein 1; *GCCX*—γ-Glutamyl Carboxylase; *KDR*—Kinase insert domain receptor; *TET2*—Tet Methylcytosine Dioxygenase 2; *FBLN2*—Fibulin 2; *PDGFD*—Platelet-Derived Growth Factor D.

Study (Reference)	Genes	Study Design	Sample	Ethnicity	Method	Reference Genome
Lane et al. 2000, [101]	*BMPR2*	Case-level data	Cases: $n = 8$ PPH kindreds for candidate gene mutational analysis	Not stated	TS	H.sapiens mRNA for BMPR-II: Genbank Z48923
Thomson 2000, [102]	*BMPR2*	Case-level data	Cases: $n = 50$ PPH	Not stated	TS	Not stated
Trembath et al. 2001, [103]	*ACVRL1*	Case-level data	Cases: 5 kindreds plus 1 individual patient with HHT, including $n = 10$ cases with PH	Not stated	TS	Not stated
Chaouat 2004, [104]	*ENG*	Case-level data	Case: $n=1$ HHT, PPH with history of anorexigen use	Not stated	TS	Not stated
Harrison et al. 2005, [105]	*ACVRL1, ENG*	Case-level data	Cases: $n = 18$ I/APAH	Not stated	TS	Not stated
Shintani et al. 2009, [106]	*SMAD9 (SMAD8)*	Case-level data	Cases: $n = 23$ IPAH	Japanese	TS	Not stated
Nasim et al. 2011, [54]	*SMAD1, SMAD4, SMAD9*	Case-level data	Cases: $n = 324$ IPAH/APAH/CTEPH; Controls: $n = 1584$	European & Japanese	TS	Not stated
Austin et al. 2012, [74]	*CAV1*	Case-level data	Cases: 3-generation family, 6 with PAH; Additional cohort: $n = 260$ unrelated I/HPAH cases; Controls: $n = 1000$	European	WES	GRCh37
Ma et al. 2013, [107]	*KCNK3*	Case-level data	Cases: Family in which multiple members had PAH	Not stated	WES	GRCh37
Kerstjens-Frederikse et al. 2013, [108]	*TBX4*	Case-level data	Cases: $n = 20$ childhood-onset I/HPAH; $n = 49$ adult-onset I/HPAH; $n = 23$ SPS	Not stated	TS	Not stated
Eyries et al. 2014, [109]	*EIF2AK4*	Case-level data	Cases: $n = 13$ PVOD families	Not stated	WES	GRCh37
Best et al. 2017, [110]	*EIF2AK4*	Case-level data	Cases: $n = 81$ I/HPAH	Not stated	TS	Not stated

Table 2. *Cont.*

Study (Reference)	Genes	Study Design	Sample	Ethnicity	Method	Reference Genome
Hadinnapola et al. 2017, [111]	*EIF2AK4*	Case-control data	Cases: n = 880 I/FPAH, PVOD/PCH; Controls: n = 7134 non-PAH controls and their relatives recruited to NBR	European: 84.6%	WGS	GRCh37
Gräf et al. 2018, [24]	*GDF2, SOX17, ATP13A3, AQP1*	Case-control data	Cases: n = 1048 I/F/PAH, PVOD/PCH; Controls: n = 7979 non-PAH controls and their relatives recruited to NBR	European: 84.6%	WGS	GRCh37
Zhu et al. 2018, [75]	*SOX17*	Case-control data	Cases: n = 256 I/FPAH-CHD; Additional cohort: n = 413 I/FPAH screened for rare variants in SOX17; Controls: n = 7509 gnomAD	Not stated	WES	GRCh37
Hiraide et al. 2018, [112]	*SOX17*	Case-level data	Cases: n = 12 IPAH and 12 family members; Additional cohort: n = 128 I/HPAH screened for *SOX17* mutations	Japanese: 100%	WES	Not stated
Bohnen et al. 2018, [113]	*ABCC8*	Case-control data	Cases: n = 913; Controls: n = 33,369 European adults from ExAC & n = 49,630 Europeans from the Regeneron-Geisinger DiscovEHR study	Not stated	WES, WGS	GRCh37
Wang et al. 2019, [26]	*GDF2*	Case-control data	Cases: n = 331 IPAH; Controls: n = 10,508 from available reference data sets	East Asian: 100%	WES, WGS	GRCh37
Eyries et al. 2019, [114]	*BMP10*	Case-level data	Cases: n = 268 I/HPAH, PVOD/PCH	European: > 90%	TS	GRCh37
Hodgson et al. 2020, [115]	*BMP10*	Case-level data	Cases: n = 1048 I/FPAH, PVOD/PCH	European: 84.6%	WGS	GRCh37
Zhu et al. 2019, [25]	*KLK1, GCCX*	Case-control data	Cases: n = 2572 Group 1 PAH; Controls: n = 12,771	European: 72%	WES	GRCh38
Rhodes et al. 2019, [116]	*HLA-DPA1/DPB1 SOX17 enhancer*	Case-control data	Cases: n = 2085 cases; Controls: n = 9659	European: 100%	WGS	GRCh37
Swietlik et al. 2019, [30]	*KDR*	Case-control data	Cases: n = 1122 PAH; Controls: n = 11,889 non-PAH NBR	European: 84%	WGS	GRCh37
Eyries et al. 2020, [117]	*KDR*	Case-level data	Cases: n = 311 unrelated PAH	Not stated	TS	Not stated
Potus et al. 2020, [55]	*TET2*	Case-control data	Cases: n = 2572; Controls: n = 7509 non-Finnish European subjects from gnomAD	European: 72%	WES	GRCh38
Zhu et al. 2020, [118]	*FBLN2, PDGFD*	Case-control data	Cases: n = 4175; Controls: n = 18,819 from SPARK and gnomAD cohorts	European: 54.5%	WES	GRCh38

4.3. Studies

4.3.1. Rare Genetic Variation

Over the last two decades, forward genetics approaches have associated PAH with numerous genes (Table 2); the level of evidence supporting the causal role of these genes, however, is variable and depends on multiple factors (Table 3). PAH is considered to be a monogenic condition transmitted in autosomal dominant fashion with incomplete penetrance. Heterozygous germline mutations in *BMPR2*, a member of the TGF-β superfamily, are the most common genetic cause of PAH [101,119], accounting for over 80% of familial PAH, and approximately 25% of idiopathic PAH [24]. Additional mutations within the TGF-β/BMP signalling pathway, such as activin A like type 1 (*ACVRL1*), endoglin (*ENG*), SMAD family members (*SMAD1, SMAD4, SMAD9*), caveolin 1 (*CAV1*), growth differentiation factor 2 (*GDF2*), loss of function variants in channelopathy genes, potassium two pore domain channel subfamily K member 3 (*KCNK3*), ATP binding cassette subfamily C member 8 (*ABCC8*), ATPase type 13A3 (*ATP13A3*), variants within developmental transcription factors, SRY-box transcription factor 17 (*SOX17*) and T-box transcription factor 4 (*TBX4*), and newly reported risk genes, γ-glutamyl carboxylase (*GGCX*), kallikrein 1 (*KLK1*) [25], kinase insert domain receptor (*KDR*) [30], fibulin 2 (*FBLN2*) and platelet-derived growth factor D (*PDGFD*) [118], have all been identified as individually rare causes of PAH. Infrequent cases of autosomal recessive transmission in *KCNK3* [120] and *GDF2* [121] have been associated with early disease onset and severe phenotype. A subtype of PAH, PVOD/PCH is linked to biallelic mutations in *EIF2AK4* [109].

Table 3. Supporting evidence for the role of risk genes in PAH pathogenesis. Abbreviations: (+) indicates that the paper provides information in favour of the role of the given gene in the pathogenesis of PAH, (−) indicates that the paper does not provide support for the role of the given gene in the pathogenesis of PAH.

	Forward Genetics				**Reverse Genetics**	
Gene	**Case-Control Data**	**Case-Level Data**	**Segregation Data**	**Functional Aberration**	**Disease Model**	**Rescue**
MOI: Autosomal Dominant						
BMPR2	(+) [24,25]	(+) [101,114]	(+) [101,122]	(+) [123–130]	Animal: (+) [123,126,128–131] Cell culture: (+) [124,129,130]	(+) [128–131] (+)
ACVRL1	(+) [24]	(+) [25,103,105,114]	(+) [103]			
ENG	(−) [24,25]	(+) [24,75,104]	(+) [104]		Animal: (−) [132]	
SMAD9	(−) [24,25]	(+) [24,25,75,106,114]		(+) [54,106,133]	Animal: (+) [134] Cell culture: (+) [54,106,133]	(+) [133,135]
SMAD1	(−) [24,25]	(+) [54], (−) [24,25]		(+) [54,136]	Animal: (+) [136] Cell culture: (+) [54]	
SMAD4	(−) [24,25]	(+) [25,54], (−) [24]		(+) [54], (−) [54]	Cell culture: (+) [54]	
CAV1	(+) [74]	(+) [25,75], (−) [24]		(+) [137]	Animal: (+) [138,139] Cell culture: (+) [137]	(+) [137]
TBX4	(+) [24,140]	(+) [75,108,114, 141,142]	(+) [143,144]			
KCNK3		(+) [24,25,107]	(+) [107]	(+) [107,145]	Animal: (+) [146] Cell culture: (+) [107,145]	(+) [107]
ATP13A3	(+) [24]			(+) [147,148]	Animal: (+) [147,148]	
AQP1	(+) [24]			(+) [149,150] (−) [149]	Animal: (+) [149,151] Cell culture: (+) [150]	(+) [149]
GDF2	(+) [24–26]	(+) [114]	(+) [115]	(+) [26,115]	Animal: (−) [152] Cell culture: (+) [26,115]	
SOX17	(+) [24,75]	(+) [112]		(+) [153,154]	Animal: (+) [153] Cell culture: (+) [154]	
ABCC8	(+) [113]		(+) [113]	(+) [113]	Cell culture: (+) [113]	(+) [113]
BMP10		(+) [114,115]				
GGCX	(+) [25]					

Table 3. *Cont.*

	Forward Genetics				Reverse Genetics	
Gene	Case-Control Data	Case-Level Data	Segregation Data	Functional Aberration	Disease Model	Rescue
KLK1	(+) [25]					
KDR	(+) [30]		(+) [30,117]	(+) [155]	Animal: (+) [155]	
FBLN2	(+) [118]					
PDGFD	(+) [118]					
TET2	(+) [55]			(+) [55]	Animal: (+) [55]	(+) [55]
BMPR1A		(+) [25,75]		(+) [156–159]	Animal: (+) [156,157,159]	(+) [156]
BMPR1B		(+) [24,25,75]		(+) [160]	Cell culture: (+) [160]	
TOPBP1		(+) [24]				
THBS1				(+) [161]	Cell culture: (+) [161]	(+) [161]
KCNA5		(+) [162]		(+) [163]	Cell culture: (+) [163]	(+) [163]
MOI: Autosomal Recessive						
EIF2AK4	(+) [111]	(+) [114]	(+) [109,110]	(+) [109]	Animal: (+) [164]	
Common Variation						
enhancer near SOX17	(+) [116]		(+) [116]	(+) [116]		
locus within HLA-DPA1/DPB1	(+) [116]					
CBLN2	(+) [165] (−) [116]					
SIRT3		(+) [166]	(+) [166]	(+) [166]	Animal: (+) [166,167]	
UCP2				(+) [168]	Animal: (+) [167,169]	
EDN1		(+) [170]	(+) [171]			
AGTR1	(+) [34]					
TOPBP1		(+) [172]		(+) [172]	Cell culture: (+) [172]	
Endostatin	(+) [173]					
TRPC6				(+) [174]	Cell culture: (+) [174]	

Autosomal Dominant Mode of Inheritance

TGF-ß Pathway

In 2000, the International Primary Pulmonary Hypertension (PPH) Consortium demonstrated that familial PAH (FPAH) is caused by mutations in *BMPR2*, located on chromosome 2, encoding a TGF-β type II receptor [101]. They established a panel of eight kindreds, in which at least two members had the typical manifestations of PAH. Sequence variants were detected in seven probands; these variants, including two frameshift, two nonsense and three missense mutations, were distributed across the gene and each of the amino acid substitutions occurred at a highly conserved and functionally important site of the *BMPR2* protein. They observed segregation of the mutations with the disease phenotype in seven of the eight families studied. As control subjects, they screened 150 normal chromosomes from the same population and 64 normal chromosomes from ethnically diverse subjects and observed no *BMPR2* mutations [101]. The predicted functional impact of these mutations, their segregation with the phenotype, and the absence of these variants in healthy controls provided strong support for the role of *BMPR2* and the TGF-β signalling pathway in the pathobiology of PAH. The role of *BMPR2* mutations has been subsequently reported in IPAH. Thomson et al. [102] investigated *BMPR2* gene mutations in 50 unrelated IPAH patients with no family history of the disease. In 13 patients (26%), 11 novel heterozygous mutations in *BMPR2* were identified, these included three missense, three nonsense and five frameshift. They also sequenced both parents for five of the 13 probands;

paternal transmission was observed for three families, whereas the remaining two mutations arose spontaneously. *BMPR2* mutations were not observed in 150 normal chromosomes [102]. Screening of other disease subtypes revealed *BMPR2* mutations among patients with PAH associated with congenital heart disease (PAH-CHD) [175] and PVOD [176].

Large cohort studies have proved useful in defining the relative contribution of *BMPR2* mutations in various PAH subtypes. Gräf et al. [24] reported rare heterozygous *BMPR2* mutations in 160 of 1048 PAH cases (15.3%); the frequency of *BMPR2* mutations in FPAH, IPAH and anorexigen-exposed PAH were 75.9%, 12.2% and 8.3%, respectively. Fourteen percent of *BMPR2* mutations resulted in the deletion of larger protein-coding regions, ranging from 5 kb to 3.8 Mb in size. Additionally, 52% of the observed *BMPR2* mutations were newly identified in their study [24], suggesting that nearly two decades after the first *BMPR2* mutation was identified, the use of WGS has allowed for closer study of *BMPR2*, including large deletions around the *BMPR2* locus, and the TGF-β pathway. Another large-scale study in a more heterogeneous group of patients (Group 1 PAH) [25] reported *BMPR2* mutations in 180 of 2572 cases (7%); the frequency of *BMPR2* mutations in FPAH and IPAH patients were 62.4% and 9.3%, respectively. Taken together, over 600 distinct mutations in *BMPR2* have been identified in PAH patients [24,25,177–179] of which around 70–80% are identified in FPAH and 10–20% in IPAH [180].

Importantly, impaired *BMPR2* signalling was shown to be a universal feature of PAH and pointed towards other key members of the canonical *BMPR2* signalling pathway as potential culprits for the disease [181–183].

Mutations in *ACVRL1* and *ENG* have been reported in PAH patients and in patients with PH in association with hereditary hemorrhagic telangiectasia (HHT). HHT is a rare autosomal dominant genetic disorder characterised by arteriovenous malformations and multiple telangiectasias [184]; it is frequently linked to defects in *ACVRL1* and *ENG* and as HHT and PAH may co-present in families, suggests a common molecular aetiology [103–105]. Of note, PH secondary to high cardiac output from arteriovenous fistulas is much more common in HHT, and such phenocopies, if unrecognised, may introduce significant bias to the studies [185]. Conversely, I/HPAH associated with *ACVRL1* and *ENG* can occur without clinical features of HHT [105,185,186], as the latter shows age-related penetrance. In a large case-control study, which employed deep phenotyping prior to association analysis, *ACVRL1* was associated with HPAH [30] but fell just below the cut-off for significance when studied in unselected patients with Group 1 PAH [25].

Two studies using targeted sequencing of *BMPR2* signalling intermediates provided further evidence supporting the role of this pathway in the pathogenesis of PAH. Shintani et al. [106] screened 23 patients with IPAH for mutations in *ENG*, *SMAD1*, *SMAD2*, *SMAD3*, *SMAD4*, *SMAD5*, *SMAD6* and *SMAD9* (*SMAD8*) and identified a nonsense mutation in *SMAD9* in a child who was diagnosed at eight years of age and his unaffected father. The results of immunoblotting and co-immunoprecipitation assays indicated that the *SMAD9* mutant disturbs the downstream signalling of TGF-β/BMP. In a later study, *SMAD1*, *SMAD4*, *SMAD5* and *SMAD9* were screened by direct sequencing in a cohort of 324 PAH cases (188 IPAH and 136 anorexigen-induced PAH) [54]. Four gene defects in three genes were observed. A novel missense variant in *SMAD1* was observed in an IPAH patient, a predicted splice-site mutation and a missense variant in *SMAD4* were observed in two IPAH patients and a novel missense variant in *SMAD9* was observed in a patient of Japanese origin. These four variants were absent in the 960 European and 284 French control samples and the *SMAD9* variant was excluded from the panel of 340 Japanese controls. A case-control study using WGS detected two cases harbouring protein-truncating variants in *SMAD1*, of which one co-existed with a protein-truncating variant in *BMPR2*, and eight *SMAD9* variants, two of which co-occurred with protein-truncating variants in *BMPR2* and *GDF2*; statistical analysis did not reveal significant association with studied phenotypes [30]. In another large cohort study (*n* = 2572 cases, 72% European), deleterious variants were observed in *SMAD1* (two cases), *SMAD4* (two cases) and *SMAD9* (13 cases) but were not statistically significant [25]. Taken together, these findings demonstrate that variations within the *SMAD* family have a small effect size, suggesting that a second genetic or environmental hit is needed, or that they perturb other non-investigated pathways.

Besides *BMPR2* mutations, *CAV1* mutations are a rare cause of PAH. Variants in *CAV1* were initially implicated in PAH pathogenesis by exome sequencing of a three-generation family with autosomal dominant HPAH who were negative for established variants in the TGF-β family [74]. They identified a frameshift mutation in *CAV1*; all PAH patients and several unaffected family members carried the *CAV1* mutation, suggesting incomplete penetrance. Subsequent evaluation of an additional 62 unrelated HPAH and 198 IPAH patients identified an independent de novo *CAV1* mutation in a child with IPAH. Two separate studies have since identified *CAV1* mutations in PAH patients; the first identified a novel heterozygous frameshift mutation in an adult PAH patient with a paediatric-onset daughter who died at nine years old [140], and the second identified deleterious variants in 10 patients with I/F/APAH, with three related cases carrying the same likely gene damaging mutation [25]. WGS in an I/HPAH cohort did not detect deleterious variants in *CAV1* [24]. Whilst these findings highlight the importance of caveolae in the homeostasis of the pulmonary vasculature, the link between *CAV1* mutations and PAH requires further study.

Bone morphogenetic protein (BMP)-9 (encoded by *GDF2*) and *BMP10* are ligands involved in TGF-β signalling pathway. Wang et al. [121] identified a novel homozygous nonsense mutation in the *GDF2* gene in a five-year-old Hispanic child with severe PAH. Genetic testing revealed that both parents were heterozygous for the same mutation, indicating that the child inherited the *GDF2* mutant allele from each parent. This study was the first to report a novel homozygous nonsense mutation in *GDF2* in an IPAH patient, suggestive of the causative role of *GDF2* mutations in PAH. Further evidence came from the NBR study which identified associations between rare heterozygous missense ($n = 7$) and frameshift variants ($n = 1$) in adult-onset IPAH (88% European) [24]; additionally, Hodgson et al. [115] identified two patients with large deletions encompassing the *GDF2* locus and several neighbouring genes.

The identification of *GDF2* mutations has since been independently replicated in a Chinese cohort [26]. Wang et al. [26] performed an exome-wide gene-based burden analysis on two independent case-control studies. The discovery analysis, containing 251 IPAH patients, identified rare heterozygous mutations in *BMPR2* (49 cases), *ACVRL1* (15 cases), *TBX4* (10 cases), *SMAD1* (two cases), *BMPR1B* (one case), *KCNK3* (one case) and *SMAD9* (one case). In a gene-based burden analysis (cases: $n = 251$; controls: $n = 1884$), only three genes (*BMPR2*, *GDF2* and *ACVRL1*) had an exome-wide significant enrichment of mutations in IPAH cases when compared to healthy controls. *GDF2* mutations were identified in 17 cases (6.8%) and ranked second to *BMPR2* (56 cases, 22.3%). To validate the risk effects of *GDF2*, they performed WES in an independent replication cohort of 80 IPAH cases and in a second gene-based burden analysis (cases: $n = 80$; controls: $n = 8624$), *BMPR2*, *GDF2* and *ACVRL1* were again identified as the top three disease-associated genes. Within this analysis, five additional *GDF2* heterozygous mutations were identified. Among the 331 IPAH patients, they identified 22 cases carrying 21 distinct rare heterozygous mutations in *GDF2*, only two of which had been reported previously [24], accounting for 6.7% of IPAH cases. An independent cohort confirmed the genome-wide association of *GDF2* among 1832 PAH and 812 IPAH cases of European ancestry [25]; twenty-four *GDF2* variants were observed in 28 cases, only two of which had been reported previously, and 75% of these occurred in IPAH cases.

Additionally, gene panel sequencing of 263 PAH patients (180 IPAH, 11 FPAH, 13 drug and toxin-induced PAH and 59 sporadic PVOD) revealed two (1.2%) *BMP9* mutations in adult PAH cases [114]; due to the close similarity of *BMP9* and *BMP10* (a close paralogue of *BMP9* that encodes an activating ligand for *ACVRL1*), the *BMP10* gene was also included in the capture design. Two mutations were identified in *BMP10*, a truncating mutation and a predicted loss of function variant were identified in two severely affected IPAH patients. Two rare missense variants in *BMP10* were identified in patients with IPAH in an independent cohort [115]. These results emphasise the role of *GDF2* in the pathobiology of PAH and suggest *BMP10* might act as a predisposing risk factor.

Channelopathies

Channelopathies are a group of diseases caused by dysfunction of ion channels localised in cellular membranes and organelles. These diseases include, but are not limited to, cardiac, respiratory, neurological and endocrine disorders. LoF variants in channelopathy genes have also been reported in PAH.

The first channelopathy described in PAH was caused by a genetic defect in *KCNK3* in patients with familial and sporadic PAH [107]. *KCNK3* belongs to a family of mammalian potassium channels and encodes for a two-pore potassium channel which is expressed in pulmonary artery smooth muscle cells (PASMCs); this channel plays a role in the regulation of resting membrane potential and pulmonary vascular tone and vascular remodelling [120]. In studying a family in which multiple members had PAH, Ma et al. [107] identified a novel heterozygous missense variant in *KCNK3* as a disease-causing candidate gene within the family. WES was used to study an additional 10 probands with FPAH and two novel heterozygous *KCNK3* variants were identified and segregated with the disease. In addition, three novel *KCNK3* variants were identified in 230 patients with IPAH. These five variants were predicted to be damaging. In summary, *KCNK3* mutations were identified in three of 93 unrelated patients (3.2%) with FPAH and in three of 230 patients (1.3%) with IPAH [107].

In another study using targeted sequencing, two *KCNK3* mutations were observed in three patients from two families. One of these mutations, a homozygous missense variant in *KCNK3*, was identified in a patient belonging to a consanguineous Romani family; his affected mother and asymptomatic father were carriers of the same *KCNK3* mutation. This is the first report of a young patient with severe PAH carrying a homozygous mutation in *KCNK3* [120]. Of note, in the Ma et al. [107] study, as the pedigree suggested an autosomal dominant mode of inheritance, homozygous variants were excluded from the analysis. The two biggest case-control studies, reported by Gräf et al. [24] and Zhu et al. [25], identified heterozygous KCNK3 mutations in only four (0.4%) and three (0.1%) cases, respectively, and did not show statistically significant associations.

Conversely, Gräf et al. [24] detected statistically significant enrichment of rare deleterious variants in two new channel genes: *ATP13A3* and *AQP1*. Utilising a rigorous case-control comparison using a tiered search for variants, they searched for high-impact protein-truncating variants (PTVs) overrepresented in cases and identified a higher frequency of PTVs in *ATP13A3* (six cases). *ATP13A3* is a poorly characterised member of the P-type ATPase family of proteins that transport a variety of cations across membranes [187]; *ATP13A3* is thought to play a role in polyamine transport [188]. Within PAH cases, Gräf et al. [24] identified three heterozygous frameshift variants, two stop gain, two splice region variants and four heterozygous likely pathogenic missense variants in *ATP13A3*. These variants were predicted to lead to loss of ATPase catalytic activity, and to destabilise the conformation of the catalytic domain; six variants were predicted to cause protein truncation, suggesting that loss of function of *ATP13A3* contributes to PAH pathogenesis.

Within the same study, SKAT-O analyses revealed a significant association with rare variants in *AQP1* [24]; *AQP1* ranked second in their combined rare PTV and missense variant case-control analysis. Along with statistical evidence, familial segregation of *AQP1* variants with the phenotype was shown in three families [24]. *AQP1* belongs to the aquaporins family, a family of water-specific membrane channel proteins that facilitate water transport in response to osmotic gradients [189]. Zhu et al. [25] identified seven cases with *ATP13A3* rare deleterious variants. However, *ATP13A3* and *AQP1* failed to reach genome-wide significance in their study and additionally, *AQP1* was not among the expanded list of genes with $p \leq 0.001$ for either the whole cohort or the IPAH subset.

Interestingly, a mutation in ATPase Na^+/K^+ transporting subunit α 2 (*ATP1A2*), previously associated with familial hemiplegic migraine (FHM), was reported in a 24-year old male with IPAH and history of FHM [190]. Genetic analysis of the proband and two siblings (one with FHM) revealed a nucleotide substitution in the coding sequence of the *ATP1A2* gene for both the proband and the affected sibling [190]. A co-occurrence of PAH and FHM supports the hypothesis of a potential common pathophysiological link; several studies have reported the presence and activity of the α2-subunit of the

Na$^+$/K$^+$-ATPase in pulmonary vascular smooth muscle cells [191,192], and the decrease in expression and/or activity of different types of K$^+$ channels in PASMCs of IPAH patients [193,194], and Montani et al. [190] suggested that mutations in *ATP1A2* may contribute to pulmonary arterial remodelling through the disturbance of intracellular Ca^{2+} and K$^+$ concentrations.

Mutations in *ABCC8* have recently been identified as a potential second potassium channelopathy in PAH [113]. *ABCC8* encodes SUR1 (sulfonylurea receptor 1), a regulatory subunit of the ATP-sensitive potassium channel; *ABCC8* is highly expressed in the human brain and endocrine pancreas and moderately expressed in human lungs [195]. Mutations in *ABCC8* have previously been related to type II diabetes mellitus and congenital hyperinsulinism [196]. In exome-sequencing a cohort of 99 paediatric- and 134 adult-onset Group 1 PAH patients (182 IPAH and 52 HPAH), Bohnen et al. [113] identified a de novo heterozygous predicted deleterious missense variant in *ABCC8* in a child with IPAH. All individuals within this cohort and the second cohort of 680 adult-onset PAH patients (NBR study) were screened for rare or novel variants in *ABCC8*. Eleven heterozygous predicted damaging *ABCC8* variants were identified, seven of these were observed in the original cohort, including one familial case. In a study of 2572 PAH cases, Zhu et al. [25] identified rare deleterious variants in newly reported risk genes and nearly two-thirds of these variants were in *ABCC8* (26 variants in 29 IPAH/APAH patients). In a more recent study, utilising a custom NGS targeted sequencing panel of 21 genes, Lago-Docampo et al. [197] identified 11 rare variants in *ABCC8* within a cohort of 624 paediatric and adult patients from the Spanish PAH registry. To date, *ABCC8* variants have been identified in patients with IPAH, FPAH, PAH-CHD and APAH and account for ~0.5–1.7% of cases.

Transcription Factors

An emerging category of HPAH is the one that can be labelled as a disorder of transcriptional regulation. Interestingly, variants within developmental transcription factors are enriched in paediatric patients.

The first transcription factor implicated in the pathogenesis of PAH was *TBX4*. *TBX4*, expressed in the atrium of the heart, limbs, and the mesenchyme of the lung and trachea, encodes a transcription factor in the T-box gene family [198]. Deletions and LoF mutations in *TBX4* cause a variety of developmental lung disorders [199] and have been identified as a prominent risk factor in small patella syndrome (SPS) [200], childhood-onset PAH [108] and more recently, persistent pulmonary hypertension in neonates [201]. In a 2013 study incorporating array-comparative hybridisation and direct sequencing, three *TBX4* mutations and three novel *TBX4* microdeletions were detected in six out of 20 children with I/HPAH and interestingly, features of SPS were detected in all living *TBX4* mutation carriers [108].

Zhu et al. [140] performed exome sequencing on a cohort of 155 paediatric- and 257 adult-onset PAH patients. Within 13 probands (12 paediatric- and one adult-onset), they identified 13 likely pathogenic/predicted highly deleterious *TBX4* variants; eight of these variants were inherited from an unaffected parent, whereas one was de novo. This pattern is consistent with the incomplete penetrance observed for *BMPR2* mutation carriers [122]. Similar frequencies of rare, deleterious *BMPR2* mutations were observed in paediatric- and adult-onset I/FPAH patients; however, there was significant enrichment of rare, predicted deleterious *TBX4* mutations in paediatric- (10 of 130 patients) compared with adult-onset (0 of 178 patients) IPAH patients. In comparison to *BMPR2* mutation carriers, *TBX4* carriers had a 20-year younger age of onset, with a mean age of onset of 28.2 ± 15.4 years and 7.9 ± 9.0, respectively. After *BMPR2* mutations (10%), variants in *TBX4* (7.7%) conferred the highest degree of genetic risk of paediatric-onset IPAH [140]. Similar estimates of *BMPR2* (12.5% of I/FPAH patients) and *TBX4* mutation carriers (7.5% of I/FPAH patients) were observed in a study of 66 paediatric patients [141]. Additionally, in a 2019 study, examining 263 PAH and PVOD/PCH patients (paediatric and adult cases), *TBX4* mutations were the second most frequent mutations after *BMPR2* in both paediatric and adult cases [114].

Indicative of bimodal age distribution, pathogenic *TBX4* variants have also been reported in adult-onset PAH. Gräf et al. [24] identified deleterious heterozygous rare variants in 14 cases, Navas Tejedor et al. [120], in a Spanish cohort of 136 adult-onset PAH patients, identified three pathogenic mutations, and Kerstjens-Frederikse et al. [108], in a much smaller adult cohort ($n = 49$), detected a rare *TBX4* mutation. In a more recent study, 448 index patients were screened for PAH predisposing genes; 20 patients (nine childhood-onset) from 17 unrelated families carried heterozygous mutations in the *TBX4* gene, bringing the frequency of *TBX4* mutations in France to 6% and 3% in childhood- and adult-onset PAH, respectively [142]. Within this cohort, SPS was present in 80% of cases [142] and interestingly, all patients showed decreased DLCO and 87% had parenchymal abnormalities. These findings suggest that *TBX4* mutations may occur with or without skeletal abnormalities and whilst such mutations are mainly associated with childhood-onset PAH, the prevalence of PAH in adult *TBX4* mutation carriers could be up to 3%, depending on the population studied.

A recent study of 1038 PAH patients observed that rare heterozygous variants in *SOX17* were significantly overrepresented in the I/HPAH cohort [24]. *SOX17* is a member of the conserved SOX family of transcription factors. These transcription factors play a pivotal role in cardiovascular development and figure prominently in the aetiology of human vascular disease; they are involved in the regulation of embryonic development, the determination of cell fate and participate in vasculogenesis and remodelling [202]. In the NBR study [24], deleterious variants in *SOX17* were detected in less than 1% of the studied population and were characterised by younger age at diagnosis; familial segregation was shown in one patient.

To identify novel genetic causes of PAH-CHD, Zhu et al. [75] performed WES in 256 PAH-CHD patients; the cohort included 15 familial and 241 sporadic cases. Fifty-six percent of the cohort were of European ancestry and 26% were Hispanic; most cases (56%) had an age of onset < 18 years and so were categorised as paediatric-onset. The cohort was screened for 11 known risk genes for PAH and 253 candidate risk genes for CHD; PAH risk variants were identified in only 6.4% of sporadic PAH-CHD cases and four of the 15 familial cases. They performed a case-control gene-based association test of rare deleterious variants comparing European cases and controls (gnomAD: $n = 7509$ non-Finnish Europeans) and identified *SOX17* as a novel PAH-CHD candidate risk gene [75]. They estimated that rare deleterious variants in *SOX17* contributed to approximately 3% of European PAH-CHD patients. Following this discovery, they screened for *SOX17* variants in non-European cases and an additional cohort of PAH patients without CHD ($n = 413$) and identified five additional rare variants in the PAH-CHD cohort and three additional rare variants in the I/HPAH cohort [75]. In this second cohort, rare deleterious variants in *SOX17* were observed in 0.7% of cases. In total, 13 patients across the two cohorts were observed to have rare deleterious *SOX17* variants; nine of these had paediatric-onset PAH, suggesting these variants may be enriched in paediatric patients.

A Japanese study also demonstrated familial segregation of *SOX17* variants [112]. This study whole-exome sequenced 12 patients with PAH, 12 asymptomatic family members and 128 index cases and identified *SOX17* mutations in four PAH patients (three of these had congenital heart defects, i.e., atrial septal defect or patent ductus arteriosus) and one asymptomatic family member. Interestingly, the same heterozygous missense mutation in *SOX17* (c.397C) was observed in a Japanese patient [112] and in a patient with PAH from the NBR study [24], suggesting that this base position may be a pan-ethnic mutational hot spot. Taken together, these data strongly implicate *SOX17* as a new risk gene for PAH-CHD and suggest that this gene has a pleiotropic effect. Replication analyses in other PAH cohorts, with specific PAH subclasses, are needed to confirm the precise role of *SOX17*.

New Genes

In recent years, new risk genes for PAH have emerged from large WES and WGS studies. Zhu et al. [25] performed targeted gene sequencing alongside WES in a large cohort from the National Biological Sample and Data Repository for PAH (US PAH Biobank: $n = 2572$). Despite screening for 11 established PAH risk genes and seven recently reported risk genes, they failed to identify rare deleterious variants in known risk genes for 86% of the PAH Biobank cases (Group 1 PAH). They performed gene-based case-control association analysis and to prevent confounding by genetic ancestry, only participants of European ancestry (cases: n = 1832; controls: n = 7509 gnomAD WGS subjects and n = 5262 unaffected parents from the Pediatric Cardiac Genomics Consortium) were included. Using a variable threshold method, they identified two genes that exceeded the Bonferroni-corrected cut-off for significance: *BMPR2* and *KLK1*. *KLK1*, which encodes kallikrein 1, also known as tissue kallikrein, has not previously been associated with pulmonary hypertension.

The analysis was then repeated using 812 European IPAH cases and significant associations were observed for *BMPR2*, *KLK1* and *GGCX*. *GGCX* encodes γ-glutamyl carboxylase and has previously been implicated in coagulation factor deficiencies and vascular calcification [203], but again, never in PAH. In a final analysis, the entire PAH Biobank cohort was screened for rare deleterious variants in *KLK1* and *GGCX*; twelve cases carried *KLK1* variants (all European), whereas 28 cases carried *GGCX* variants (19 European, six African, three Hispanic), accounting for ~0.4% and ~0.9% of PAH Biobank cases, respectively. Carriers of *KLK1* and *GGCX* had a later mean age of onset and relatively moderate disease phenotype compared to *BMPR2* carriers. Both *KLK1* and *GGCX*, expressed in the lung and vascular tissues, play an important role in vascular haemodynamics and inflammation. Whilst Zhu et al. [25] identified *KLK1* and *GGCX* as new candidate risk genes for IPAH, suggesting new pathogenic mechanisms outside of the TGF-β/BMP signalling pathway, further research needs to be conducted to better understand these findings, especially in larger cohorts of similar phenotypic characteristics.

In a recent large-scale analysis utilising the 13,037 participants enrolled in the NBR study, of which 1148 patients were recruited to the PAH domain, we discovered *KDR* as a novel PAH candidate gene [24,30] utilising the Bayesian model comparison method, BeviMed [64], and deep phenotype data. Under an autosomal dominant mode of inheritance, high impact variants in *KDR* were associated with a significantly reduced KCO (transfer coefficient for carbon monoxide) and older age at diagnosis [30]. Six ultra-rare high impact variants in *KDR* were identified in the study cohort; four of these were in unrelated PAH cases, one in a relative and one nonsense variant was identified in a non-PAH control subject. The latter variant appeared late in the protein sequence and hence might not impair protein structure. To seek further evidence for *KDR* as a new candidate gene for PAH, we analysed subjects recruited to two cohorts with similar phenotypic characteristics (US PAH Biobank: n = 2572; Columbia University Medical Center: n = 440); four additional individuals harbouring rare high impact *KDR* variants were identified, with one variant identified in both cohorts. A combined analysis of both cohorts confirmed the association of *KDR* with PAH. Further evidence came from a French cohort of 311 PAH patients prospectively analysed by targeted panel sequencing (which included *KDR*) [117]. Two index cases with severe PAH, from two different families, were found to carry LoF mutations in *KDR*, providing further genetic evidence for considering *KDR* as a newly identified PAH-causing gene. Across both studies [30,117], there are now three reported familial cases with a distinct phenotype in which LoF variants in *KDR* segregate with PAH and significantly reduced KCO.

Power to detect novel genotype–phenotype associations can be increased by merging existing datasets, aligning sequences to the most recent genome assembly, using the most up to date reference variome, and improved variant classification tools, as well as updated clinical phenotypes. Such an approach was recently taken by a large international consortium of 4241 PAH cases from three cohorts (US PAH Biobank: n = 2572; Columbia University Medical Center: n = 469; NBR: n = 1134). Of the available 4175 sequenced exomes, most cases (92.6%) were adult-onset, with 54.6% IPAH, 34.8% APAH, 5.9% FPAH and 4.6% other; 74.5% of the cohort was European [118]. Gene-based case-control association analysis in unrelated participants of European ancestry was performed, using 2789 cases and 18,819 controls taken from the Simons Foundation Powering Autism Research for Knowledge (SPARK) cohort and gnomAD, before screening the whole cohort, including non-Europeans, for rare deleterious variants in associated genes. Statistical analyses revealed that rare predicted deleterious variants in seven genes were significantly associated with IPAH, including three established PAH risk genes (*BMPR2*, *GDF2*, and *TBX4*), two recently identified candidate genes (*SOX17* and *KDR*) and two new candidate genes (*FBLN2* and *PDGFD*).

Both new candidate genes have known functions in vasculogenesis and remodelling but have not previously been implicated in PAH. In total, they identified seven cases with *FBLN2* variants and ten cases with *PDGFD* variants, accounting for 0.26% and 0.35% of IPAH cases, respectively; most of these were of European ancestry and all were adult-onset, except for one paediatric *PDGFD* variant carrier. Analysis of single-cell RNAseq data showed that *FBLN2* and *PDGFD* have similar expression patterns to well-known PAH risk genes [118]. Whilst this provides additional support and mechanistic insight for the new genes, as with the discovery of *KLK1* and *GGCX*, variants within *FBLN2* and *PDGFD* require independent validation.

Autosomal Recessive Mode of Inheritance

PVOD/PCH has recently been reclassified as an ultra-rare form of Group 1 PAH [18]. PVOD and PCH often show significant phenotypic overlap. Indeed, 73% of patients diagnosed with PVOD are found to have capillary proliferation and 80% of patients with PCH demonstrate typical venous and arterial changes [204], and are, therefore, referred to as PVOD/PCH. Clinically, PVOD/PCH is characterised by early-onset, significantly reduced DLCO and patchy centrilobular ground-glass opacities, septal lines and lymph node enlargement seen on high-resolution computed tomography. The disease outcome is dismal, with rapid progression and frequent pulmonary oedema in response to PAH medication. Similarly to PAH, PVOD/PCH can present as either a sporadic or familial disease [205,206]. It was the latter that triggered a family-based study into the genetic basis of this condition. Familial linkage mapping, WES, and Sanger sequencing were employed and identified biallelic *EIF2AK4* mutations in affected siblings. Subsequently, biallelic *EIF2AK4* mutations were also identified in 25% of sporadic PVOD cases [109], 11.1% of HPAH cases (one of nine cases) [110] and 1.04% of I/HPAH cases (nine of 864 cases) [111]. Harbouring *EIF2AK4* mutations confer a poor prognosis irrespective of clinical diagnosis and importantly, radiological assessments were unable to distinguish reliably between PVOD/PCH patients and patients with IPAH [111].

4.3.2. Common Genetic Variation

Complex pathobiology, low penetrance, heterogeneous phenotype and variable disease trajectory allow for common sequence variation that contributes to PAH risk and natural history. In HPAH, several common genetic variants were shown to impact the disease. Firstly, *BMPR2* mutations are a significant source of sex-related bias in disease penetrance (42% in females vs 14% in males) [122]; one explanation for this could be gene polymorphisms involved in estrogen metabolism [207]. Females harbouring deleterious variants in *BMPR2* show a significant reduction in *CYP1B1* gene expression and as a consequence, a lower 2-hydroxyestrone to 16α-hydroxyestrone ratio leading to activation of mitogenic pathways [207]. Moreover, direct estrogen receptor α binding to *BMPR2* promoter results in reduced *BMPR2* gene expression in females and may contribute to the increased prevalence of PAH [208]. Secondly, variation in *BMPR2* expression in HPAH, caused by nonsense-mediated decay positive (NMD$^+$) *BMPR2* mutations, is conditional upon individual polymorphisms of the wild type (WT) allele [209]. Thirdly, a SNP in *TGFβ1* modulates age at disease onset in patients harbouring *BMPR2* mutations [210].

In IPAH, a SNP (rs11246020) in the Sirtuin3 (*SIRT3*), mitochondrial deacetylase, in either homozygote or heterozygote fashion, was associated with increased acetylation of mitochondrial proteins compared to the IPAH patients or disease comparator group with the WT genotype [166]. *SIRT3*, by deacetylating and thus activating multiple enzymes and electron transport chain complexes, plays a significant role in cell bioenergetics, and its polymorphism has been associated with susceptibility to metabolic syndrome [211] and IPAH, but not APAH [166]. Similarly, uncoupling protein 2 (*UCP2*), shown to conduct calcium from the endoplasmic reticulum (ER) to mitochondria and suppressing mitochondrial function, has been implicated in the pathogenesis of PAH [168].

These findings have important therapeutic implications as common variants in *SIRT3* and *UCP2* predicted response to dichloroacetate, pyruvate dehydrogenase kinase inhibitor, in a PAH phase 2 clinical trial [167]. Renin–angiotensin–aldosterone (RAA) system has been implicated in the pathogenesis of both systemic and pulmonary hypertension. For example, polymorphisms in the gene encoding angiotensin-converting enzyme (ACE) have been associated with IPAH [212] and with diaphragmatic hernia with persistent pulmonary hypertension [213]. Likewise, common variation in angiotensin II type 1 receptor (*AGTR1*) was associated with age at diagnosis in PAH. These findings indicate that RAA might be a therapeutic target [34].

Considering the important role of vasoactive, angiogenic and angiostatic substances in the pathogenesis and therapy of PAH, studies investigating the impact of common variation on gene expression of these substances are warranted. Recently, Villar et al. [170] found a recurrent SNP (rs397751713) in the promoter region of the endothelin-1 (*END1*) gene that had important regulatory consequences in both IPAH and APAH patients; rs397751713, which consists of an adenine deletion, allows transcription factors (Peroxisome proliferator-activated receptor γ (PPARG) and Kruppel Like Factor 4 (KLF4)) to bind to the promoter. Both of these transcription factors are linked to PAH pathogenesis and have been suggested as potential therapeutic targets [214–217]. In a similar fashion, a polymorphism in a potent angiostatic factor, endostatin, Collagen Type XVIII α 1 Chain (*Col18a1*), was studied. A SNP (rs12483377) in *Col18a1* was observed at an increased frequency in PAH patients (MAF 21.6) relative to published controls (MAF 7.5), or patients with scleroderma without PAH (MAF 12.5). Rs12483377 encodes uncharged amino acid asparagine (N) at residue 104 in place of negatively charged aspartic acid (D); carriers of single minor allele A (genotype AG), had lower serum endostatin levels than those who carried WT (genotype GG) and showed better survival even after adjusting for confounders [173].

Along with studies looking at polymorphisms in a particular gene or its promoter region, two GWAS have been carried out in PAH. The first consisted of a discovery and validation cohort totalling 625 patients (*sans BMPR2* mutations) and 1525 healthy individuals. A genome-wide significant association was found at Cerebellin 2 Precursor (*CBLN2*) locus; rs2217560-G allele was more frequent in cases than in controls in both the discovery and validation cohorts with a combined odds ratio (OR) for I/HPAH of 1.97 [95% CI 1.59;2.45] (*p*-value 7.47×10^{-7}). Although the rs2217560 genotype did not correlate with *CBLN2* mRNA expression in monocytes, higher *CBLN2* mRNA levels were seen in lung tissue explanted from PAH patients than in control lung samples. Further analysis showed that increasing concentrations of *CBLN2* lead to inhibition of proliferation of PASMCs [165].

To date, the largest PAH GWAS of multiple international I/HPAH cohorts, totalling 2085 cases and 9659 controls, identified two novel loci associated with PAH: an enhancer near *SOX17* (rs10103692, OR 1.80 (95% CI 1.55;2.08), $p = 5.13 \times 10^{-15}$), and a locus within *HLA-DPA1/DPB1* (rs2856830, OR 1.56 (95% CI 1.42;1.71), $p = 7.65 \times 10^{-20}$) [116]. CRISPR-mediated inhibition of the enhancer reduced *SOX17* expression; this finding corroborates and extends the previous discovery of the association of rare variants in *SOX17* with PAH [24]. The *HLA-DPA1/DPB1* rs2856830 genotype was not only associated with I/HPAH but also with a beneficial effect on survival (CC genotype 13.5 vs TT genotype 6.97 years) irrespective of baseline disease severity, age and sex [116]. The latter study did not replicate previous results at genome-wide significance.

4.4. Limitations, Challenges and Future Directions

The major limitations in PAH genetic studies originate from either their study design or the genetic methods used. The former is mostly limited by sample size and population structure and features in GWAS and RVAS alike, the latter is impacted by sequencing technology, genome assembly and analysis pipelines. Additionally, the variable quality of reporting may obscure the validity of the results. Several approaches may accelerate causal variant discovery. Firstly, international efforts to set up platforms and governance, thereby reducing the barriers to genotype and phenotype data harmonisation, across studies and countries, along with regulatory standards, compatible with the international legislative landscape, aid variant discovery. Secondly, the application of computational and/or intermediate phenotypes may further increase power to detect new risk genes. Likewise, researching isolated populations resulting from recent bottlenecks (i.e., Icelanders, Ashkenazi Jews) have shown promising results in other rare diseases. As for genetic methods, long-read technologies will improve significantly and with improvements to the error rate, long-read alignment to a pan-genome graph may identify additional genetic variants. Expansion of resources for allele frequency estimation (i.e., gnomAD), encompassing diverse ethnicities and variant classes, including structural variants, will improve variant interpretation. Studying gene sets that are likely to be enriched for disease-associated loci (i.e., implicated by GWAS, or expressed in disease relevant tissues) and extending RVAS to non-coding regions (a natural step given GWAS findings) is likely to further increase discovery yield.

5. Reverse Genetics

5.1. Concepts

Reverse genetic techniques have been used extensively in the context of PAH. These experimental approaches involve targeted modifications to specific genes in order to analyse phenotypic impact (Figure 1, process 3) and are critical in validating in silico predictions. Most deleterious mutations found in PAH patients are LoF, suggesting a knockout or knockdown approach to analysing the phenotypic effect of a mutation. The main benefit of direct gene mutagenesis is the permanence and potentially ubiquitous effect of that alteration. In comparison, gene knockdown is a transient change that can be created after the developmental stage in the animal life cycle. This is particularly useful for understanding LoF in genes that are embryonically lethal [123]. Conditional and inducible systems are also excellent means of circumventing this issue, permitting temporally and spatially controlled gene disruption [55,125,128,218]. Inducible systems display higher efficiency and limited side effects compared to stably-expressed mutations and have the added benefit of reversibility [219].

5.2. Methodology

Techniques for genetic modification have permitted targeted genetic analysis using in vitro and in vivo models of PAH. Modification of a specific gene, through mutagenesis or knockout, can be carried out using a wide range of tools, including Zinc-finger nucleases (ZFNs) [220], transcription activator-like effector nucleases (TALENs) [221], and CRISPR/Cas9 [222]. Similarly, transient genetic changes can be achieved through gene expression knockdown, such as RNAi [223], or using conditional/inducible expression systems, such as Cre/LoxP and Flp/FRT sites or the Tet inducible system [224]. Generating models with specific mutations is also critical for the identification of therapeutic targets and drug testing. Specific techniques and their utility in animal models have been extensively reviewed [224–226].

5.2.1. In Vitro

As PAH is a disease of the pulmonary circulation, most studies seek to model genetic changes in physiologically relevant cell lines. This includes pulmonary artery endothelial cells (PAECs), PASMCs, blood outgrowth endothelial cells (BOECs) [227] and human umbilical vein endothelial cells (HUVECs). PAH is characterised by molecular and cellular deviations in pulmonary vascular function [228,229], including hyperproliferation [124,129,130,149], impaired migration [127,130,150] and aberrant apoptosis [124,230]. Traditionally, research has focused on using 2D cellular monolayers for analysis of these features; however, with rapidly evolving technologies, future analysis may utilise multi-layer, 3D models such as organoids [231] and hydrogels [232] to better replicate the pathobiology of PAH.

5.2.2. In Vivo

Understanding the pathobiology of PAH is dependent upon robust animal models for functional analysis and exploration of potential therapeutics. To achieve this aim, a wide range of animal models have been developed across species, including pig, sheep, dog and most commonly rodents [233].

The most widely accepted rat models for PAH are pharmacologically induced, namely monocrotaline (MCT) and Sugen 5416-hypoxia (SuHx) models [234]. MCT is a toxin that induces damage to PAECs, causes RV and pulmonary arterial medial hypertrophy, and alters pulmonary artery pressures [235,236]. In the SuHx rat model, SU5416, a Vascular Endothelial Growth Factor Receptor (VEGFR) antagonist that promotes endothelial cell apoptosis and smooth muscle cell (SMC) proliferation [237], is administered followed by a period of hypoxia [234]. SU5416 requires hypoxia to induce severe PAH, causing vascular changes that model human IPAH [237]. Both models induce non-PAH related symptoms, including alveolar oedema, pulmonary vein occlusion, acute lung injury, liver toxicity and emphysema [238,239].

The chronic hypoxia mouse model is also widely used in PAH research, with mice housed at 10% oxygen for a variable length of time [234,240,241]. While mice display changes in pulmonary artery pressure, they display limited vascular remodelling, which consists of muscularization of vessels and medial hypertrophy of muscular resistance vessels [242]. Alternatively, pulmonary artery banding (PAB) has been used in various species to increase pulmonary artery pressures and induce RV hypertrophy. The main benefit of this technique is the lack of pharmacologic or environmental modification of the animal [243]. In practicality, the popularity of each model varies across PAH research groups; a recent poll showed that two-thirds of groups used two or more models [244], suggesting that no one model truly replicates the disease.

In comparison, genetic models tend to be more favourable as they are free from the systemic effects of pharmacological models. Several heterozygous *BMPR2* mutant models have been developed. While some models have shown elevated mPAP and PVR compared to WT littermates [123], others did not replicate these findings [245,246]. Alternate studies have shown the development of pulmonary vascular lesions [126] and often, an environment hit may be required for development of PAH [131,245,246]. More pronounced phenotypes are seen in homozygous deletions; as *BMPR2* is critical for embryogenesis, conditional deletion in pulmonary endothelial cells has been used and showed incomplete penetrance of the mutation, with a subset of mice displaying elevated RVSP, RV hypertrophy and inflammation [125].

The following genetic models have also been developed: *KDR* (cell-specific conditional deletion of *Kdr* in mice) [155], *AQP1* (*Aqp1* null mice [151] and *Aqp1* with a COOH-terminal tail deletion [149]), *TET2* (conditional, haematopoietic heterozygous and homozygous deletion in mice, generated using the Vav1-Cre system) [55], *BMPR1a* (several conditional knockout mice models) [156,157,159], *SOX17* (conditional deletion in mice, using *Demo1-Cre* in descendants of the splanchnic mesenchyme) [153], *UCP2* (knockout mouse model created using a plasmid vector carrying modified *Ucp2*) [169], *CAV1* (modified mouse cav-1 targeting construct) [138], *EIF2AK4* (modified mouse *Gcn2* targeting construct) [164], *SIRT3* (heterozygous and homozygous mice, generated using targeted embryonic stem cell clones harbouring a mutated *SIRT3*) [166], *ATP13A3* (generation of a protein truncating mutation in mouse *Atp13a3* using CRISPR/Cas9) [147], *SMAD9* (Disrupted *Smad8* allele generated in mice using LoxP/cre and Frt sites) [134], *KCNK3* (heterozygous and homozygous rats harbouring a CRISPR/Cas9-generated exonic, inframe deletion) [146] and *SMAD1* (mice with *Smad1* deletion in endothelial cells or SMCs, generated using L1Cre or Tagln-Cre, respectively) [136].

In humans, the severity of disease corresponds to the extent of pulmonary vascular changes and impairment of RV function. However, rodent models may display milder and isolated symptoms of PAH and, thus, require thorough characterisation to validate the model. Echocardiography is useful for the assessment of RV function but estimates of pulmonary pressures are inferior to direct measurements obtained through RHC [234]. Alternatively, cardiac magnetic resonance imaging is another noninvasive tool for assessment of RV function; it supersedes echocardiography in its ability to produce high quality, three-dimensional images from which the RV can be directly measured. However, it also has limited availability and bears higher running costs [247]. Close-chested RHC is preferred over open-chested RHC, as it is less invasive and preserves the negative intrathoracic pressure associated with breathing. Open-chested RHC is usually a terminal procedure, thus preventing longitudinal assessment of rodents; however, it also permits a more comprehensive assessment of haemodynamic characteristics and analysis using pressure–volume loops. Tissue harvesting permits further characterisation of the heart and lungs, including evaluation of RV hypertrophy [234].

It is important that measures are taken to recognise and limit bias in experimentation and phenotyping. This includes ensuring randomisation of groups when carrying out tests that require a control and treatment group, such as selecting mice to be pharmacological models or therapeutic treatment. Additionally, mice should be matched for all possible qualities, including strain, age and sex. During phenotype assessment, researchers should aim to avoid implicit bias by using blinded observers during all procedures. Furthermore, the data collected must be comprehensive and relevant for PAH, including RHC data and histological analysis [248].

5.3. Studies

5.3.1. Rare Genetic Variation

Autosomal Dominant Mode of Inheritance

TGF-β Family

BMPR2, a receptor in the TGF-β pathway, is the most commonly mutated gene in PAH. BMP ligands induce signalling through binding with *BMPR2* and *ALK* 1, 2, 3 or 6 to induce downstream SMAD-mediated transcription [249]. This pathway plays important roles in cell proliferation, differentiation, migration, apoptosis and inflammation, which are often dysregulated in PAH. Alongside *BMPR2*, several pathway components have been implicated in disease; this includes the ALK and SMAD proteins, as well as coreceptors, such as endoglin [250]. Reverse genetics techniques have been used extensively on these genes to understand their role in PAH and their interaction with other members of the pathway. *BMPR2* has been extensively analysed using the methods described; under normal conditions it is highly expressed in vascular ECs, with lower levels of expression in SMC; however, reduced levels have been reported in the lungs of PAH patients [181]. In the endothelium, *BMP9* and *BMP10* induce binding of *BMPR2* and *ALK1*, along with co-receptor endoglin, to induce downstream signalling. Such signalling is often disrupted in HPAH patients that harbour mutations in *BMPR2* [251]. However, disruption of BMP-mediated signalling is also apparent in IPAH patients, despite the absence of *BMPR2* mutation [252].

Gene disruption and knockout techniques have been key to elucidating the role of *BMPR2* in pathogenesis. Such techniques have revealed dysregulation of endothelial nitric oxide synthase (eNOS) in cell culture models of PAH; siRNA-mediated knockdown of *BMPR2* induced endothelial dysfunction in human PAECs, caused by a reduction in BMP2/4-mediated production of the vasodilator nitric oxide by eNOS [127]. In addition, reduced *BMPR2* has been associated with overactive glycolysis [130] and hyperproliferation [124].

Furthermore, a key feature of PAH is incomplete penetrance in *BMPR2* mutation carriers. This has been replicated in mice carrying conditional knockout mutations of *Bmpr2* in pulmonary endothelial cells, with only a proportion of mice developing haemodynamic symptoms of PAH [125]. In addition, it has been shown that *BMPR2* mutations disrupt RV function, with heterozygous mice displaying impaired hypertrophy and lipid accumulation in the RV [128]. Moreover, EndoMT has been suggested to be a characteristic of PAH in both animal models and patients [253,254]. This includes a heterozygous *BMPR2* mutant rat model, which displayed overexpression of two molecular markers of EndoMT. Such changes were also seen in PAH patients when analysing mRNA and protein level changes in lung and artery tissue [129]. These in vivo models have therapeutic utility, having been used to test treatments that include rapamycin [129] and metformin [128] to alleviate symptoms.

The experimental techniques discussed have been highly important in identifying other members of the TGF-β pathway that are implicated in PAH pathogenesis. This includes the Smad proteins, for which functional evidence is limited. Functional analysis of variants detected in *SMAD1*, *SMAD4* and *SMAD9*, have shown impaired signalling in *Smad4* and *Smad9*. However, variants only have a moderate impact on phenotype, suggesting further genetic or environmental hits may be required to induce PAH [54]. Furthermore, analysis of *CAV1* mutations have revealed impaired protein trafficking in mouse embryonic fibroblasts carrying a disrupted copy of *CAV1*. Furthermore, patient fibroblast cells displayed reductions in both *CAV1* and associated accessory proteins [137]. Thrombospondin 1 (*THBS1)* mutants, which regulate TGFβ function, have been found to have impaired ability to activate TGFβ, and permit proliferation of PASMCs. Additionally, an electrophoretic mobility shift assay (EMSA) was used to analyse the binding of biotin labelled oligonucleotides with two transcription factors, MAZ and SP1, showing a reduction in efficiency [174].

As previously discussed, *GDF2*, encoding the ligand BMP9, is critical in *BMPR2*-mediated signalling in endothelial cells. Functional studies are particularly important in validating in silico predictions regarding the pathogenicity of certain variants. Hodgson et al. [115] analysed seven missense mutations in *GDF2*, which had been previously associated with PAH and predicted pathogenic in silico. When analysed in vitro, all variants displayed dysfunctional processing, secretion or protein stability. In addition, functional studies found a predicted benign mutation to be deleterious. Similarly, association analysis in IPAH patients of Chinese Han ethnicity identified *GDF2* mutations [26]. Six mutations were selected for functional studies, using mutant GDF2 plasmids transfected into HEK293E cells/PAECs. Mutants displayed reduced *BMP9* levels, reduced BMP activity and increased apoptosis [26].

There is some evidence implicating *BMPR1A*, which encodes the type I receptor ALK3, in PAH. A hypoxia mouse model harbouring a mosaic deletion of *Bmpr1a*, generated through conditional deletion of the gene in SMCs using the TRE-Cre system, displayed proximal artery stiffening caused by excess collagen. This may lead to constricted vessels and impairment of right ventricular function [255]. Furthermore, two independent studies have implicated reduced *Bmpr1a* expression with EndoMT and shown that the phenotype may be rescued by attenuating expression of TGFBR2, which is upregulated in the mice models [156,157]. Additionally, functional analysis of *BMPR1B* mutations found in PAH patients found changes in transcriptional activity, which resulted in increased SMAD8 phosphorylation [160].

Channelopathies

Alongside TGFβ-family members, various channel proteins have been linked to pathogenesis. Functional analysis was conducted on six mutations reported in *KCNK3*, all of which resulted in LoF. Patch clamp variants showed a reduction in current, with rescue of channel function, in a subset of variants by administration of a phospholipase inhibitor, ONO-RS-082 [107]. In addition, Bohnen et al. [145] introduced a heterozygous, single amino acid variant into *KCNK3* using site-directed mutagenesis and transfected hPASMCs with this mutant via a pcDNA3.1 plasmid. The physiological impact of this mutation was assessed using whole-patch clamp tests to observe changes to membrane potential and current, with experiments demonstrating loss of channel function in mutant cells compared to WT *KCNK3*. More limited evidence is available for the role of *ATP13A3*, a P-type ATPase, in PAH; siRNA-mediated knockdown of expression resulted in a reduction in cell proliferation and increased apoptosis, as well as the loss of endothelial integrity in hPAECs [230]. Furthermore, mutant *ATP13A3* mice, carrying a protein-truncating variant introduced using CRISPR/Cas9, demonstrated haemodynamic measurements indicative of PAH. This included reduced pulmonary artery acceleration time (PAAT) and increased RVSP, with rats also exhibiting reduced polyamine levels in their lungs [147].

AQP1 has been associated with cell migration and vascularity. *AQP1* is highly expressed in microvascular endothelial cells, vascular smooth muscle cells and non-vascular endothelium [256], and has recently been proposed as a novel promoter of tumour angiogenesis [257]. A recent study demonstrated that depletion of *AQP1* reduced proliferation, migratory potential, and increased apoptosis of PASMCs [258]. Saadoun et al. [150] produced an *AQP1* knockout mouse model using targeted gene disruption; when implanted with melanoma cells, the developing tumours displayed impaired angiogenesis. In-depth analysis of *APQ1* function in endothelial cells, produced from mouse aortic cells, using cell migration, wound healing, proliferation and cord development assays, revealed critical roles of *AQP1* in cell migration. *AQP1* null cells showed reduced ability in all four assays. This was corroborated by the plasmid expression of *AQP1* and another water channel protein, *AQP4*, in non-endothelial cells, resulting in increased migration. Functional analysis has confirmed the role of *AQP1* in cell migration under hypoxic conditions. Yun et al. [149] used rat pulmonary-artery-derived PASMCs, infected with adenovirus vectors carrying either WT *AQP1* or *AQP1* with a 37 amino acid terminal deletion removing the COOH tail. Overexpression of *AQP1* increased levels of β-catenin and its downstream targets, including c-Myc and cyclin D1. However, loss of the COOH tail produced no such results, suggesting COOH-mediated upregulation of β-catenin. Both *AQP1* and β-catenin displayed elevated levels under hypoxic conditions. This was further corroborated by siRNA-mediated silencing of β-catenin, in cells subject to hypoxia, and consequent infection with adenoviral constructs. Cells were rescued from the migration and proliferation associated with hypoxic conditions.

Additionally, functional analysis has been conducted on *ABCC8*. Eight mutations were analysed in COS cells using patch-clamp tests and rubidium flux assays, as a measure of ion efflux. Loss of channel function was seen in all variants, with the rescue of function by the administration of the SUR1 activator drug, diazoxide [113]. Additionally, functional analysis of *KCNA5*, by overexpressing the gene in PASMCs, showed an increase in channel current that was rescued by administering either nicotine, bepridil, correolide or endothelin-1 [163]. Additional genes have been reported, but have limited functional evidence confirming their role in PAH. One such gene is *UCP2*, encoding uncoupling protein 2, which functions as a calcium channel in vascular mitochondria that mediates ion influx from the ER. Fluorescence resonance energy transfer imaging of mitochondria from resistance pulmonary arteries of mice harbouring a *Ucp2* knockout showed a reduced calcium concentration, impairing their function, and mimicking the effects of hypoxia. Impaired mitochondrial function has been associated with the hyperproliferative and anti-apoptotic symptoms seen in PH [168]. Another channel protein is TRPC6. Using Western blot analysis, TRPC6 was found to be upregulated in PASMCs obtained from IPAH patients, compared to control cells. The endothelin receptor antagonist, Bosentan, was found to act by downregulating *TRPC6* and had a more pronounced anti-proliferative effect on IPAH cells than on WT cells [259]. Alternatively, potassium channel function has been rescued using adenoviral infection of human Kv1.5, into mice exposed to chronic hypoxia for three to four weeks. Mice were alleviated of hypoxia-induced vasoconstriction, displaying reduced right ventricle and pulmonary artery medial hypertrophy, when compared to control mice subject to hypoxia [260].

Transcription Factors

TBX4 is an evolutionarily conserved transcription factor, which plays a critical role in limb development during embryogenesis [261]. Detection of a lung-specific *Tbx4* enhancer, specific to the lung mesenchyme and trachea during embryogenesis, suggest an important role in lung development [262]. It has been shown to induce myofibroblast proliferation and drive invasion of the matrix by fibroblasts, promoting the development of pulmonary fibrosis, and deletion of *Tbx4* has led to a reduction in fibrosis [262]. *TBX4* plays a role in lung development and has been linked to pulmonary fibrosis, suggesting a possible role in PAH; nonetheless, experimental evidence is required to confirm this. Similarly, *SOX17* has been shown to be critical for the development of pulmonary vasculature. A mouse model, harbouring a conditional deletion of *Sox17* in splanchnic mesenchyme descendants, displayed impaired pulmonary vascular development leading to death by three weeks of age [153]. Shih et al. [154] assessed the role of *SOX17* using siRNA-mediated knockdown of expression in PAECs, followed by angiogenesis assays that showed impaired vessel formation. Concurrently, a mutant *SOX17* cell line, generated through CRISPR/Cas9 mutagenesis of human iPSC, showed a reduction in expression of arterial genes compared to WT cells.

New Genes

More recently, a plethora of new genes have been found to harbour mutations in PAH patients; however, functional evidence for their involvement remains limited. *TET2*, involved in DNA demethylation, is a key epigenetic regulator and has been found dysregulated in PAH. Potus et al. [55] developed heterozygous and homozygous *Tet2* knockout mice models, by crossing floxed *Tet2* parents and *Vav1-cre* to excise the gene. Their phenotype was ascertained through echocardiography and RHC. Patients carrying *TET2* mutations were found to have a later age of onset compared to other PAH patients; this finding was replicated in mice. Genetic evidence has also implicated *KDR* in PAH; Winter et al. [155] developed a mouse model with a conditional deletion of the *KDR* gene, encoding VEGFR2, in endothelial cells. The mice developed mild PAH symptoms under normoxic conditions; however, these were exacerbated under hypoxia with mice displaying pulmonary vascular remodelling.

Other genes which have been associated with PAH through sequencing analysis have no functional evidence with respect to PAH; however, physiological functions reveal possible mechanistic roles in the disease. *GGCX* has also been associated with PAH using forward genetics studies but lacks any functional evidence to corroborate this. This gene encodes a protein that carboxylates glutamate residues on Vitamin-K-dependent proteins; these are vital for the activation of coagulation proteins [263], inhibiting vascular calcification and inflammation [25]. Mutations in *GGCX* have been associated with Vitamin-K-dependent clotting factors deficiency, a congenital bleeding disorder [264] and Pseudoxanthoma elasticum [265].

Additionally, *KLK1* plays a key role in cardiac and renal function, notably regulating blood pressure. Kinins are known to affect endothelial cells, especially playing a role in vasodilation, increasing vascular permeability, nitric oxide production and inflammation. Tissue kallikrein is highly expressed in the kidney, pancreas, central nervous system and blood vessels [266]. Kinins, which include bradykinin, are regulated by inactivating enzymes known as kininases, of which the ACE is the most well known [267]; ACE-inhibitors are effectively used to reduce blood pressure [268]. Indeed, hypertension has been associated with polymorphisms and deficiencies in *KLK1* expression in rat models [269,270]. While genetic variation in *KLK1* has been associated with essential hypertension in a Chinese Han cohort [271], other studies have reported conflicting evidence on its role in the condition [268]. Nonetheless, this gene and pathway offer potential therapeutic targets, having major functions in areas of dysfunction seen in PAH, demonstrated by the evidence showing that *KLK1* can be used for its vasodilative properties in the treatment of acute ischemic stroke [272].

Autosomal Recessive Mode of Inheritance

Anthony et al. [273] demonstrated that homozygous *Eif2ak4* knockout mice displayed increased levels of protein carbonylation, a symptom of oxidative damage that was induced by dietary leucine deprivation. Additionally, these mice exhibited reduced viability compared to controls, with some dying shortly after birth. Their findings suggest that the gene may elicit protective effects against oxidative damage.

5.3.2. Common Genetic Variation

Another interesting PH animal model is the *Sirt3* knockout mice, which displays higher mPAP, PVR, RV hypertrophy and reduced exercise tolerance, compared with controls. This gene has been implicated in IPAH, with reduced expression in hPASMCs obtained from patients. The severity of symptoms in the model occur in a dose dependent manner, with *Sirt3* heterozygous mice displaying milder symptoms than homozygous knockout mice. This includes the extent of muscularisation and medial wall thickness of resistance PAs, which is more severe in homozygotes [166]. However, another *Sirt3*KO mouse model, against a C57BL/6 background [274] failed to develop PH, which may be attributed to differences between the mouse strains used. The 129/Sv strain used by Paulin et al. [166] possess a slower metabolism compared to C57BL/6 mice, which may mean loss of *Sirt3* is more damaging than in the latter strain.

Mutations in vasodilatory and vasoconstrictive proteins have been implicated in PAH, including *END-1*. In situ hybridisation has shown *END-1* upregulation in the lungs of patients with PAH, with greatest abundance in vessels subject to disease-associated remodelling [171]. Furthermore, analysis of a SNP found in the reporter region of *END-1* using a luciferase assay has shown upregulated activity. Homozygous mutants display a 30% increase in promoter activity and loss of binding of KLF4 and PPAR γ, which have both been linked to PAH [170].

Common variation in the DNA Topoisomerase II Binding Protein 1 (*TopBP1*) has been found in IPAH patients. Immunohistochemical analysis of vascular lesions demonstrated a reduction in TopBP1 protein. Analysis of cultured pulmonary microvascular endothelial cells from IPAH lungs, using q-PCR and Western blotting, showed a reduction in TopBP1 mRNA and protein. Further analysis of cells for increased susceptibility to DNA damage, by administering hydroxyurea and staining nuclei for phosphorylated histone foci, revealed an excess of DNA strand breaks and apoptosis. However, patient cells could be rescued by transfection with a plasmid containing wild-type human *TopBP1*, displaying a reduction in DNA damage-induced apoptosis [172].

In addition, functional analysis has been completed in *SOX17*. A luciferase reporter construct was generated for each haplotype of four heterozygous *SOX17* mutations and transfected into hPAECs to assess transcriptional activity [116,275]. Comparison of each haplotype revealed a haplotype-specific difference in promoter activity in a subset of variants. Two variants showed a statistically significant decrease in promoter activity in the risk allele, compared to the non-risk allele. Evidence exists for the intricate function of the various genes implicated in PAH, including regulatory roles of *SOX17*. VEGF has been shown to upregulate *SOX17*, while the Notch pathway plays critical roles in downregulating Sox17 in the endothelium. Additionally, there is a positive feedback loop between VEGR and *SOX17* [275]. These findings support the evidence that a wide range of mutations across genes are capable of eliciting similar phenotypic impacts through large-scale changes to pathway regulation.

5.4. Limitations, Challenges and Future Directions

To conclude, reverse genetics techniques are vital components of PAH research, without which forward genetics' discoveries cannot be validated. Nonetheless, it is important that such studies are interrogated for any limitations that may affect their results. As previously mentioned, it is important to have a robust experimental design and protocol, which includes power calculations for animal studies, minimising bias and reporting all measurements in publications. Cellular models are powerful in vitro tools for understanding specific processes in PAH; however, focusing on certain cell types may produce more pronounced effects than those seen in vivo. Understanding the widespread impact of PAH mutations are best captured using animal models. However, it is important to remember that all fail to fully recapitulate the disease and that any toxic side effects of pharmacological models may confound results. Studies must choose models which are most appropriate for the hypothesis, or use multiple where necessary. Furthermore, focusing specifically on rodent models, these often capture longitudinal data through the use of terminal procedures for haemodynamic measurements. The experimental techniques discussed here are well established in the field; however, several new tools have been developed which may increase the speed and depth of information derived using reverse genetics; these include high-throughput technologies for mutagenesis and spatial gene expression analysis [276,277]. Such developments offer exciting prospects for future research in PAH.

6. Reverse Phenotyping

6.1. Concepts

The concept of reverse phenotyping (Figure 1, process 4) refers to the use of genetic marker data to refine the disease subgroup definitions [36]. The method was first used in phenotyping patients with sarcoidosis [278] and has proven successful in many other diseases characterised by high heterogeneity [279].

6.2. Theoretical Basis

Heterogeneity provides a unique challenge in the diagnosis and treatment of rare diseases, and is further complicated by the accuracy of reporting. Knowledge about the genetic makeup of the individual allows targeted treatments, thereby enhancing efficacy and decreasing the risk of potential side effects [280]. Conversely, knowledge about the phenotypic characteristics of mutation carriers is indispensable to match study design and methodological approaches to disease attributes [281].

6.3. Methodology

Studying the complexity of genotype–phenotype associations is of particular interest to medical genetics. Evidence for such studies can be retrieved from published peer-review literature and publicly available databases, and facilitated by computational tools such as the R package *VarformPDB* (Disease-Gene-Variant Relations Mining from the Public Databases and Literature), which captures and compiles the genes and variants related to a disease, a phenotype or a clinical feature from public databases including HPO (Human Phenotype Ontology), Orphanet, OMIM (Online Mendelian Inheritance in Man), ClinVar, and UniProt (Universal Protein Resource) and PubMed abstracts. Alternatively, the development of large biobanks that link rich electronic health record (EHR) data and dense genetic information have made it possible to enhance clinical characterisation of mutation carriers in a cost-effective fashion. Besides deep clinical, often longitudinal phenotyping, multi-omic approaches can be employed thanks to the concurrent collection of patients' biological samples. A number of efforts have been made to facilitate EHR-based genetic and genomic research, such as the establishment of the UK Biobank [282], the Electronic Medical Record and Genomics Network [283] or the Precision Medicine Initiative Cohort Program [284], which allow both forward genetics and reverse phenotyping applications. Mining EHRs for genetic research purposes possesses a number of advantages over classical population or family-based genetic association studies, large samples can be collected over a short period of time in a cost-effective fashion. Moreover, the robustness of EHRs allows for the investigation of a spectrum of phenotypes in a hypothesis-free manner. Familial relationships reported in EHRs have also been used for the estimation of disease heritability shown to be consistent with the literature [285].

Multiple tools are being used by the Clinical Genetics community to address the issue of reverse phenotyping when interpreting NGS data. This includes DECIPHER [286], a publicly-available database that houses a collection of case-level evidence for 36,801 individuals with genetic diseases, for comparison of phenotypic and genotypic data. Projects affiliated to DECIPHER can deposit and share patients, variants and phenotypes to invite collaboration and increase diagnostic yield. Another example (also accessible via DECIPHER) is that of Matchmaker Exchange [287] or Genematcher [288], for mining other databases based on matching genotype and phenotype data. VarSome is a data aggregator that encompasses platforms and bioinformatic tools with the aim of consolidating information required for variant analysis. VarSome draws information from 30 databases [289] and is also embedded into another web-based tool, VariantValidator, which enables validation, mapping and formatting of sequence variants using HGVS nomenclature [290].

Knowledge of age-dependent and/or reduced penetrance is vital in reverse-phenotyping as new clinical features may emerge in an individual over time. This is applicable to relatives of PAH patients found to carry presumed pathogenic variants in known disease-associated genes. A detailed evaluation of multiple family members alongside longitudinal follow-up of these individuals is vital to study the disease trajectory which has often higher definition than snapshot phenotyping at diagnosis.

Apart from mining existing databases and health records, recall by genotype (RbG) studies will be a relevant and valuable source of information regarding the biological mechanics of genotype–phenotype associations. By making use of the random allocation of alleles at conception (mendelian randomization—MR) these studies enhance the ability to draw causal inferences in population-based studies and minimise the problems related to confounders; additionally, the focus on phenotypic assessment of a specific carrier subgroup can enhance the understanding of pathobiology in a cost-effective manner. RbG studies can focus on a single variant or multiple variants. The former considers rare and large effect loci to understand the biological pathways of interest, and as a result, the groups are relatively small so the phenotype can be studied in detail, in the latter, genetic variation is used as a proxy of relevant exposure (proviso there is a credible association between genetic markers and exposure). To increase power for studies in which a single variant has a small effect size deployment of aggregate genetic risk scores can be beneficial [291].

6.4. Studies

The reversal of the usual hypothesis-driven paradigm for the refined diagnosis was first used by Grunewald and Eklund [278] who by deep clinical phenotyping of patients with Lofgren's syndrome, showed that erythema nodosum predominantly occurs in women and bilateral periarticular arthritis in men, and that the acute sarcoidosis subgroup can be further characterized according to the presence of HLA-DRB1*0301/DQB1*0201. The reverse phenotyping method has been subsequently adopted in other disease domains (Table 4) and has led to new disease discovery, such as the Koolen–DeVries syndrome, due to a 17q21.31 microdeletion involving the *KANSL1* gene [292], and the Potocki–Lupski Syndrome, due to a 17p11.2 microduplication [293,294]. Recently, a retrospective study of 111 patients with nephrotic syndrome, who underwent WES, showed that reverse phenotyping increased the diagnostic accuracy in patients referred with the diagnosis of steroid-resistant nephrotic syndrome [279].

The reverse phenotyping of *BMPR2* mutation carriers has been studied in PH patients. A large individual participant data meta-analysis found that patients with *BMPR2* mutations have earlier disease onset, worse haemodynamics, are less likely to respond to nitric oxide challenge and have lower survival when compared to those without *BMPR2* mutations [180]. The histological analysis of lungs explanted from those patients also revealed a higher degree of bronchial artery hypertrophy/dilatation, which correlated with the frequency of haemoptysis at presentation [295]. Patients with missense mutations in *BMPR2* that escape nonsense-mediated decay have more severe disease than those with truncating mutations, suggestive of a dominant-negative impact of mutated protein on downstream signalling [296]. However, missense variants in the cytoplasmic tail appear to confer less severe phenotype than other *BMPR2* variants with a later age of onset, milder haemodynamics, and more vasoreactivity [297].

Other good candidates for reverse phenotyping are patients with a mutation in transcription factors, i.e., *TBX4*, which by virtue of gene function would be associated with a more complex and heterogeneous phenotype. Patients with mutations in *TBX4* present with severe PAH associated with bronchial and parenchymal changes, low DLCO, with or without skeletal abnormalities [142], and bimodal age of onset [25]. Interestingly, the penetrance of *TBX4* mutations for skeletal abnormalities is much higher than for PAH [108]. Descriptions of paediatric subjects with mutations in *TBX4*, and likely loci nearby including *TBX2*, characterised by skeletal dysplasias, developmental delay and hearing loss have been reported in CHD and PH patients [298,299]. Interestingly, the US PAH Biobank study confirmed a causative role of *TBX4* not only across various age groups (12/266; 4.6%—paediatric-onset; 11/2345; 0.47%—adult-onset) but also across different PH phenotypes: 2 (one adult; one paediatric)/23; 8.7% PAH-CTD, 3 (two paediatric and one adult-onset)/23; 13% PAH-CHD [25]. Not all patients harbouring deleterious variants in *TBX4* fit neatly into Group 1 PH, some individuals with heterozygous mutations in *TBX4* or large deletions encompassing *TBX4* were characterised by death in infancy secondary to acinar dysplasia [300], congenital alveolar dysplasia and pulmonary hypoplasia [301]. The studies dissecting phenotype by mutation type or looking for subtle features (i.e., SPS) among patients harbouring deleterious variants in *TBX4* who were initially diagnosed with PAH are lacking.

Reverse phenotyping of patients with PTVs in *KDR* revealed mild interstitial lung disease in all index cases and one affected relative [30], which was further confirmed in an independent case report [117].

Table 4. Examples of successful studies that used reverse phenotyping. Abbreviations: *BMPR2*—Bone morphogenic protein receptor type 2; PAH—pulmonary arterial hypertension; WES—whole-exome sequencing, * denotes allele number.

Gene	Condition	Study Design	Data Sources	Results	References
HLA-DRB1* 0301/DQB1*0201	Sarcoidosis	Mix of retrospective and prospective cases with acute onset sarcoidosis	Health records	Patients with acute onset sarcoidosis carrying HLA-DRB1*0301/DQB1*0201 genotype have good prognosis, manifestations of Lofgren's syndrome differ between man and woman.	[278,302]
BMPR2	PAH	Retrospective analysis of 169 PAH patients	WES & Health records	Patients with missense mutations that escape nonsense-mediated decay have more severe disease than those with truncating mutations.	[296]
BMPR2	PAH	Retrospective analysis of 171 patients		Missense variants in the cytoplasmic tail appear to confer less severe phenotype than other *BMPR2* variants with a later age of onset, milder haemodynamics, and more vasoreactivity.	[297]
BMPR2	PAH	A large individual participant data meta-analysis	Literature review	Patients harbouring deleterious *BMPR2* mutations have earlier disease onset, worse haemodynamics, are less likely to respond to NO challenge and have lower survival when compared to those without *BMPR2* mutations.	[180]
BMPR2	PAH	Retrospective analysis of 44 PAH patients who underwent lung transplantation between 2005 and 2014	Histology, immunohisto chemistry, morphometry of explanted lungs; French registry database	*BMPR2* mutation carriers are more prone to haemoptysis; haemoptysis is closely correlated to bronchial arterial remodelling and angiogenesis; pronounced changes in the systemic vasculature correlate with increased pulmonary venous remodelling, creating a distinctive profile in PAH patients harbouring a *BMPR2* mutation.	[295]
multiple genes	Nephrotic syndrome	A retrospective analysis of all patients diagnosed with nephrotic syndrome between 2000 and 2018	WES & personalised diagnostic workflow	Reverse phenotyping after WES increased the diagnostic accuracy in patients referred with the diagnosis of steroid-resistant nephrotic syndrome.	[279]

6.5. Limitations, Challenges and Future Directions

Reverse phenotyping has the potential to shed light on genotype–phenotype associations and accelerate the recruitment of homogenous, well-defined groups to clinical trials, thereby increasing the chances of effective treatment and decreasing the risk of side effects. Its success depends on the quality of forward phenotyping, and forward and reverse genetic studies accurately reporting the phenotype of interest (use of standardised vocabulary to describe phenotypic abnormalities (i.e., HPO terms), clinical measurements (i.e., Logical Observation Identifiers (LOINC) [303]) and systematic approaches to data sharing.

In the future, national and international efforts should be directed at optimising recruitment of large homogeneous cohorts for forward genetics studies and, simultaneously, preempting any potential ethical issues related to subsequent recall by genotype studies. Similarly, recall by genotype studies will face challenges related to specific study design, i.e., when recruiting newborns or children who might not express the full phenotype by the time of diagnosis/inclusion.

7. Missing Heritability in the Postgenomic Era

As already alluded to in the introduction, missing heritability is the proportion of phenotypic variance not explained by the additive effects of known variants. There are multiple sources of missing heritability beyond the rare and common sequence variation described above (Figure 2). In the following paragraphs, we will discuss studies which contribute to a better understanding of the genetic landscape in PAH, beyond germline, rare, and common sequence variation.

7.1. Unknown Genetic Variation

Recently, large case-control studies have allowed us to discover new genes involved in the pathogenesis of PAH; all these studies, however, assumed that PAH is a monogenic condition and examined only protein-coding space. Although this permitted the discovery of more than 16 genes associated with PAH (Table 3), and increased the diagnostic yield, most PAH cases remain unexplained, suggesting additional variation in non-coding regions. It is now known that the remaining 98% of non-coding space plays a crucial role in the regulation of gene expression and requires interpretation using its 3D structure that allows regulatory elements to function across long distances [304]. Regulatory elements such as enhancers, promoters and actively transcribed genes are located in open-chromatin regions so they can be accessed by transcriptional machinery. A number of methods for the identification of non-coding regulatory elements (NCRE) have been developed over the last 20 years [305]. A recent study investigated variation in non-coding regions by researching differences in chromatin marks and gene expression in PAECs from patients with I/HPAH and controls. Using ChiP-Seq profiling of the active histone marks (H3K27ac, H3K4me1, H3K4me3) and RNA-Seq, it was shown that while PAECs from PAH patients and healthy controls are similar in terms of gene expression, they undergo large-scale remodelling of the active chromatin regions, which leads to differential gene expression in response to external stimuli. Overall, this study not only validated and expanded our knowledge regarding the genes involved in the pathogenesis of PAH, but also highlighted that steady-state expression analyses are of limited value in systems where the mechanism involves an aberrant response to stimuli [306].

So far, mostly germline mutations have been implicated in the pathogenesis of PAH but the ever-increasing age of onset may suggest a role of acquired somatic mutations in disease development. While rare somatic mutations of the *BMPR2* gene within pulmonary vasculature cells could cause or aggravate *BMPR2* haploinsufficiency, previous work has failed to find this to be true among those with a germline *BMPR2* mutation [307]. Conversely, a study by Aldred et al. [308] showed a high frequency of genetically abnormal subclones of PAEC from patients with PAH including one patient harbouring a deleterious *BMPR2* variant. Similarly, Drake et al. [309] demonstrated that a somatic deletion on chromosome 13, encompassing *SMAD9*, in PAECs from a patient with PAH associated with CHD could have contributed to the pathogenesis of PAH, and potentially also to CHD [309]. Likewise, postzygotic mutations have been implicated in the pathogenesis of HHT [310] and HHT with concomitant PAH [311,312]. Somatic mutations in *TET2* have recently been reported in PAH cases, and the phenotype was replicated in the mouse model [55].

Impaired bioenergetics is an established feature of PAH [313,314] whether its origins can be traced back to mitochondrial DNA (mtDNA) remains an open question, but a recent study revealed that mitochondrial haplogroups influence the risk of PAH and that susceptibility to PAH emerged as a result of selective enrichment of specific haplogroups upon the migration of populations out of Africa [315]. Studies into the role of mtDNA in PAH will require novel bioinformatics tools and variant prioritisation approaches to account for differences between nuclear and mitochondrial genomes [316].

7.2. Epigenetic Inheritance

Epigenetics are heritable changes in gene function that occur without a change in the sequence of DNA and include DNA methylation, histone modification and RNA interference. All three of these modifications have been shown to be involved in the development of PAH. Epigenetic changes can be heritable [317,318], de novo and may be modified by environmental factors such as diet, smoking, and drugs.

7.2.1. DNA Methylation

DNA methylation is a well-described form of epigenetic modification which involves the transfer of a methyl group to cytosine residues in dinucleotide CpG sequences of DNA resulting in gene silencing. In PAH, the best described is hypermethylation of the gene encoding superoxide dismutase 2 (*SOD2*), an enzyme involved in H_2O_2 regulation which also acts as a tumour suppressor gene. Hypermethylation of *SOD2* leads to uncontrolled cell proliferation in multiple cancers; similarly reduced expression of *SOD2* has been shown in the rat model of PH [319] and in IPAH [320]. Further evidence comes from a recent study which found rare germline and somatic mutations in the key regulator of DNA demethylation, *TET2*, in 0.39% of PAH cases [55]. The phenotype was replicated in the $Tet^{-/-}$ mouse and reversed with IL-1β blockade.

7.2.2. Histone Acetylation

Histone acetylation is the most common form of histone modification [321], with a proven impact on PAH development. In rat and human PH histone deacetylases, HDAC1 and HDAC5 concentrations are elevated and histone deacetylase inhibitor can reverse hyperproliferation of PASMCs and decrease mPAP and RV hypertrophy [322].

7.2.3. RNA Interference

A large number of non-coding RNAs (ncRNAs), including long non-coding RNAs (lncRNAs), miRNAs, circular RNAs (circRNAs), and piwi-interacting RNAs (piRNAs) are involved in the regulation of a variety of cellular processes. For example, so far, almost 27,000 lncRNAs have been identified in the human genome [323] and reported regulating gene expression via diverse mechanisms [324]. In PAH, both miRNA and lncRNAs have been studied and reported to impact cells relevant to PAH pathogenesis, including PASMCs, PAECs and fibroblasts. Regulatory effects of lncRNA in PAH range from enhanced proliferation of PASMCs (H19 [325], PAXIP-AS1 [326], MALAT1 [327], lnc-Ang362 [328], Tug1 [329], HOXA-AS3 [330], MEG3 [331], TCONS_00034812 [148], and UCA1 [332]), over suppressed proliferation and/or migration of PASMCs (MEG3 [333,334], CASC2 [335], and LnPRT [336]) to induction of EndoMT in PAECs (MALT1 [337] and GATA6-AS [338]). Significant causative links exist between multiple miRNAs and BMPR2 expression and signalling [339–341], as well as hypoxia [342]. Importantly, although attractive as a therapeutic target, miRNAs have not been shown to have the potential to reverse the disease, which is most likely due to the high degree of redundancy and "fine-tuning" rather than "switch on/off" mode of action of these molecules.

Although still limited, our knowledge of ncRNA in the pathobiology of cardiovascular diseases [343,344] including PAH, is rapidly growing. The major limitation of studies so far is that the focus has been mostly on identifying the role of preselected ncRNA with very limited effort to investigate ncRNA in a global manner accounting for multiple interactions.

7.3. Interactions

Although difficult to estimate, gene and environment covariation and gene-environment interactions may play a significant role in heritability, expressivity and penetration of the disease. The most commonly studied factors are drugs, diet, toxins, radiation and stress, and some of these factors have been implicated in the development of PAH. Definitive associations between PAH and drugs and toxins based on outbreaks, epidemiological case-control studies or large multicentre series have been confirmed for aminorexigenes, methamphetamines, dasatinib and toxic rapeseed oil, yet some others require further validation [18]. Interestingly, several studies reported that BMPR2 expression and degradation can be affected by viral proteins and cocaine [345–347]. Similarly, a number of volatile organic compounds (VOCs) identified in exhaled breath condensate in association with PAH, are exogenous [348] and may be considered pollutants due to exposure to cigarette smoke, air pollution and radiation [349]. Contribution of volatile compounds was also reported in the pathogenesis of PVOD [350].

A specific example of the environmental effects on phenotype is the foetal origins of adult diseases (FOAD) hypothesis. This hypothesis was based on the studies which reported that intrauterine exposure increased the risk of specific cardiovascular and metabolic diseases [351–353] as well as impacted on lung function [354–356] in adult life. In the area of PH, prenatal exposure to antidepressants and persistent pulmonary hypertension of the newborn (PPHN) has been a subject of multiple studies. A meta-analysis revealed that the risk for PPHN in infants exposed to SSRIs during late pregnancy is small, although significantly increased [357]. In systemic hypertension, the transmission of gut microbiota from parent to offspring was shown to influence the disease risk in adulthood [358] and similarly, bacterial translocation may contribute to PAH development [359]. Air pollution was also shown to correlate with PAH severity and outcomes [360].

With the advent of NGS, the notion of digenic or oligogenic inheritance has gained traction. HPAH has been historically considered a monogenic condition but incomplete penetrance may indicate that other germline or somatic variants are required for the disease to develop. Therefore, at least in a proportion of cases, the inheritance may be digenic or even oligogenic. In the true digenic model, both genes are required to develop the disease. Conversely, in the composite class model, a variant in one gene is sufficient to produce the phenotype, but an additional variant in a second gene impacts the disease phenotype or alters the age of onset [361]. The latter model seems to be plausible in PAH, where co-occurrence of the variants in different PAH risk genes has been reported to impact on disease onset and penetrance [362]. Patients harbouring deleterious variants in more than one PAH risk gene have been reported in case reports [162], small [363] and large cohorts of HPAH patients [24,26,113]. Finally, the phenomenon of synergistic heterozygosity, whereby the effect of susceptibility genes is enhanced by modifier genes, in which common variants may influence the disease onset and severity, has been described in PAH [210].

7.4. Population Dependent Heterogeneity

It is well recognised that although rare diseases can occur in any population, some ethnic groups are characterised by their higher incidence, i.e., sickle cell disease is more common in African, African-Americans and Mediterranean populations and similarly, Tay-Sachs disease occurs mostly in people of Ashkenazi Jewish or French Canadian ancestry. National and international PAH registries have shown differences in demographic characteristics between European and East Asian patients. East Asian IPAH cohorts resemble those of European cohorts from 40 years ago; additionally, they show more pronounced female predominance, a younger age of onset and lower comorbidity burden, and importantly, they demonstrate that significantly more cases can be explained by variation in known PAH risk genes (24% [24,25] vs. 39% [26]).

8. Summary

In this review, we have summarised how four theoretical and methodological approaches impact genetic discoveries in rare diseases, with a particular focus on PAH. These steps include: (1) forward phenotyping, which refers to clustering patients into homogenous groups likely to share genetic architecture, (2) forward genetics, which aims to identify the candidate genes responsible for the phenotype, (3) reverse genetics, which consists of experimental in vivo and in vitro targeted modifications to candidate genes in order to analyse their phenotypic impact, and (4) reverse phenotyping, which uses genetic marker data to refine phenotype definitions. Whilst these approaches have been employed in PAH research, it is important to mention that at present, the majority of patients diagnosed with PAH do not have a genetic diagnosis, a finding that indicates "missing heritability". To date, some progress has been made in addressing this issue; however, there is still a way to go and only through the use of large-scale international cohorts will studies have the power to detect novel genetic risk loci underlying PAH pathobiology and to elucidate this missing heritability.

Author Contributions: Conceptualization, E.M.S.; writing—original draft preparation, E.M.S., M.P., J.M.M., D.P., K.A.; writing—review and editing, E.M.S., M.P., J.M.M., D.P., K.A., N.W.M., S.G.; visualization, E.M.S., S.G.; supervision, E.M.S., N.W.M., S.G. All authors have read and agreed to the published version of the manuscript.

Funding: This research received no external funding.

Conflicts of Interest: The authors declare no conflict of interest.

References

1. Montserrat Moliner, A.; Waligóra, J. The European union policy in the field of rare diseases. *Public Health Genom.* **2013**, *16*, 268–277. [CrossRef] [PubMed]

2. Nguengang Wakap, S.; Lambert, D.M.; Olry, A.; Rodwell, C.; Gueydan, C.; Lanneau, V.; Murphy, D.; Le Cam, Y.; Rath, A. Estimating cumulative point prevalence of rare diseases: Analysis of the Orphanet database. *Eur. J. Hum. Genet.* **2020**, *28*, 165–173. [CrossRef] [PubMed]

3. OMIM Entry Statistics. Available online: https://www.omim.org/statistics/entry (accessed on 27 August 2020).

4. Wright, C.F.; FitzPatrick, D.R.; Firth, H.V. Paediatric genomics: Diagnosing rare disease in children. *Nat. Rev. Genet.* **2018**, *19*, 253–268. [CrossRef] [PubMed]

5. Simonneau, G.; Robbins, I.M.; Beghetti, M.; Channick, R.N.; Delcroix, M.; Denton, C.P.; Elliott, C.G.; Gaine, S.P.; Gladwin, M.T.; Jing, Z.-C.; et al. Updated clinical classification of pulmonary hypertension. *J. Am. Coll. Cardiol.* **2009**, *54*, S43–S54. [CrossRef] [PubMed]

6. Fisher, R.A. XV.—The Correlation between Relatives on the Supposition of Mendelian Inheritance. *Trans. R. Soc. Edinb.* **1919**, *52*, 399–433. [CrossRef]

7. Wright, S. The Relative Importance of Heredity and Environment in Determining the Piebald Pattern of Guinea-Pigs. *Proc. Natl. Acad. Sci. USA* **1920**, *6*, 320–332. [CrossRef]

8. Tenesa, A.; Haley, C.S. The heritability of human disease: Estimation, uses and abuses. *Nat. Rev. Genet.* **2013**, *14*, 139–149. [CrossRef]

9. Manolio, T.A.; Collins, F.S.; Cox, N.J.; Goldstein, D.B.; Hindorff, L.A.; Hunter, D.J.; McCarthy, M.I.; Ramos, E.M.; Cardon, L.R.; Chakravarti, A.; et al. Finding the missing heritability of complex diseases. *Nature* **2009**, *461*, 747–753. [CrossRef]

10. Maher, B. Personal genomes: The case of the missing heritability. *Nature* **2008**, *456*, 18–21. [CrossRef]

11. Génin, E. Missing heritability of complex diseases: Case solved? *Hum. Genet.* **2020**, *139*, 103–113. [CrossRef]

12. WSPH, Geneva. 1973. Available online: http://www.wsphassociation.org/wp-content/uploads/2019/03/Programma-WSPH-Geneva-1973.pdf (accessed on 25 November 2020).

13. WSPH, Evian. 1998. Available online: http://www.wsphassociation.org/wp-content/uploads/2019/04/Primary-Pulmonary-Hypertension-Evian-1998.pdf (accessed on 25 November 2020).

14. Simonneau, G.; Galiè, N.; Rubin, L.J.; Langleben, D.; Seeger, W.; Domenighetti, G.; Gibbs, S.; Lebrec, D.; Speich, R.; Beghetti, M.; et al. Clinical classification of pulmonary hypertension. *J. Am. Coll. Cardiol.* **2004**, *43*, 5S–12S. [CrossRef]

15. Galiè, N.; Hoeper, M.M.; Humbert, M.; Torbicki, A.; Vachiery, J.-L.; Barbera, J.A.; Beghetti, M.; Corris, P.; Gaine, S.; Gibbs, J.S.; et al. Guidelines for the diagnosis and treatment of pulmonary hypertension: The Task Force for the Diagnosis and Treatment of Pulmonary Hypertension of the European Society of Cardiology (ESC) and the European Respiratory Society (ERS), endorsed by the International Society of Heart and Lung Transplantation (ISHLT). *Eur. Heart J.* **2009**, *30*, 2493–2537. [CrossRef] [PubMed]

16. Updated Clinical Classification of Pulmonary Hypertension. *J. Am. Coll. Cardiol.* **2013**, *62*, D34–D41. [CrossRef] [PubMed]

17. Galiè, N.; Humbert, M.; Vachiery, J.-L.; Gibbs, S.; Lang, I.; Torbicki, A.; Simonneau, G.; Peacock, A.; Vonk Noordegraaf, A.; Beghetti, M.; et al. 2015 ESC/ERS Guidelines for the diagnosis and treatment of pulmonary hypertension: The Joint Task Force for the Diagnosis and Treatment of Pulmonary Hypertension of the European Society of Cardiology (ESC) and the European Respiratory Society (ERS): Endorsed by: Association for European Paediatric and Congenital Cardiology (AEPC), International Society for Heart and Lung Transplantation (ISHLT). *Eur. Respir. J.* **2015**, *46*, 903–975. [CrossRef] [PubMed]

18. Simonneau, G.; Montani, D.; Celermajer, D.S.; Denton, C.P.; Gatzoulis, M.A.; Krowka, M.; Williams, P.G.; Souza, R. Haemodynamic definitions and updated clinical classification of pulmonary hypertension. *Eur. Respir. J.* **2019**, *53*. [CrossRef]

19. Kato, G.J.; Piel, F.B.; Reid, C.D.; Gaston, M.H.; Ohene-Frempong, K.; Krishnamurti, L.; Smith, W.R.; Panepinto, J.A.; Weatherall, D.J.; Costa, F.F.; et al. Sickle cell disease. *Nat. Rev. Dis. Primers* **2018**, *4*, 18010. [CrossRef]

20. Peron, A.; Au, K.S.; Northrup, H. Genetics, genomics, and genotype-phenotype correlations of TSC: Insights for clinical practice. *Am. J. Med. Genet. C Semin. Med. Genet.* **2018**, *178*, 281–290. [CrossRef]

21. Bravo-Gil, N.; González-Del Pozo, M.; Martín-Sánchez, M.; Méndez-Vidal, C.; Rodríguez-de la Rúa, E.; Borrego, S.; Antiñolo, G. Unravelling the genetic basis of simplex Retinitis Pigmentosa cases. *Sci. Rep.* **2017**, *7*, 41937. [CrossRef]

22. Chiurazzi, P.; Pirozzi, F. Advances in understanding - genetic basis of intellectual disability. *F1000Res* **2016**, *5*. [CrossRef]

23. Southgate, L.; Machado, R.D.; Gräf, S.; Morrell, N.W. Molecular genetic framework underlying pulmonary arterial hypertension. *Nat. Rev. Cardiol.* **2020**, *17*, 85–95. [CrossRef]

24. Gräf, S.; Haimel, M.; Bleda, M.; Hadinnapola, C.; Southgate, L.; Li, W.; Hodgson, J.; Liu, B.; Salmon, R.M.; Southwood, M.; et al. Identification of rare sequence variation underlying heritable pulmonary arterial hypertension. *Nat. Commun.* **2018**, *9*, 1416. [CrossRef] [PubMed]

25. Zhu, N.; Pauciulo, M.W.; Welch, C.L.; Lutz, K.A.; Coleman, A.W.; Gonzaga-Jauregui, C.; Wang, J.; Grimes, J.M.; Martin, L.J.; He, H.; et al. Novel risk genes and mechanisms implicated by exome sequencing of 2572 individuals with pulmonary arterial hypertension. *Genome Med.* **2019**, *11*, 69. [CrossRef]

26. Wang, X.-J.; Lian, T.-Y.; Jiang, X.; Liu, S.-F.; Li, S.-Q.; Jiang, R.; Wu, W.-H.; Ye, J.; Cheng, C.-Y.; Du, Y.; et al. Germline BMP9 mutation causes idiopathic pulmonary arterial hypertension. *Eur. Respir. J.* **2019**, *53*, 1801609. [CrossRef] [PubMed]

27. Dweik, R.A.; Rounds, S.; Erzurum, S.C.; Archer, S.; Fagan, K.; Hassoun, P.M.; Hill, N.S.; Humbert, M.; Kawut, S.M.; Krowka, M.; et al. An official American Thoracic Society Statement: Pulmonary hypertension phenotypes. *Am. J. Respir. Crit. Care Med.* **2014**, *189*, 345–355. [CrossRef] [PubMed]

28. Sutendra, G.; Michelakis, E.D. The metabolic basis of pulmonary arterial hypertension. *Cell Metab.* **2014**, *19*, 558–573. [CrossRef] [PubMed]

29. Buyske, S.; Yang, G.; Matise, T.C.; Gordon, D. When a Case Is Not a Case: Effects of Phenotype Misclassification on Power and Sample Size Requirements for the Transmission Disequilibrium Test with Affected Child Trios. *Hum. Hered.* **2009**, *67*, 287–292. [CrossRef]

30. Swietlik, E.M.; Greene, D.; Zhu, N.; Megy, K.; Cogliano, M.; Rajaram, S.; Pandya, D.; Tilly, T.; Lutz, K.A.; Welch, C.C.L.; et al. Reduced transfer coefficient of carbon monoxide in pulmonary arterial hypertension implicates rare protein-truncating variants in KDR. *bioRxiv* **2020**. [CrossRef]

31. Pullabhatla, V.; Roberts, A.L.; Lewis, M.J.; Mauro, D.; Morris, D.L.; Odhams, C.A.; Tombleson, P.; Liljedahl, U.; Vyse, S.; Simpson, M.A.; et al. De novo mutations implicate novel genes in systemic lupus erythematosus. *Hum. Mol. Genet.* **2018**, *27*, 421–429. [CrossRef]

32. Husson, T.; Duboc, J.-B.; Quenez, O.; Charbonnier, C.; Rotharmel, M.; Cuenca, M.; Jegouzo, X.; Richard, A.-C.; Frebourg, T.; Deleuze, J.-F.; et al. Identification of potential genetic risk factors for bipolar disorder by whole-exome sequencing. *Transl. Psychiatry* **2018**, *8*, 268. [CrossRef]

33. Bjørnland, T.; Bye, A.; Ryeng, E.; Wisløff, U.; Langaas, M. Powerful extreme phenotype sampling designs and score tests for genetic association studies. *Stat. Med.* **2018**, *37*, 4234–4251. [CrossRef]

34. Chung, W.K.; Deng, L.; Carroll, J.S.; Mallory, N.; Diamond, B.; Rosenzweig, E.B.; Barst, R.J.; Morse, J.H. Polymorphism in the angiotensin II type 1 receptor (AGTR1) is associated with age at diagnosis in pulmonary arterial hypertension. *J. Heart Lung Transplant.* **2009**, *28*, 373–379. [CrossRef] [PubMed]

35. Li, D.; Lewinger, J.P.; Gauderman, W.J.; Murcray, C.E.; Conti, D. Using extreme phenotype sampling to identify the rare causal variants of quantitative traits in association studies. *Genet. Epidemiol.* **2011**, *35*, 790–799. [CrossRef] [PubMed]

36. Schulze, T.G.; McMahon, F.J. Defining the phenotype in human genetic studies: Forward genetics and reverse phenotyping. *Hum. Hered.* **2004**, *58*, 131–138. [CrossRef] [PubMed]

37. Sweatt, A.J.; Hedlin, H.K.; Balasubramanian, V.; Hsi, A.; Blum, L.K.; Robinson, W.H.; Haddad, F.; Hickey, P.M.; Condliffe, R.; Lawrie, A.; et al. Discovery of Distinct Immune Phenotypes Using Machine Learning in Pulmonary Arterial Hypertension. *Circ. Res.* **2019**, *124*, 904–919. [CrossRef] [PubMed]

38. Hoeper, M.M.; Pausch, C.; Grünig, E.; Klose, H.; Staehler, G.; Huscher, D.; Pittrow, D.; Olsson, K.M.; Vizza, C.D.; Gall, H.; et al. Idiopathic pulmonary arterial hypertension phenotypes determined by cluster analysis from the COMPERA registry. *J. Heart Lung Transplant.* **2020**. [CrossRef] [PubMed]

39. Köhler, S.; Carmody, L.; Vasilevsky, N.; Jacobsen, J.O.B.; Danis, D.; Gourdine, J.-P.; Gargano, M.; Harris, N.L.; Matentzoglu, N.; McMurry, J.A.; et al. Expansion of the Human Phenotype Ontology (HPO) knowledge base and resources. *Nucleic Acids Res.* **2019**, *47*, D1018–D1027. [CrossRef] [PubMed]

40. Köhler, S.; Schulz, M.H.; Krawitz, P.; Bauer, S.; Dölken, S.; Ott, C.E.; Mundlos, C.; Horn, D.; Mundlos, S.; Robinson, P.N. Clinical diagnostics in human genetics with semantic similarity searches in ontologies. *Am. J. Hum. Genet.* **2009**, *85*, 457–464. [CrossRef]

41. Greene, D.; Richardson, S.; Turro, E. ontologyX: A suite of R packages for working with ontological data. *Bioinformatics* **2017**, *33*, 1104–1106. [CrossRef]

42. Peters, J.L.; Cnudde, F.; Gerats, T. Forward genetics and map-based cloning approaches. *Trends Plant Sci.* **2003**, *8*, 484–491. [CrossRef]

43. Ionita-Laza, I.; Lee, S.; Makarov, V.; Buxbaum, J.D.; Lin, X. Family-based association tests for sequence data, and comparisons with population-based association tests. *Eur. J. Hum. Genet.* **2013**, *21*, 1158–1162. [CrossRef]

44. Risch, N.; Merikangas, K. The Future of Genetic Studies of Complex Human Diseases. *Science* **1996**, *273*, 1516–1517. [CrossRef] [PubMed]

45. Marchini, J.; Donnelly, P.; Cardon, L.R. Genome-wide strategies for detecting multiple loci that influence complex diseases. *Nat. Genet.* **2005**, *37*, 413–417. [CrossRef] [PubMed]

46. Petrovski, S.; Todd, J.L.; Durheim, M.T.; Wang, Q.; Chien, J.W.; Kelly, F.L.; Frankel, C.; Mebane, C.M.; Ren, Z.; Bridgers, J.; et al. An Exome Sequencing Study to Assess the Role of Rare Genetic Variation in Pulmonary Fibrosis. *Am. J. Respir. Crit. Care Med.* **2017**, *196*, 82–93. [CrossRef] [PubMed]

47. Adzhubei, I.A.; Schmidt, S.; Peshkin, L.; Ramensky, V.E.; Gerasimova, A.; Bork, P.; Kondrashov, A.S.; Sunyaev, S.R. A method and server for predicting damaging missense mutations. *Nat. Methods* **2010**, *7*, 248–249. [CrossRef] [PubMed]

48. Sim, N.-L.; Kumar, P.; Hu, J.; Henikoff, S.; Schneider, G.; Ng, P.C. SIFT web server: Predicting effects of amino acid substitutions on proteins. *Nucleic Acids Res.* **2012**, *40*, W452–W457. [CrossRef]

49. Ioannidis, N.M.; Rothstein, J.H.; Pejaver, V.; Middha, S.; McDonnell, S.K.; Baheti, S.; Musolf, A.; Li, Q.; Holzinger, E.; Karyadi, D.; et al. REVEL: An Ensemble Method for Predicting the Pathogenicity of Rare Missense Variants. *Am. J. Hum. Genet.* **2016**, *99*, 877–885. [CrossRef]

50. Cooper, G.M.; Stone, E.A.; Asimenos, G.; NISC Comparative Sequencing Program; Green, E.D.; Batzoglou, S.; Sidow, A. Distribution and intensity of constraint in mammalian genomic sequence. *Genome Res.* **2005**, *15*, 901–913. [CrossRef]

51. Pollard, K.S.; Hubisz, M.J.; Rosenbloom, K.R.; Siepel, A. Detection of nonneutral substitution rates on mammalian phylogenies. *Genome Res.* **2010**, *20*, 110–121. [CrossRef]

52. Siepel, A.; Bejerano, G.; Pedersen, J.S.; Hinrichs, A.S.; Hou, M.; Rosenbloom, K.; Clawson, H.; Spieth, J.; Hillier, L.W.; Richards, S.; et al. Evolutionarily conserved elements in vertebrate, insect, worm, and yeast genomes. *Genome Res.* **2005**, *15*, 1034–1050. [CrossRef]

53. Kircher, M.; Witten, D.M.; Jain, P.; O'Roak, B.J.; Cooper, G.M.; Shendure, J. A general framework for estimating the relative pathogenicity of human genetic variants. *Nat. Genet.* **2014**, *46*, 310–315. [CrossRef]

54. Nasim, M.T.; Ogo, T.; Ahmed, M.; Randall, R.; Chowdhury, H.M.; Snape, K.M.; Bradshaw, T.Y.; Southgate, L.; Lee, G.J.; Jackson, I.; et al. Molecular genetic characterization of SMAD signaling molecules in pulmonary arterial hypertension. *Hum. Mutat.* **2011**, *32*, 1385–1389. [CrossRef]

55. Potus, F.; Pauciulo, M.W.; Cook, E.K.; Zhu, N.; Hsieh, A.; Welch, C.L.; Shen, Y.; Tian, L.; Lima, P.; Mewburn, J.; et al. Novel Mutations and Decreased Expression of the Epigenetic Regulator TET2 in Pulmonary Arterial Hypertension. *Circulation* **2020**, *141*, 1986–2000. [CrossRef]

56. Povysil, G.; Petrovski, S.; Hostyk, J.; Aggarwal, V.; Allen, A.S.; Goldstein, D.B. Rare-variant collapsing analyses for complex traits: Guidelines and applications. *Nat. Rev. Genet.* **2019**, *20*, 747–759. [CrossRef] [PubMed]

57. Asimit, J.L.; Day-Williams, A.G.; Morris, A.P.; Zeggini, E. ARIEL and AMELIA: Testing for an accumulation of rare variants using next-generation sequencing data. *Hum. Hered.* **2012**, *73*, 84–94. [CrossRef] [PubMed]

58. Morris, A.P.; Zeggini, E. An evaluation of statistical approaches to rare variant analysis in genetic association studies. *Genet. Epidemiol.* **2010**, *34*, 188–193. [CrossRef] [PubMed]

59. Li, B.; Leal, S.M. Methods for detecting associations with rare variants for common diseases: Application to analysis of sequence data. *Am. J. Hum. Genet.* **2008**, *83*, 311–321. [CrossRef]

60. Madsen, B.E.; Browning, S.R. A groupwise association test for rare mutations using a weighted sum statistic. *PLoS Genet.* **2009**, *5*, e1000384. [CrossRef]

61. Ionita-Laza, I.; Buxbaum, J.D.; Laird, N.M.; Lange, C. A New Testing Strategy to Identify Rare Variants with Either Risk or Protective Effect on Disease. *PLoS Genet.* **2011**, *7*, e1001289. [CrossRef]

62. Wu, M.C.; Lee, S.; Cai, T.; Li, Y.; Boehnke, M.; Lin, X. Rare-variant association testing for sequencing data with the sequence kernel association test. *Am. J. Hum. Genet.* **2011**, *89*, 82–93. [CrossRef]

63. Lee, S.; Emond, M.J.; Bamshad, M.J.; Barnes, K.C.; Rieder, M.J.; Nickerson, D.A.; NHLBI GO Exome Sequencing Project—ESP Lung Project Team; Christiani, D.C.; Wurfel, M.M.; Lin, X. Optimal unified approach for rare-variant association testing with application to small-sample case-control whole-exome sequencing studies. *Am. J. Hum. Genet.* **2012**, *91*, 224–237. [CrossRef]

64. Greene, D.; NIHR BioResource; Richardson, S.; Turro, E. A Fast Association Test for Identifying Pathogenic Variants Involved in Rare Diseases. *Am. J. Hum. Genet.* **2017**, *101*, 104–114. [CrossRef] [PubMed]

65. Eilbeck, K.; Quinlan, A.; Yandell, M. Settling the score: Variant prioritization and Mendelian disease. *Nat. Rev. Genet.* **2017**, *18*, 599–612. [CrossRef] [PubMed]

66. Cacheiro, P.; Muñoz-Fuentes, V.; Murray, S.A.; Dickinson, M.E.; Bucan, M.; Nutter, L.M.J.; Peterson, K.A.; Haselimashhadi, H.; Flenniken, A.M.; Morgan, H.; et al. Human and mouse essentiality screens as a resource for disease gene discovery. *Nat. Commun.* **2020**, *11*, 655. [CrossRef] [PubMed]

67. Richards, S.; Aziz, N.; Bale, S.; Bick, D.; Das, S.; Gastier-Foster, J.; Grody, W.W.; Hegde, M.; Lyon, E.; Spector, E.; et al. Standards and guidelines for the interpretation of sequence variants: A joint consensus recommendation of the American College of Medical Genetics and Genomics and the Association for Molecular Pathology. *Genet. Med.* **2015**, *17*, 405–424. [CrossRef] [PubMed]

68. Rehm, H.L.; Berg, J.S.; Brooks, L.D.; Bustamante, C.D.; Evans, J.P.; Landrum, M.J.; Ledbetter, D.H.; Maglott, D.R.; Martin, C.L.; Nussbaum, R.L.; et al. ClinGen—The Clinical Genome Resource. *N. Engl. J. Med.* **2015**, *372*, 2235–2242. [CrossRef] [PubMed]

69. Goswami, R.S.; Harada, S. An Overview of Molecular Genetic Diagnosis Techniques. *Curr. Protoc. Hum. Genet.* **2020**, *105*, e97. [CrossRef] [PubMed]

70. Bean, L.J.H.; Funke, B.; Carlston, C.M.; Gannon, J.L.; Kantarci, S.; Krock, B.L.; Zhang, S.; Bayrak-Toydemir, P. Diagnostic gene sequencing panels: From design to report—A technical standard of the American College of Medical Genetics and Genomics (ACMG). *Genet. Med.* **2020**, *22*, 453–461. [CrossRef]

71. Song, J.; Eichstaedt, C.A.; Viales, R.R.; Benjamin, N.; Harutyunova, S.; Fischer, C.; Grünig, E.; Hinderhofer, K. Identification of genetic defects in pulmonary arterial hypertension by a new gene panel diagnostic tool. *Clin. Sci.* **2016**, *130*, 2043–2052. [CrossRef]

72. Rabbani, B.; Tekin, M.; Mahdieh, N. The promise of whole-exome sequencing in medical genetics. *J. Hum. Genet.* **2014**, *59*, 5–15. [CrossRef]

73. Retterer, K.; Juusola, J.; Cho, M.T.; Vitazka, P.; Millan, F.; Gibellini, F.; Vertino-Bell, A.; Smaoui, N.; Neidich, J.; Monaghan, K.G.; et al. Clinical application of whole-exome sequencing across clinical indications. *Genet. Med.* **2016**, *18*, 696–704. [CrossRef]

74. Austin, E.D.; Ma, L.; LeDuc, C.; Berman Rosenzweig, E.; Borczuk, A.; Phillips, J.A., 3rd; Palomero, T.; Sumazin, P.; Kim, H.R.; Talati, M.H.; et al. Whole exome sequencing to identify a novel gene (caveolin-1) associated with human pulmonary arterial hypertension. *Circ. Cardiovasc. Genet.* **2012**, *5*, 336–343. [CrossRef] [PubMed]

75. Zhu, N.; Welch, C.L.; Wang, J.; Allen, P.M.; Gonzaga-Jauregui, C.; Ma, L.; King, A.K.; Krishnan, U.; Rosenzweig, E.B.; Ivy, D.D.; et al. Rare variants in SOX17 are associated with pulmonary arterial hypertension with congenital heart disease. *Genome Med.* **2018**, *10*, 691. [CrossRef] [PubMed]

76. Zare, F.; Dow, M.; Monteleone, N.; Hosny, A.; Nabavi, S. An evaluation of copy number variation detection tools for cancer using whole exome sequencing data. *BMC Bioinform.* **2017**, *18*, 286. [CrossRef] [PubMed]

77. Turro, E.; Astle, W.J.; Megy, K.; Gräf, S.; Greene, D.; Shamardina, O.; Allen, H.L.; Sanchis-Juan, A.; Frontini, M.; Thys, C.; et al. Whole-genome sequencing of patients with rare diseases in a national health system. *Nature* **2020**, *583*, 96–102. [CrossRef] [PubMed]

78. National Human Genome Research Institute Home. Available online: https://www.genome.gov (accessed on 17 October 2020).

79. Schwarze, K.; Buchanan, J.; Fermont, J.M.; Dreau, H.; Tilley, M.W.; Taylor, J.M.; Antoniou, P.; Knight, S.J.L.; Camps, C.; Pentony, M.M.; et al. The complete costs of genome sequencing: A microcosting study in cancer and rare diseases from a single center in the United Kingdom. *Genet. Med.* **2020**, *22*, 85–94. [CrossRef] [PubMed]

80. Barozzi, C.; Galletti, M.; Tomasi, L.; De Fanti, S.; Palazzini, M.; Manes, A.; Sazzini, M.; Galiè, N. A Combined Targeted and Whole Exome Sequencing Approach Identified Novel Candidate Genes Involved in Heritable Pulmonary Arterial Hypertension. *Sci. Rep.* **2019**, *9*, 753. [CrossRef]

81. Pulmonary Arterial Hypertension (Version 2.5). Available online: https://panelapp.genomicsengland.co.uk/panels/193/ (accessed on 14 October 2020).

82. Kim, J.; Weber, J.A.; Jho, S.; Jang, J.; Jun, J.; Cho, Y.S.; Kim, H.-M.; Kim, H.; Kim, Y.; Chung, O.; et al. KoVariome: Korean National Standard Reference Variome database of whole genomes with comprehensive SNV, indel, CNV, and SV analyses. *Sci. Rep.* **2018**, *8*, 5677. [CrossRef]

83. Finer, S.; Martin, H.C.; Khan, A.; Hunt, K.A.; MacLaughlin, B.; Ahmed, Z.; Ashcroft, R.; Durham, C.; MacArthur, D.G.; McCarthy, M.I.; et al. Cohort Profile: East London Genes & Health (ELGH), a community-based population genomics and health study in British Bangladeshi and British Pakistani people. *Int. J. Epidemiol.* **2020**, *49*, 20–21i. [CrossRef]

84. BioBank Japan (BBJ). Available online: http://biobankjp.org (accessed on 14 October 2020).

85. IBD BioResource—Translating Today's Science into Tomorrow's Treatments. Available online: https://www.ibdbioresource.nihr.ac.uk (accessed on 14 October 2020).

86. Genetic Links to Anxiety and Depression Study—GLAD Study. Available online: https://gladstudy.org.uk (accessed on 14 October 2020).

87. Eating Disorders Genetics Initiative. Available online: https://edgiuk.org/ (accessed on 14 October 2020).

88. Flannick, J.; Beer, N.L.; Bick, A.G.; Agarwala, V.; Molnes, J.; Gupta, N.; Burtt, N.P.; Florez, J.C.; Meigs, J.B.; Taylor, H.; et al. Assessing the phenotypic effects in the general population of rare variants in genes for a dominant Mendelian form of diabetes. *Nat. Genet.* **2013**, *45*, 1380–1385. [CrossRef]

89. Castel, S.E.; Cervera, A.; Mohammadi, P.; Aguet, F.; Reverter, F.; Wolman, A.; Guigo, R.; Iossifov, I.; Vasileva, A.; Lappalainen, T. Modified penetrance of coding variants by cis-regulatory variation contributes to disease risk. *Nat. Genet.* **2018**, *50*, 1327–1334. [CrossRef]

90. International Human Genome Sequencing Consortium. Initial sequencing and analysis of the human genome. *Nature* **2001**, *409*, 860–921.

91. International Human Genome Sequencing Consortium Finishing the euchromatic sequence of the human genome. *Nature* **2004**, *431*, 931–945. [CrossRef]

92. Guo, Y.; Dai, Y.; Yu, H.; Zhao, S.; Samuels, D.C.; Shyr, Y. Improvements and impacts of GRCh38 human reference on high throughput sequencing data analysis. *Genomics* **2017**, *109*, 83–90. [CrossRef] [PubMed]

93. Pan, B.; Kusko, R.; Xiao, W.; Zheng, Y.; Liu, Z.; Xiao, C.; Sakkiah, S.; Guo, W.; Gong, P.; Zhang, C.; et al. Similarities and differences between variants called with human reference genome HG19 or HG38. *BMC Bioinform.* **2019**, *20*, 17–29. [CrossRef]

94. Schneider, V.A.; Graves-Lindsay, T.; Howe, K.; Bouk, N.; Chen, H.-C.; Kitts, P.A.; Murphy, T.D.; Pruitt, K.D.; Thibaud-Nissen, F.; Albracht, D.; et al. Evaluation of GRCh38 and de novo haploid genome assemblies demonstrates the enduring quality of the reference assembly. *Genome Res.* **2017**, *27*, 849–864. [CrossRef]

95. Church, D.M.; Schneider, V.A.; Steinberg, K.M.; Schatz, M.C.; Quinlan, A.R.; Chin, C.-S.; Kitts, P.A.; Aken, B.; Marth, G.T.; Hoffman, M.M.; et al. Extending reference assembly models. *Genome Biol.* **2015**, *16*, 13. [CrossRef]

96. Lappalainen, T.; Scott, A.J.; Brandt, M.; Hall, I.M. Genomic Analysis in the Age of Human Genome Sequencing. *Cell* **2019**, *177*, 70–84. [CrossRef]

97. Li, R.; Li, Y.; Zheng, H.; Luo, R.; Zhu, H.; Li, Q.; Qian, W.; Ren, Y.; Tian, G.; Li, J.; et al. Building the sequence map of the human pan-genome. *Nat. Biotechnol.* **2010**, *28*, 57–63. [CrossRef]

98. Paten, B.; Novak, A.M.; Eizenga, J.M.; Garrison, E. Genome graphs and the evolution of genome inference. *Genome Res.* **2017**, *27*, 665–676. [CrossRef]

99. Rakocevic, G.; Semenyuk, V.; Lee, W.-P.; Spencer, J.; Browning, J.; Johnson, I.J.; Arsenijevic, V.; Nadj, J.; Ghose, K.; Suciu, M.C.; et al. Fast and accurate genomic analyses using genome graphs. *Nat. Genet.* **2019**, *51*, 354–362. [CrossRef]

100. Ballouz, S.; Dobin, A.; Gillis, J.A. Is it time to change the reference genome? *Genome Biol.* **2019**, *20*, 159. [CrossRef] [PubMed]

101. International PPH Consortium; Lane, K.B.; Machado, R.D.; Pauciulo, M.W.; Thomson, J.R.; Phillips, J.A., 3rd; Loyd, J.E.; Nichols, W.C.; Trembath, R.C. Heterozygous germline mutations in BMPR2, encoding a TGF-beta receptor, cause familial primary pulmonary hypertension. *Nat. Genet.* **2000**, *26*, 81–84. [CrossRef] [PubMed]

102. Thomson, J.R. Sporadic primary pulmonary hypertension is associated with germline mutations of the gene encoding BMPR-II, a receptor member of the TGF-beta family. *J. Med Genet.* **2000**, *37*, 741–745. [CrossRef]

103. Trembath, R.C.; Thomson, J.R.; Machado, R.D.; Morgan, N.V.; Atkinson, C.; Winship, I.; Simonneau, G.; Galie, N.; Loyd, J.E.; Humbert, M.; et al. Clinical and molecular genetic features of pulmonary hypertension in patients with hereditary hemorrhagic telangiectasia. *N. Engl. J. Med.* **2001**, *345*, 325–334. [CrossRef]

104. Chaouat, A. Endoglin germline mutation in a patient with hereditary haemorrhagic telangiectasia and dexfenfluramine associated pulmonary arterial hypertension. *Thorax* **2004**, *59*, 446–448. [CrossRef] [PubMed]

105. Harrison, R.E.; Berger, R.; Haworth, S.G.; Tulloh, R.; Mache, C.J.; Morrell, N.W.; Aldred, M.A.; Trembath, R.C. Transforming growth factor-beta receptor mutations and pulmonary arterial hypertension in childhood. *Circulation* **2005**, *111*, 435–441. [CrossRef]

106. Shintani, M.; Yagi, H.; Nakayama, T.; Saji, T.; Matsuoka, R. A new nonsense mutation of SMAD8 associated with pulmonary arterial hypertension. *J. Med. Genet.* **2009**, *46*, 331–337. [CrossRef] [PubMed]

107. Ma, L.; Roman-Campos, D.; Austin, E.D.; Eyries, M.; Sampson, K.S.; Soubrier, F.; Germain, M.; Trégouët, D.-A.; Borczuk, A.; Rosenzweig, E.B.; et al. A novel channelopathy in pulmonary arterial hypertension. *N. Engl. J. Med.* **2013**, *369*, 351–361. [CrossRef]

108. Kerstjens-Frederikse, W.S.; Bongers, E.M.H.F.; Roofthooft, M.T.R.; Leter, E.M.; Douwes, J.M.; Van Dijk, A.; Vonk-Noordegraaf, A.; Dijk-Bos, K.K.; Hoefsloot, L.H.; Hoendermis, E.S.; et al. TBX4 mutations (small patella syndrome) are associated with childhood-onset pulmonary arterial hypertension. *J. Med. Genet.* **2013**, *50*, 500–506. [CrossRef]

109. Eyries, M.; Montani, D.; Girerd, B.; Perret, C.; Leroy, A.; Lonjou, C.; Chelghoum, N.; Coulet, F.; Bonnet, D.; Dorfmüller, P.; et al. EIF2AK4 mutations cause pulmonary veno-occlusive disease, a recessive form of pulmonary hypertension. *Nat. Genet.* **2014**, *46*, 65–69. [CrossRef]

110. Best, D.H.; Sumner, K.L.; Smith, B.P.; Damjanovich-Colmenares, K.; Nakayama, I.; Brown, L.M.; Ha, Y.; Paul, E.; Morris, A.; Jama, M.A.; et al. EIF2AK4 Mutations in Patients Diagnosed with Pulmonary Arterial Hypertension. *Chest* **2017**, *151*, 821–828. [CrossRef] [PubMed]

111. Hadinnapola, C.; Bleda, M.; Haimel, M.; Screaton, N.; Swift, A.; Dorfmüller, P.; Preston, S.D.; Southwood, M.; Hernandez-Sanchez, J.; Martin, J.; et al. Phenotypic Characterization of Mutation Carriers in a Large Cohort of Patients Diagnosed Clinically with Pulmonary Arterial Hypertension. *Circulation* **2017**, *136*, 2022–2033. [CrossRef] [PubMed]

112. Hiraide, T.; Kataoka, M.; Suzuki, H.; Aimi, Y.; Chiba, T.; Kanekura, K.; Satoh, T.; Fukuda, K.; Gamou, S.; Kosaki, K. SOX17 Mutations in Japanese Patients with Pulmonary Arterial Hypertension. *Am. J. Respir. Crit. Care Med.* **2018**, *198*, 1231–1233. [CrossRef]

113. Bohnen, M.S.; Ma, L.; Zhu, N.; Qi, H.; McClenaghan, C.; Gonzaga-Jauregui, C.; Dewey, F.E.; Overton, J.D.; Reid, J.G.; Shuldiner, A.R.; et al. Loss-of-Function ABCC8 Mutations in Pulmonary Arterial Hypertension. *Circ. Genom. Precis. Med.* **2018**, *11*, e002087. [CrossRef]

114. Eyries, M.; Montani, D.; Nadaud, S.; Girerd, B.; Levy, M.; Bourdin, A.; Trésorier, R.; Chaouat, A.; Cottin, V.; Sanfiorenzo, C.; et al. Widening the landscape of heritable pulmonary hypertension mutations in paediatric and adult cases. *Eur. Respir. J.* **2019**, *53*. [CrossRef]

115. Hodgson, J.; Swietlik, E.M.; Salmon, R.M.; Hadinnapola, C.; Nikolic, I.; Wharton, J.; Guo, J.; Liley, J.; Haimel, M.; Bleda, M.; et al. Characterization of Mutations and Levels of BMP9 and BMP10 in Pulmonary Arterial Hypertension. *Am. J. Respir. Crit. Care Med.* **2020**, *201*, 575–585. [CrossRef]

116. Rhodes, C.J.; Batai, K.; Bleda, M.; Haimel, M.; Southgate, L.; Germain, M.; Pauciulo, M.W.; Hadinnapola, C.; Aman, J.; Girerd, B.; et al. Genetic determinants of risk in pulmonary arterial hypertension: International genome-wide association studies and meta-analysis. *Lancet Respir Med.* **2019**, *7*, 227–238. [CrossRef]

117. Eyries, M.; Montani, D.; Girerd, B.; Favrolt, N.; Riou, M.; Faivre, L.; Manaud, G.; Perros, F.; Gräf, S.; Morrell, N.W.; et al. Familial pulmonary arterial hypertension by KDR heterozygous loss of function. *Eur. Respir. J.* **2020**, *55*, 1902165. [CrossRef]

118. Zhu, N.; Swietlik, E.M.; Welch, C.L.; Pauciulo, M.W.; Hagen, J.J.; Zhou, X.; Guo, Y.; Karten, J.; Pandya, D.; Tilly, T.; et al. Rare variant analysis of 4,241 pulmonary arterial hypertension cases from an international consortium implicate FBLN2, PDGFD and rare de novo variants in PAH. *bioRxiv* **2020**. [CrossRef]

119. Deng, Z.; Morse, J.H.; Slager, S.L.; Cuervo, N.; Moore, K.J.; Venetos, G.; Kalachikov, S.; Cayanis, E.; Fischer, S.G.; Barst, R.J.; et al. Familial primary pulmonary hypertension (gene PPH1) is caused by mutations in the bone morphogenetic protein receptor-II gene. *Am. J. Hum. Genet.* **2000**, *67*, 737–744. [CrossRef]

120. Navas Tejedor, P.; Tenorio Castaño, J.; Palomino Doza, J.; Arias Lajara, P.; Gordo Trujillo, G.; López Meseguer, M.; Román Broto, A.; Lapunzina Abadía, P.; Escribano Subía, P. An homozygous mutation in KCNK3 is associated with an aggressive form of hereditary pulmonary arterial hypertension. *Clin. Genet.* **2017**, *91*, 453–457. [CrossRef] [PubMed]

121. Wang, G.; Fan, R.; Ji, R.; Zou, W.; Penny, D.J.; Varghese, N.P.; Fan, Y. Novel homozygous BMP9 nonsense mutation causes pulmonary arterial hypertension: A case report. *BMC Pulm. Med.* **2016**, *16*, 17. [CrossRef] [PubMed]

122. Larkin, E.K.; Newman, J.H.; Austin, E.D.; Hemnes, A.R.; Wheeler, L.; Robbins, I.M.; West, J.D.; Phillips, J.A., 3rd; Hamid, R.; Loyd, J.E. Longitudinal analysis casts doubt on the presence of genetic anticipation in heritable pulmonary arterial hypertension. *Am. J. Respir. Crit. Care Med.* **2012**, *186*, 892–896. [CrossRef] [PubMed]

123. Beppu, H.; Ichinose, F.; Kawai, N.; Jones, R.C.; Yu, P.B.; Zapol, W.M.; Miyazono, K.; Li, E.; Bloch, K.D. BMPR-II heterozygous mice have mild pulmonary hypertension and an impaired pulmonary vascular remodeling response to prolonged hypoxia. *Am. J. Physiol.-Lung Cell. Mol. Physiol.* **2004**, *287*, L1241–L1247. [CrossRef]

124. Yang, X.; Long, L.; Southwood, M.; Rudarakanchana, N.; Upton, P.D.; Jeffery, T.K.; Atkinson, C.; Chen, H.; Trembath, R.C.; Morrell, N.W. Dysfunctional Smad Signaling Contributes to Abnormal Smooth Muscle Cell Proliferation in Familial Pulmonary Arterial Hypertension. *Circ. Res.* **2005**, *96*, 1053–1063. [CrossRef]

125. Hong, K.-H.; Lee, Y.J.; Lee, E.; Park, S.O.; Han, C.; Beppu, H.; Li, E.; Raizada, M.K.; Bloch, K.D.; Oh, S.P. Genetic ablation of the BMPR2 gene in pulmonary endothelium is sufficient to predispose to pulmonary arterial hypertension. *Circulation* **2008**, *118*, 722–730. [CrossRef]

126. West, J.; Harral, J.; Lane, K.; Deng, Y.; Ickes, B.; Crona, D.; Albu, S.; Stewart, D.; Fagan, K. Mice expressing BMPR2R899X transgene in smooth muscle develop pulmonary vascular lesions. *Am. J. Physiol. Lung Cell. Mol. Physiol.* **2008**, *295*, L744–L755. [CrossRef]

127. Gangopahyay, A.; Oran, M.; Bauer, E.M.; Wertz, J.W.; Comhair, S.A.; Erzurum, S.C.; Bauer, P.M. Bone morphogenetic protein receptor II is a novel mediator of endothelial nitric-oxide synthase activation. *J. Biol. Chem.* **2011**, *286*, 33134–33140. [CrossRef]

128. Hemnes, A.R.; Brittain, E.L.; Trammell, A.W.; Fessel, J.P.; Austin, E.D.; Penner, N.; Maynard, K.B.; Gleaves, L.; Talati, M.; Absi, T.; et al. Evidence for right ventricular lipotoxicity in heritable pulmonary arterial hypertension. *Am. J. Respir. Crit. Care Med.* **2014**, *189*, 325–334. [CrossRef]

129. Ranchoux, B.; Antigny, F.; Rucker-Martin, C.; Hautefort, A.; Péchoux, C.; Bogaard, H.J.; Dorfmüller, P.; Remy, S.; Lecerf, F.; Planté, S.; et al. Endothelial-to-mesenchymal transition in pulmonary hypertension. *Circulation* **2015**, *131*, 1006–1018. [CrossRef]

130. Caruso, P.; Dunmore, B.J.; Schlosser, K.; Schoors, S.; Dos Santos, C.; Perez-Iratxeta, C.; Lavoie, J.R.; Zhang, H.; Long, L.; Flockton, A.R.; et al. Identification of MicroRNA-124 as a Major Regulator of Enhanced Endothelial Cell Glycolysis in Pulmonary Arterial Hypertension via PTBP1 (Polypyrimidine Tract Binding Protein) and Pyruvate Kinase M2. *Circulation* **2017**, *136*, 2451–2467. [CrossRef] [PubMed]

131. Tian, W.; Jiang, X.; Sung, Y.K.; Shuffle, E.; Wu, T.-H.; Kao, P.N.; Tu, A.B.; Dorfmüller, P.; Cao, A.; Wang, L.; et al. Phenotypically Silent Bone Morphogenetic Protein Receptor 2 Mutations Predispose Rats to Inflammation-Induced Pulmonary Arterial Hypertension by Enhancing the Risk for Neointimal Transformation. *Circulation* **2019**, *140*, 1409–1425. [CrossRef] [PubMed]

132. Gore, B.; Izikki, M.; Mercier, O.; Dewachter, L.; Fadel, E.; Humbert, M.; Dartevelle, P.; Simonneau, G.; Naeije, R.; Lebrin, F.; et al. Key role of the endothelial TGF-β/ALK1/endoglin signaling pathway in humans and rodents pulmonary hypertension. *PLoS ONE* **2014**, *9*, e100310. [CrossRef] [PubMed]

133. Drake, K.M.; Zygmunt, D.; Mavrakis, L.; Harbor, P.; Wang, L.; Comhair, S.A.; Erzurum, S.C.; Aldred, M.A. Altered MicroRNA Processing in Heritable Pulmonary Arterial Hypertension. *Am. J. Respir. Crit. Care Med.* **2011**, *184*, 1400–1408. [CrossRef]

134. Huang, Z.; Wang, D.; Ihida-Stansbury, K.; Jones, P.L.; Martin, J.F. Defective pulmonary vascular remodeling in Smad8 mutant mice. *Hum. Mol. Genet.* **2009**, *18*, 2791–2801. [CrossRef]

135. Drake, K.M.; Dunmore, B.J.; McNelly, L.N.; Morrell, N.W.; Aldred, M.A. Correction of nonsense BMPR2 and SMAD9 mutations by ataluren in pulmonary arterial hypertension. *Am. J. Respir. Cell Mol. Biol.* **2013**, *49*, 403–409. [CrossRef]

136. Han, C.; Hong, K.-H.; Kim, Y.H.; Kim, M.-J.; Song, C.; Kim, M.J.; Kim, S.-J.; Raizada, M.K.; Oh, S.P. SMAD1 deficiency in either endothelial or smooth muscle cells can predispose mice to pulmonary hypertension. *Hypertension* **2013**, *61*, 1044–1052. [CrossRef]

137. Copeland, C.A.; Han, B.; Tiwari, A.; Austin, E.D.; Loyd, J.E.; West, J.D.; Kenworthy, A.K. A disease-associated frameshift mutation in caveolin-1 disrupts caveolae formation and function through introduction of a de novo ER retention signal. *Mol. Biol. Cell* **2017**, *28*, 3095–3111. [CrossRef]

138. Zhao, Y.-Y.; Liu, Y.; Stan, R.-V.; Fan, L.; Gu, Y.; Dalton, N.; Chu, P.-H.; Peterson, K.; Ross, J., Jr.; Chien, K.R. Defects in caveolin-1 cause dilated cardiomyopathy and pulmonary hypertension in knockout mice. *Proc. Natl. Acad. Sci. USA* **2002**, *99*, 11375–11380. [CrossRef]

139. Drab, M.; Verkade, P.; Elger, M.; Kasper, M.; Lohn, M.; Lauterbach, B.; Menne, J.; Lindschau, C.; Mende, F.; Luft, F.C.; et al. Loss of caveolae, vascular dysfunction, and pulmonary defects in caveolin-1 gene-disrupted mice. *Science* **2001**, *293*, 2449–2452. [CrossRef]

140. Zhu, N.; Gonzaga-Jauregui, C.; Welch, C.L.; Ma, L.; Qi, H.; King, A.K.; Krishnan, U.; Rosenzweig, E.B.; Ivy, D.D.; Austin, E.D.; et al. Exome Sequencing in Children with Pulmonary Arterial Hypertension Demonstrates Differences Compared with Adults. *Circ Genom Precis Med* **2018**, *11*, e001887. [CrossRef] [PubMed]

141. Levy, M.; Eyries, M.; Szezepanski, I.; Ladouceur, M.; Nadaud, S.; Bonnet, D.; Soubrier, F. Genetic analyses in a cohort of children with pulmonary hypertension. *Eur. Respir. J.* **2016**, *48*, 1118–1126. [CrossRef] [PubMed]

142. Thoré, P.; Girerd, B.; Jaïs, X.; Savale, L.; Ghigna, M.-R.; Eyries, M.; Levy, M.; Ovaert, C.; Servettaz, A.; Guillaumot, A.; et al. Phenotype and outcome of pulmonary arterial hypertension patients carrying a TBX4 mutation. *Eur. Respir. J.* **2020**, *55*, 1902340. [CrossRef] [PubMed]

143. Jansen, S.M.A.; van den Heuvel, L.; Meijboom, L.J.; Alsters, S.I.M.; Vonk Noordegraaf, A.; Houweling, A.; Bogaard, H.J. Correspondence regarding "T-box protein 4 mutation causing pulmonary arterial hypertension and lung disease": A single-centre case series. *Eur. Respir. J.* **2020**, *55*, 1902272. [CrossRef]

144. Hernandez-Gonzalez, I.; Tenorio, J.; Palomino-Doza, J.; Martinez Meñaca, A.; Morales Ruiz, R.; Lago-Docampo, M.; Valverde Gomez, M.; Gomez Roman, J.; Enguita Valls, A.B.; Perez-Olivares, C.; et al. Clinical heterogeneity of Pulmonary Arterial Hypertension associated with variants in TBX4. *PLoS ONE* **2020**, *15*, e0232216. [CrossRef]

145. Bohnen, M.S.; Roman-Campos, D.; Terrenoire, C.; Jnani, J.; Sampson, K.J.; Chung, W.K.; Kass, R.S. The Impact of Heterozygous KCNK3 Mutations Associated with Pulmonary Arterial Hypertension on Channel Function and Pharmacological Recovery. *J. Am. Heart Assoc.* **2017**, *6*, e006465. [CrossRef]

146. Lambert, M.; Capuano, V.; Boet, A.; Tesson, L.; Bertero, T.; Nakhleh, M.K.; Remy, S.; Anegon, I.; Pechoux, C.; Hautefort, A.; et al. Characterization of Kcnk3 -Mutated Rat, a Novel Model of Pulmonary Hypertension. *Circ. Res.* **2019**, *125*, 678–695. [CrossRef]

147. Legchenko, E.; Liu, B.; West, J.; Vangheluwe, P.; Upton, P.; Morrell, N. Protein truncating mutations in ATP13A3 promote pulmonary arterial hypertension in mice. *ERJ Open Res.* **2020**, *6*, 83.

148. Liu, Y.; Sun, Z.; Zhu, J.; Xiao, B.; Dong, J.; Li, X. LncRNA-TCONS_00034812 in cell proliferation and apoptosis of pulmonary artery smooth muscle cells and its mechanism. *J. Cell. Physiol.* **2018**, *233*, 4801–4814. [CrossRef]

149. Yun, X.; Jiang, H.; Lai, N.; Wang, J.; Shimoda, L.A. Aquaporin 1-mediated changes in pulmonary arterial smooth muscle cell migration and proliferation involve β-catenin. *Am. J. Physiol. Lung Cell. Mol. Physiol.* **2017**, *313*, L889–L898. [CrossRef]

150. Saadoun, S.; Papadopoulos, M.C.; Hara-Chikuma, M.; Verkman, A.S. Impairment of angiogenesis and cell migration by targeted aquaporin-1 gene disruption. *Nature* **2005**, *434*, 786–792. [CrossRef] [PubMed]

151. Ma, T.; Yang, B.; Gillespie, A.; Carlson, E.J.; Epstein, C.J.; Verkman, A.S. Severely impaired urinary concentrating ability in transgenic mice lacking aquaporin-1 water channels. *J. Biol. Chem.* **1998**, *273*, 4296–4299. [CrossRef] [PubMed]

152. Tu, L.; Desroches-Castan, A.; Mallet, C.; Guyon, L.; Cumont, A.; Phan, C.; Robert, F.; Thuillet, R.; Bordenave, J.; Sekine, A.; et al. Selective BMP-9 Inhibition Partially Protects Against Experimental Pulmonary Hypertension. *Circ. Res.* **2019**, *124*, 846–855. [CrossRef] [PubMed]

153. Lange, A.W.; Haitchi, H.M.; LeCras, T.D.; Sridharan, A.; Xu, Y.; Wert, S.E.; James, J.; Udell, N.; Thurner, P.J.; Whitsett, J.A. Sox17 is required for normal pulmonary vascular morphogenesis. *Dev. Biol.* **2014**, *387*, 109–120. [CrossRef] [PubMed]

154. Shih, J.; Liu, B.; Huang, C.; Haimel, M.; Bleda, M.; Gräf, S.; Rana, A.; Upton, P.D.; Morrell, N.W. SOX17 Deficiency Impairs Tube Network Formation Through the Reduction of Arterial Identity in Pulmonary Arterial Hypertension. *Am. J. Respir. Crit. Care Med.* **2020**, *201*, A2366.

155. Winter, M.-P.; Sharma, S.; Altmann, J.; Seidl, V.; Panzenböck, A.; Alimohammadi, A.; Zelniker, T.; Redwan, B.; Nagel, F.; Santer, D.; et al. Interruption of vascular endothelial growth factor receptor 2 signaling induces a proliferative pulmonary vasculopathy and pulmonary hypertension. *Basic Res. Cardiol.* **2020**, *115*, 58. [CrossRef]

156. Lee, H.-W.; Adachi, T.; Park, S.; Kowalski, P.; Anderson, D.; Simons, M.; Jin, S.W.; Chun, H.J. Bmpr1a and Its Role in Endomt in Pulmonary Arterial Hypertension. *Circulation* **2018**, *138*, A17070.

157. Lee, H.W.; Adachi, T.; Park, S.; Pak, B.; Kowalski, P.; Choi, W.; Clapham, K.; Lee, A.; Anderson, D.; Simons, M.; et al. Abstract 14616: Loss of Endothelial BMPR1A Results in Pulmonary Hypertension Through Endothelial to Mesenchymal Transition. *Circulation* **2019**, *140*, A14616. [CrossRef]

158. Du, L.; Sullivan, C.C.; Chu, D.; Cho, A.J.; Kido, M.; Wolf, P.L.; Yuan, J.X.-J.; Deutsch, R.; Jamieson, S.W.; Thistlethwaite, P.A. Signaling molecules in nonfamilial pulmonary hypertension. *N. Engl. J. Med.* **2003**, *348*, 500–509. [CrossRef]

159. El-Bizri, N.; Wang, L.; Merklinger, S.L.; Guignabert, C.; Desai, T.; Urashima, T.; Sheikh, A.Y.; Knutsen, R.H.; Mecham, R.P.; Mishina, Y.; et al. Smooth muscle protein 22alpha-mediated patchy deletion of Bmpr1a impairs cardiac contractility but protects against pulmonary vascular remodeling. *Circ. Res.* **2008**, *102*, 380–388. [CrossRef]

160. Chida, A.; Shintani, M.; Nakayama, T.; Furutani, Y.; Hayama, E.; Inai, K.; Saji, T.; Nonoyama, S.; Nakanishi, T. Missense mutations of the BMPR1B (ALK6) gene in childhood idiopathic pulmonary arterial hypertension. *Circ. J.* **2012**, *76*, 1501–1508. [CrossRef] [PubMed]

161. Yu, Y.; Keller, S.H.; Remillard, C.V.; Safrina, O.; Nicholson, A.; Zhang, S.L.; Jiang, W.; Vangala, N.; Landsberg, J.W.; Wang, J.-Y.; et al. A functional single-nucleotide polymorphism in the TRPC6 gene promoter associated with idiopathic pulmonary arterial hypertension. *Circulation* **2009**, *119*, 2313–2322. [CrossRef] [PubMed]

162. Wang, G.; Knight, L.; Ji, R.; Lawrence, P.; Kanaan, U.; Li, L.; Das, A.; Cui, B.; Zou, W.; Penny, D.J.; et al. Early onset severe pulmonary arterial hypertension with "two-hit" digenic mutations in both BMPR2 and KCNA5 genes. *Int. J. Cardiol.* **2014**, *177*, e167–e169. [CrossRef] [PubMed]

163. Remillard, C.V.; Tigno, D.D.; Platoshyn, O.; Burg, E.D.; Brevnova, E.E.; Conger, D.; Nicholson, A.; Rana, B.K.; Channick, R.N.; Rubin, L.J.; et al. Function of Kv1.5 channels and genetic variations of KCNA5 in patients with idiopathic pulmonary arterial hypertension. *Am. J. Physiol. Cell Physiol.* **2007**, *292*, C1837–C1853. [CrossRef]

164. Zhang, P.; McGrath, B.C.; Reinert, J.; Olsen, D.S.; Lei, L.; Gill, S.; Wek, S.A.; Vattem, K.M.; Wek, R.C.; Kimball, S.R.; et al. The GCN2 eIF2alpha kinase is required for adaptation to amino acid deprivation in mice. *Mol. Cell. Biol.* **2002**, *22*, 6681–6688. [CrossRef] [PubMed]

165. Germain, M.; Eyries, M.; Montani, D.; Poirier, O.; Girerd, B.; Dorfmüller, P.; Coulet, F.; Nadaud, S.; Maugenre, S.; Guignabert, C.; et al. Genome-wide association analysis identifies a susceptibility locus for pulmonary arterial hypertension. *Nat. Genet.* **2013**, *45*, 518–521. [CrossRef]

166. Paulin, R.; Dromparis, P.; Sutendra, G.; Gurtu, V.; Zervopoulos, S.; Bowers, L.; Haromy, A.; Webster, L.; Provencher, S.; Bonnet, S.; et al. Sirtuin 3 deficiency is associated with inhibited mitochondrial function and pulmonary arterial hypertension in rodents and humans. *Cell Metab.* **2014**, *20*, 827–839. [CrossRef]

167. Michelakis, E.D.; Gurtu, V.; Webster, L.; Barnes, G.; Watson, G.; Howard, L.; Cupitt, J.; Paterson, I.; Thompson, R.B.; Chow, K.; et al. Inhibition of pyruvate dehydrogenase kinase improves pulmonary arterial hypertension in genetically susceptible patients. *Sci. Transl. Med.* **2017**, *9*. [CrossRef]

168. Dromparis, P.; Paulin, R.; Sutendra, G.; Qi, A.C.; Bonnet, S.; Michelakis, E.D. Uncoupling protein 2 deficiency mimics the effects of hypoxia and endoplasmic reticulum stress on mitochondria and triggers pseudohypoxic pulmonary vascular remodeling and pulmonary hypertension. *Circ. Res.* **2013**, *113*, 126–136. [CrossRef]

169. Zhang, C.Y.; Baffy, G.; Perret, P.; Krauss, S.; Peroni, O.; Grujic, D.; Hagen, T.; Vidal-Puig, A.J.; Boss, O.; Kim, Y.B.; et al. Uncoupling protein-2 negatively regulates insulin secretion and is a major link between obesity, beta cell dysfunction, and type 2 diabetes. *Cell* **2001**, *105*, 745–755. [CrossRef]

170. Baloira Villar, A.; Valverde, D.; Lago, M. Functional study of polymorphisms in the promoter region of the endothelin-1 gene in Pulmonary Arterial Hypertension. *Eur. Respir. J.* **2019**, *54*, PA5043.

171. Giaid, A.; Yanagisawa, M.; Langleben, D.; Michel, R.P.; Levy, R.; Shennib, H.; Kimura, S.; Masaki, T.; Duguid, W.P.; Stewart, D.J. Expression of Endothelin-1 in the Lungs of Patients with Pulmonary Hypertension. *N. Engl. J. Med.* **1993**, *328*, 1732–1739. [CrossRef] [PubMed]

172. De Jesus Perez, V.A.; Yuan, K.; Lyuksyutova, M.A.; Dewey, F.; Orcholski, M.E.; Shuffle, E.M.; Mathur, M.; Yancy, L., Jr.; Rojas, V.; Li, C.G.; et al. Whole-exome sequencing reveals TopBP1 as a novel gene in idiopathic pulmonary arterial hypertension. *Am. J. Respir. Crit. Care Med.* **2014**, *189*, 1260–1272. [CrossRef] [PubMed]

173. Damico, R.; Kolb, T.M.; Valera, L.; Wang, L.; Housten, T.; Tedford, R.J.; Kass, D.A.; Rafaels, N.; Gao, L.; Barnes, K.C.; et al. Serum endostatin is a genetically determined predictor of survival in pulmonary arterial hypertension. *Am. J. Respir. Crit. Care Med.* **2015**, *191*, 208–218. [CrossRef]

174. Maloney, J.P.; Stearman, R.S.; Bull, T.M.; Calabrese, D.W.; Tripp-Addison, M.L.; Wick, M.J.; Broeckel, U.; Robbins, I.M.; Wheeler, L.A.; Cogan, J.D.; et al. Loss-of-function thrombospondin-1 mutations in familial pulmonary hypertension. *Am. J. Physiol. Lung Cell. Mol. Physiol.* **2012**, *302*, L541–L554. [CrossRef]

175. Pfarr, N.; Fischer, C.; Ehlken, N.; Becker-Grünig, T.; López-González, V.; Gorenflo, M.; Hager, A.; Hinderhofer, K.; Miera, O.; Nagel, C.; et al. Hemodynamic and genetic analysis in children with idiopathic, heritable, and congenital heart disease associated pulmonary arterial hypertension. *Respir. Res.* **2013**, *14*, 3. [CrossRef]

176. Runo, J.R.; Vnencak-Jones, C.L.; Prince, M.; Loyd, J.E.; Wheeler, L.; Robbins, I.M.; Lane, K.B.; Newman, J.H.; Johnson, J.; Nichols, W.C.; et al. Pulmonary veno-occlusive disease caused by an inherited mutation in bone morphogenetic protein receptor II. *Am. J. Respir. Crit. Care Med.* **2003**, *167*, 889–894. [CrossRef]

177. Machado, R.D.; Southgate, L.; Eichstaedt, C.A.; Aldred, M.A.; Austin, E.D.; Best, D.H.; Chung, W.K.; Benjamin, N.; Elliott, C.G.; Eyries, M.; et al. Pulmonary Arterial Hypertension: A Current Perspective on Established and Emerging Molecular Genetic Defects. *Hum. Mutat.* **2015**, *36*, 1113–1127. [CrossRef]

178. Machado, R.D.; Eickelberg, O.; Elliott, C.G.; Geraci, M.W.; Hanaoka, M.; Loyd, J.E.; Newman, J.H.; Phillips, J.A., 3rd; Soubrier, F.; Trembath, R.C.; et al. Genetics and genomics of pulmonary arterial hypertension. *J. Am. Coll. Cardiol.* **2009**, *54*, S32–S42. [CrossRef]

179. Machado, R.D.; Aldred, M.A.; James, V.; Harrison, R.E.; Patel, B.; Schwalbe, E.C.; Gruenig, E.; Janssen, B.; Koehler, R.; Seeger, W.; et al. Mutations of the TGF-beta type II receptor BMPR2 in pulmonary arterial hypertension. *Hum. Mutat.* **2006**, *27*, 121–132. [CrossRef]

180. Evans, J.D.W.; Girerd, B.; Montani, D.; Wang, X.-J.; Galiè, N.; Austin, E.D.; Elliott, G.; Asano, K.; Grünig, E.; Yan, Y.; et al. BMPR2 mutations and survival in pulmonary arterial hypertension: An individual participant data meta-analysis. *Lancet Respir. Med.* **2016**, *4*, 129–137. [CrossRef]

181. Atkinson, C.; Stewart, S.; Upton, P.D.; Machado, R.; Thomson, J.R.; Trembath, R.C.; Morrell, N.W. Primary pulmonary hypertension is associated with reduced pulmonary vascular expression of type II bone morphogenetic protein receptor. *Circulation* **2002**, *105*, 1672–1678. [CrossRef] [PubMed]

182. Lavoie, J.R.; Ormiston, M.L.; Perez-Iratxeta, C.; Courtman, D.W.; Jiang, B.; Ferrer, E.; Caruso, P.; Southwood, M.; Foster, W.S.; Morrell, N.W.; et al. Proteomic analysis implicates translationally controlled tumor protein as a novel mediator of occlusive vascular remodeling in pulmonary arterial hypertension. *Circulation* **2014**, *129*, 2125–2135. [CrossRef] [PubMed]

183. Chen, N.-Y.; D Collum, S.; Luo, F.; Weng, T.; Le, T.-T.; M Hernandez, A.; Philip, K.; Molina, J.G.; Garcia-Morales, L.J.; Cao, Y.; et al. Macrophage bone morphogenic protein receptor 2 depletion in idiopathic pulmonary fibrosis and Group III pulmonary hypertension. *Am. J. Physiol. Lung Cell. Mol. Physiol.* **2016**, *311*, L238–L254. [CrossRef]

184. Donaldson, J.W.; McKeever, T.M.; Hall, I.P.; Hubbard, R.B.; Fogarty, A.W. The UK prevalence of hereditary haemorrhagic telangiectasia and its association with sex, socioeconomic status and region of residence: A population-based study. *Thorax* **2014**, *69*, 161–167. [CrossRef]

185. Harrison, R.E.; Flanagan, J.A.; Sankelo, M.; Abdalla, S.A.; Rowell, J.; Machado, R.D.; Elliott, C.G.; Robbins, I.M.; Olschewski, H.; McLaughlin, V.; et al. Molecular and functional analysis identifies ALK-1 as the predominant cause of pulmonary hypertension related to hereditary haemorrhagic telangiectasia. *J. Med. Genet.* **2003**, *40*, 865–871. [CrossRef]

186. Fujiwara, M.; Yagi, H.; Matsuoka, R.; Akimoto, K.; Furutani, M.; Imamura, S.-I.; Uehara, R.; Nakayama, T.; Takao, A.; Nakazawa, M.; et al. Implications of mutations of activin receptor-like kinase 1 gene (ALK1) in addition to bone morphogenetic protein receptor II gene (BMPR2) in children with pulmonary arterial hypertension. *Circ. J.* **2008**, *72*, 127–133. [CrossRef]

187. Kwasnickacrawford, D.; Carson, A.; Roberts, W.; Summers, A.; Rehnstrom, K.; Jarvela, I.; Scherer, S. Characterization of a novel cation transporter ATPase gene (ATP13A4) interrupted by 3q25–q29 inversion in an individual with language delay. *Genomics* **2005**, *86*, 182–194. [CrossRef]

188. Madan, M.; Patel, A.; Skruber, K.; Geerts, D.; Altomare, D.A.; Iv, O.P. ATP13A3 and caveolin-1 as potential biomarkers for difluoromethylornithine-based therapies in pancreatic cancers. *Am. J. Cancer Res.* **2016**, *6*, 1231–1252.

189. Sui, H.; Han, B.G.; Lee, J.K.; Walian, P.; Jap, B.K. Structural basis of water-specific transport through the AQP1 water channel. *Nature* **2001**, *414*, 872–878. [CrossRef]

190. Montani, D.; Girerd, B.; Günther, S.; Riant, F.; Tournier-Lasserve, E.; Magy, L.; Maazi, N.; Guignabert, C.; Savale, L.; Sitbon, O.; et al. Pulmonary arterial hypertension in familial hemiplegic migraine with ATP1A2 channelopathy. *Eur. Respir. J.* **2014**, *43*, 641–643. [CrossRef] [PubMed]

191. Firth, A.L.; Remillard, C.V.; Platoshyn, O.; Fantozzi, I.; Ko, E.A.; Yuan, J.X.-J. Functional ion channels in human pulmonary artery smooth muscle cells: Voltage-dependent cation channels. *Pulm. Circ.* **2011**, *1*, 48–71. [CrossRef] [PubMed]

192. Makino, A.; Firth, A.L.; Yuan, J.X.-J. Endothelial and smooth muscle cell ion channels in pulmonary vasoconstriction and vascular remodeling. *Compr. Physiol.* **2011**, *1*, 1555–1602. [CrossRef] [PubMed]

193. Burg, E.D.; Remillard, C.V.; Yuan, J.X.-J. Potassium channels in the regulation of pulmonary artery smooth muscle cell proliferation and apoptosis: Pharmacotherapeutic implications. *Br. J. Pharmacol.* **2008**, *153* (Suppl. 1), S99–S111. [CrossRef]

194. Lambert, M.; Capuano, V.; Olschewski, A.; Sabourin, J.; Nagaraj, C.; Girerd, B.; Weatherald, J.; Humbert, M.; Antigny, F. Ion Channels in Pulmonary Hypertension: A Therapeutic Interest? *Int. J. Mol. Sci.* **2018**, *19*, 3162. [CrossRef]

195. Babenko, A.P.; Polak, M.; Cavé, H.; Busiah, K.; Czernichow, P.; Scharfmann, R.; Bryan, J.; Aguilar-Bryan, L.; Vaxillaire, M.; Froguel, P. Activating mutations in the ABCC8 gene in neonatal diabetes mellitus. *N. Engl. J. Med.* **2006**, *355*, 456–466. [CrossRef]

196. Flanagan, S.E.; Clauin, S.; Bellanné-Chantelot, C.; de Lonlay, P.; Harries, L.W.; Gloyn, A.L.; Ellard, S. Update of mutations in the genes encoding the pancreatic beta-cell KATPchannel subunits Kir6.2 (KCNJ11) and sulfonylurea receptor 1 (ABCC8) in diabetes mellitus and hyperinsulinism. *Hum. Mutat.* **2009**, *30*, 170–180. [CrossRef]

197. Lago-Docampo, M.; Tenorio, J.; Hernández-González, I.; Pérez-Olivares, C.; Escribano-Subías, P.; Pousada, G.; Baloira, A.; Arenas, M.; Lapunzina, P.; Valverde, D. Characterization of rare ABCC8 variants identified in Spanish pulmonary arterial hypertension patients. *Sci. Rep.* **2020**, *10*. [CrossRef]

198. Arora, R.; Metzger, R.J.; Papaioannou, V.E. Multiple roles and interactions of Tbx4 and Tbx5 in development of the respiratory system. *PLoS Genet.* **2012**, *8*, e1002866. [CrossRef]

199. Karolak, J.A.; Vincent, M.; Deutsch, G.; Gambin, T.; Cogné, B.; Pichon, O.; Vetrini, F.; Mefford, H.C.; Dines, J.N.; Golden-Grant, K.; et al. Complex Compound Inheritance of Lethal Lung Developmental Disorders Due to Disruption of the TBX-FGF Pathway. *Am. J. Hum. Genet.* **2019**, *104*, 213–228. [CrossRef]

200. Bongers, E.M.H.F.; Duijf, P.H.G.; van Beersum, S.E.M.; Schoots, J.; Van Kampen, A.; Burckhardt, A.; Hamel, B.C.J.; Losan, F.; Hoefsloot, L.H.; Yntema, H.G.; et al. Mutations in the human TBX4 gene cause small patella syndrome. *Am. J. Hum. Genet.* **2004**, *74*, 1239–1248. [CrossRef] [PubMed]

201. Galambos, C.; Mullen, M.P.; Shieh, J.T.; Schwerk, N.; Kielt, M.J.; Ullmann, N.; Boldrini, R.; Stucin-Gantar, I.; Haass, C.; Bansal, M.; et al. Phenotype characterisation of TBX4 mutation and deletion carriers with neonatal and paediatric pulmonary hypertension. *Eur. Respir. J.* **2019**, *54*, 1801965. [CrossRef] [PubMed]

202. Francois, M.; Koopman, P.; Beltrame, M. SoxF genes: Key players in the development of the cardio-vascular system. *Int. J. Biochem. Cell Biol.* **2010**, *42*, 445–448. [CrossRef] [PubMed]

203. De Vilder, E.Y.G.; Debacker, J.; Vanakker, O.M. GGCX-Associated Phenotypes: An Overview in Search of Genotype-Phenotype Correlations. *Int. J. Mol. Sci.* **2017**, *18*, 240. [CrossRef] [PubMed]

204. Lantuéjoul, S.; Sheppard, M.N.; Corrin, B.; Burke, M.M.; Nicholson, A.G. Pulmonary veno-occlusive disease and pulmonary capillary hemangiomatosis: A clinicopathologic study of 35 cases. *Am. J. Surg. Pathol.* **2006**, *30*, 850–857. [CrossRef]

205. Davies, P.; Reid, L. Pulmonary veno-occlusive disease in siblings: Case reports and morphometric study. *Hum. Pathol.* **1982**, *13*, 911–915. [CrossRef]

206. Voordes, C.G.; Kuipers, J.R.; Elema, J.D. Familial pulmonary veno-occlusive disease: A case report. *Thorax* **1977**, *32*, 763–766. [CrossRef]

207. Austin, E.D.; Cogan, J.D.; West, J.D.; Hedges, L.K.; Hamid, R.; Dawson, E.P.; Wheeler, L.A.; Parl, F.F.; Loyd, J.E.; Phillips, J.A., 3rd. Alterations in oestrogen metabolism: Implications for higher penetrance of familial pulmonary arterial hypertension in females. *Eur. Respir. J.* **2009**, *34*, 1093–1099. [CrossRef]

208. Austin, E.D.; Hamid, R.; Hemnes, A.R.; Loyd, J.E.; Blackwell, T.; Yu, C.; Phillips, J.A., III; Gaddipati, R.; Gladson, S.; Gu, E.; et al. BMPR2 expression is suppressed by signaling through the estrogen receptor. *Biol. Sex Differ.* **2012**, *3*, 6. [CrossRef]

209. Hamid, R.; Cogan, J.D.; Hedges, L.K.; Austin, E.; Phillips, J.A., 3rd; Newman, J.H.; Loyd, J.E. Penetrance of pulmonary arterial hypertension is modulated by the expression of normal BMPR2 allele. *Hum. Mutat.* **2009**, *30*, 649–654. [CrossRef]

210. Phillips, J.A., 3rd; Poling, J.S.; Phillips, C.A.; Stanton, K.C.; Austin, E.D.; Cogan, J.D.; Wheeler, L.; Yu, C.; Newman, J.H.; Dietz, H.C.; et al. Synergistic heterozygosity for TGFbeta1 SNPs and BMPR2 mutations modulates the age at diagnosis and penetrance of familial pulmonary arterial hypertension. *Genet. Med.* **2008**, *10*, 359–365. [CrossRef] [PubMed]

211. Hirschey, M.D.; Shimazu, T.; Jing, E.; Grueter, C.A.; Collins, A.M.; Aouizerat, B.; Stančáková, A.; Goetzman, E.; Lam, M.M.; Schwer, B.; et al. SIRT3 deficiency and mitochondrial protein hyperacetylation accelerate the development of the metabolic syndrome. *Mol. Cell* **2011**, *44*, 177–190. [CrossRef] [PubMed]

212. Abraham, W.T.; Raynolds, M.V.; Badesch, D.B.; Wynne, K.M.; Groves, B.M.; Roden, R.L.; Robertson, A.D.; Lowes, B.D.; Zisman, L.S.; Voelkel, N.F.; et al. Angiotensin-converting enzyme DD genotype in patients with primary pulmonary hypertension: Increased frequency and association with preserved haemodynamics. *J. Renin-Angiotensin-Aldosterone Syst.* **2003**, *4*, 27–30. [CrossRef] [PubMed]

213. Solari, V.; Puri, P. Genetic polymorphisms of angiotensin system genes in congenital diaphragmatic hernia associated with persistent pulmonary hypertension. *J. Pediatr. Surg.* **2004**, *39*, 302–306. [CrossRef] [PubMed]

214. Sutliff, R.L.; Kang, B.-Y.; Michael Hart, C. PPARγ as a potential therapeutic target in pulmonary hypertension. *Ther. Adv. Respir. Dis.* **2010**, *4*, 143–160. [CrossRef] [PubMed]

215. Green, D.E.; Sutliff, R.L.; Michael Hart, C. Is Peroxisome Proliferator-Activated Receptor Gamma (PPARγ) a Therapeutic Target for the Treatment of Pulmonary Hypertension? *Pulm. Circ.* **2011**, *1*, 33–47. [CrossRef] [PubMed]

216. Malenfant, S.; Neyron, A.-S.; Paulin, R.; Potus, F.; Meloche, J.; Provencher, S.; Bonnet, S. Signal Transduction in the Development of Pulmonary Arterial Hypertension. *Pulm. Circ.* **2013**, *3*, 278–293. [CrossRef]

217. Shatat, M.A.; Tian, H.; Zhang, R.; Tandon, G.; Hale, A.; Fritz, J.S.; Zhou, G.; Martínez-González, J.; Rodríguez, C.; Champion, H.C.; et al. Endothelial Krüppel-Like Factor 4 Modulates Pulmonary Arterial Hypertension. *Am. J. Respir. Cell Mol. Biol.* **2014**, *50*, 647–653. [CrossRef]

218. Crosby, A.; Toshner, M.R.; Southwood, M.R.; Soon, E.; Dunmore, B.J.; Groves, E.; Moore, S.; Wright, P.; Ottersbach, K.; Bennett, C.; et al. Hematopoietic stem cell transplantation alters susceptibility to pulmonary hypertension in Bmpr2-deficient mice. *Pulm. Circ.* **2018**, *8*. [CrossRef]

219. Kallunki, T.; Barisic, M.; Jäättelä, M.; Liu, B. How to Choose the Right Inducible Gene Expression System for Mammalian Studies? *Cells* **2019**, *8*, 796. [CrossRef]

220. Urnov, F.D.; Rebar, E.J.; Holmes, M.C.; Steve Zhang, H.; Gregory, P.D. Genome editing with engineered zinc finger nucleases. *Nat. Rev. Genet.* **2010**, *11*, 636–646. [CrossRef]

221. Joung, J.K.; Keith Joung, J.; Sander, J.D. TALENs: A widely applicable technology for targeted genome editing. *Nat. Rev. Mol. Cell Biol.* **2013**, *14*, 49–55. [CrossRef]

222. Ran, F.A.; Hsu, P.D.; Wright, J.; Agarwala, V.; Scott, D.A.; Zhang, F. Genome engineering using the CRISPR-Cas9 system. *Nat. Protoc.* **2013**, *8*, 2281–2308. [CrossRef]

223. Kim, D.H.; Rossi, J.J. RNAi mechanisms and applications. *Biotech.* **2008**, *44*, 613–616. [CrossRef]

224. Lewandoski, M. Conditional control of gene expression in the mouse. *Nat. Rev. Genet.* **2001**, *2*, 743–755. [CrossRef]

225. Van der Weyden, L.; White, J.K.; Adams, D.J.; Logan, D.W. The mouse genetics toolkit: Revealing function and mechanism. *Genome Biol.* **2011**, *12*, 224. [CrossRef]

226. Van Boxtel, R.; Cuppen, E. Rat traps: Filling the toolbox for manipulating the rat genome. *Genome Biol.* **2010**, *11*, 217. [CrossRef]

227. Toshner, M.; Voswinckel, R.; Southwood, M.; Al-Lamki, R.; Howard, L.S.G.; Marchesan, D.; Yang, J.; Suntharalingam, J.; Soon, E.; Exley, A.; et al. Evidence of dysfunction of endothelial progenitors in pulmonary arterial hypertension. *Am. J. Respir. Crit. Care Med.* **2009**, *180*, 780–787. [CrossRef]

228. Morrell, N.W.; Adnot, S.; Archer, S.L.; Dupuis, J.; Jones, P.L.; MacLean, M.R.; McMurtry, I.F.; Stenmark, K.R.; Thistlethwaite, P.A.; Weissmann, N.; et al. Cellular and molecular basis of pulmonary arterial hypertension. *J. Am. Coll. Cardiol.* **2009**, *54*, S20–S31. [CrossRef]

229. Humbert, M.; Guignabert, C.; Bonnet, S.; Dorfmüller, P.; Klinger, J.R.; Nicolls, M.R.; Olschewski, A.J.; Pullamsetti, S.S.; Schermuly, R.T.; Stenmark, K.R.; et al. Pathology and pathobiology of pulmonary hypertension: State of the art and research perspectives. *Eur. Respir. J.* **2019**, *53*. [CrossRef]

230. Liu, B.; Haimel, M.; Bleda, M.; Li, W.; Gräf, S.; Upton, P.D.; Morrell, N.W. S42 Characterizing ATP13A3 loss of function in pulmonary arterial hypertension (PAH). *Fundam. Mech. Pulm. Arter. Hypertens.* **2018**.

231. Tian, L.; Gao, J.; Garcia, I.M.; Chen, H.J.; Castaldi, A.; Chen, Y.-W. Human pluripotent stem cell-derived lung organoids: Potential applications in development and disease modeling. *Wiley Interdiscip. Rev. Dev. Biol.* **2020**, e399. [CrossRef]

232. D'Amico, R.W.; Faley, S.; Shim, H.-N.; Prosser, J.R.; Agrawal, V.; Bellan, L.M.; West, J.D. Pulmonary Vascular Platform Models the Effects of Flow and Pressure on Endothelial Dysfunction in BMPR2 Associated Pulmonary Arterial Hypertension. *Int. J. Mol. Sci.* **2018**, *19*, 2561. [CrossRef]

233. Nogueira-Ferreira, R.; Faria-Costa, G.; Ferreira, R.; Henriques-Coelho, T. Animal models for the study of pulmonary hypertension: Potential and limitations. *Cardiol. Cardiovasc. Med.* **2016**, *1*, 1–22. [CrossRef]

234. Ma, Z.; Mao, L.; Rajagopal, S. Hemodynamic Characterization of Rodent Models of Pulmonary Arterial Hypertension. *J. Vis. Exp.* **2016**. [CrossRef]

235. Gomez-Arroyo, J.G.; Farkas, L.; Alhussaini, A.A.; Farkas, D.; Kraskauskas, D.; Voelkel, N.F.; Bogaard, H.J. The monocrotaline model of pulmonary hypertension in perspective. *Am. J. Physiol. Lung Cell. Mol. Physiol.* **2012**, *302*, L363–L369. [CrossRef]

236. Akhavein, F.; St-Michel, E.J.; Seifert, E.; Rohlicek, C.V. Decreased left ventricular function, myocarditis, and coronary arteriolar medial thickening following monocrotaline administration in adult rats. *J. Appl. Physiol.* **2007**, *103*, 287–295. [CrossRef]

237. Taraseviciene-stewart, L.; Kasahara, Y.; Alger, L.; Hirth, P.; Mc Mahon, G.; Waltenberger, J.; Voelkel, N.F.; Tuder, R.M. Inhibition of the VEGF receptor 2 combined with chronic hypoxia causes cell death-dependent pulmonary endothelial cell proliferation and severe pulmonary hypertension. *FASEB J.* **2001**, *15*, 427–438. [CrossRef]

238. Kojonazarov, B.; Hadzic, S.; Ghofrani, H.A.; Grimminger, F.; Seeger, W.; Weissmann, N.; Schermuly, R.T. Severe Emphysema in the SU5416/Hypoxia Rat Model of Pulmonary Hypertension. *Am. J. Respir. Crit. Care Med.* **2019**, *200*, 515–518. [CrossRef]

239. Mancuso, M.R.; Davis, R.; Norberg, S.M.; O'Brien, S.; Sennino, B.; Nakahara, T.; Yao, V.J.; Inai, T.; Brooks, P.; Freimark, B.; et al. Rapid vascular regrowth in tumors after reversal of VEGF inhibition. *J. Clin. Investig.* **2006**, *116*, 2610–2621. [CrossRef]

240. Tabima, D.M.; Hacker, T.A.; Chesler, N.C. Measuring right ventricular function in the normal and hypertensive mouse hearts using admittance-derived pressure-volume loops. *Am. J. Physiol. Heart Circ. Physiol.* **2010**, *299*, H2069–H2075. [CrossRef]

241. Luo, F.; Wang, X.; Luo, X.; Li, B.; Zhu, D.; Sun, H.; Tang, Y. Invasive Hemodynamic Assessment for the Right Ventricular System and Hypoxia-Induced Pulmonary Arterial Hypertension in Mice. *J. Vis. Exp.* **2019**. [CrossRef]

242. Stenmark, K.R.; Meyrick, B.; Galie, N.; Mooi, W.J.; McMurtry, I.F. Animal models of pulmonary arterial hypertension: The hope for etiological discovery and pharmacological cure. *Am. J. Physiol. Lung Cell. Mol. Physiol.* **2009**, *297*, L1013–L1032. [CrossRef]

243. Akazawa, Y.; Okumura, K.; Ishii, R.; Slorach, C.; Hui, W.; Ide, H.; Honjo, O.; Sun, M.; Kabir, G.; Connelly, K.; et al. Pulmonary artery banding is a relevant model to study the right ventricular remodeling and dysfunction that occurs in pulmonary arterial hypertension. *J. Appl. Physiol.* **2020**, *129*, 238–246. [CrossRef] [PubMed]

244. Lawrie, A. A report on the use of animal models and phenotyping methods in pulmonary hypertension research. *Pulm. Circ.* **2014**, *4*, 2–9. [CrossRef]

245. Song, Y.; Jones, J.E.; Beppu, H.; Keaney, J.F., Jr.; Loscalzo, J.; Zhang, Y.-Y. Increased susceptibility to pulmonary hypertension in heterozygous BMPR2-mutant mice. *Circulation* **2005**, *112*, 553–562. [CrossRef]

246. Long, L.; MacLean, M.R.; Jeffery, T.K.; Morecroft, I.; Yang, X.; Rudarakanchana, N.; Southwood, M.; James, V.; Trembath, R.C.; Morrell, N.W. Serotonin increases susceptibility to pulmonary hypertension in BMPR2-deficient mice. *Circ. Res.* **2006**, *98*, 818–827. [CrossRef]

247. Peacock, A.J.; Vonk Noordegraaf, A. Cardiac magnetic resonance imaging in pulmonary arterial hypertension. *Eur. Respir. Rev.* **2013**, *22*, 526–534. [CrossRef]

248. Provencher, S.; Archer, S.L.; Ramirez, F.D.; Hibbert, B.; Paulin, R.; Boucherat, O.; Lacasse, Y.; Bonnet, S. Standards and Methodological Rigor in Pulmonary Arterial Hypertension Preclinical and Translational Research. *Circ. Res.* **2018**, *122*, 1021–1032. [CrossRef]

249. Rol, N.; Kurakula, K.B.; Happé, C.; Bogaard, H.J.; Goumans, M.-J. TGF-β and BMPR2 Signaling in PAH: Two Black Sheep in One Family. *Int. J. Mol. Sci.* **2018**, *19*, 2585. [CrossRef]

250. Ten Dijke, P.; Goumans, M.-J.; Pardali, E. Endoglin in angiogenesis and vascular diseases. *Angiogenesis* **2008**, *11*, 79–89. [CrossRef]

251. Upton, P.D.; Davies, R.J.; Trembath, R.C.; Morrell, N.W. Bone morphogenetic protein (BMP) and activin type II receptors balance BMP9 signals mediated by activin receptor-like kinase-1 in human pulmonary artery endothelial cells. *J. Biol. Chem.* **2009**, *284*, 15794–15804. [CrossRef]

252. Barnes, J.W.; Kucera, E.T.; Tian, L.; Mellor, N.E.; Dvorina, N.; Baldwin, W.W.; Aldred, M.A.; Farver, C.F.; Comhair, S.A.A.; Aytekin, M.; et al. Bone Morphogenic Protein Type 2 Receptor Mutation-Independent Mechanisms of Disrupted Bone Morphogenetic Protein Signaling in Idiopathic Pulmonary Arterial Hypertension. *Am. J. Respir. Cell Mol. Biol.* **2016**, *55*, 564–575. [CrossRef]

253. Good, R.B.; Gilbane, A.J.; Trinder, S.L.; Denton, C.P.; Coghlan, G.; Abraham, D.J.; Holmes, A.M. Endothelial to Mesenchymal Transition Contributes to Endothelial Dysfunction in Pulmonary Arterial Hypertension. *Am. J. Pathol.* **2015**, *185*, 1850–1858. [CrossRef]

254. Liu, T.; Zou, X.-Z.; Huang, N.; Ge, X.-Y.; Yao, M.-Z.; Liu, H.; Zhang, Z.; Hu, C.-P. miR-27a promotes endothelial-mesenchymal transition in hypoxia-induced pulmonary arterial hypertension by suppressing BMP signaling. *Life Sci.* **2019**, *227*, 64–73. [CrossRef]

255. Vanderpool, R.R.; El-Bizri, N.; Rabinovitch, M.; Chesler, N.C. Patchy deletion of Bmpr1a potentiates proximal pulmonary artery remodeling in mice exposed to chronic hypoxia. *Biomech. Model. Mechanobiol.* **2013**, *12*, 33–42. [CrossRef]

256. Tie, L.; Wang, D.; Shi, Y.; Li, X. Aquaporins in Cardiovascular System. *Adv. Exp. Med. Biol.* **2017**, *969*, 105–113. [CrossRef]

257. Clapp, C.; Martínez de la Escalera, G. Aquaporin-1: A novel promoter of tumor angiogenesis. *Trends Endocrinol. Metab.* **2006**, *17*, 1–2. [CrossRef]

258. Schuoler, C.; Haider, T.J.; Leuenberger, C.; Vogel, J.; Ostergaard, L.; Kwapiszewska, G.; Kohler, M.; Gassmann, M.; Huber, L.C.; Brock, M. Aquaporin 1 controls the functional phenotype of pulmonary smooth muscle cells in hypoxia-induced pulmonary hypertension. *Basic Res. Cardiol.* **2017**, *112*, 30. [CrossRef]

259. Kunichika, N.; Landsberg, J.W.; Yu, Y.; Kunichika, H.; Thistlethwaite, P.A.; Rubin, L.J.; Yuan, J.X.-J. Bosentan inhibits transient receptor potential channel expression in pulmonary vascular myocytes. *Am. J. Respir. Crit. Care Med.* **2004**, *170*, 1101–1107. [CrossRef]

260. Pozeg, Z.I.; Michelakis, E.D.; McMurtry, M.S.; Thébaud, B.; Wu, X.-C.; Dyck, J.R.B.; Hashimoto, K.; Wang, S.; Moudgil, R.; Harry, G.; et al. In vivo gene transfer of the O2-sensitive potassium channel Kv1.5 reduces pulmonary hypertension and restores hypoxic pulmonary vasoconstriction in chronically hypoxic rats. *Circulation* **2003**, *107*, 2037–2044. [CrossRef]

261. Sheeba, C.J.; Logan, M.P.O. The Roles of T-Box Genes in Vertebrate Limb Development. *Curr. Top. Dev. Biol.* **2017**, *122*, 355–381. [CrossRef]

262. Zhang, W.; Menke, D.B.; Jiang, M.; Chen, H.; Warburton, D.; Turcatel, G.; Lu, C.-H.; Xu, W.; Luo, Y.; Shi, W. Spatial-temporal targeting of lung-specific mesenchyme by a Tbx4 enhancer. *BMC Biol.* **2013**, *11*, 111. [CrossRef]

263. Rieder, M.J.; Reiner, A.P.; Rettie, A.E. Gamma-glutamyl carboxylase (GGCX) tagSNPs have limited utility for predicting warfarin maintenance dose. *J. Thromb. Haemost.* **2007**, *5*, 2227–2234. [CrossRef]

264. Napolitano, M.; Mariani, G.; Lapecorella, M. Hereditary combined deficiency of the vitamin K-dependent clotting factors. *Orphanet J. Rare Dis.* **2010**, *5*, 21. [CrossRef]

265. Li, Q.; Schurgers, L.J.; Smith, A.C.M.; Tsokos, M.; Uitto, J.; Cowen, E.W. Co-existent pseudoxanthoma elasticum and vitamin K-dependent coagulation factor deficiency: Compound heterozygosity for mutations in the GGCX gene. *Am. J. Pathol.* **2009**, *174*, 534–540. [CrossRef]

266. Moreau, M.E.; Garbacki, N.; Molinaro, G.; Brown, N.J.; Marceau, F.; Adam, A. The Kallikrein-Kinin System: Current and Future Pharmacological Targets. *J. Pharmacol. Sci.* **2005**, *99*, 6–38. [CrossRef]

267. Carretero, O.A.; Scicli, A.G. The renal kallikrein-kinin system in human and in experimental hypertension. *Klin. Wochenschr.* **1978**, *56* (Suppl. 1), 113–125. [CrossRef]

268. Madeddu, P.; Emanueli, C.; El-Dahr, S. Mechanisms of disease: The tissue kallikrein-kinin system in hypertension and vascular remodeling. *Nat. Clin. Pract. Nephrol.* **2007**, *3*, 208–221. [CrossRef]

269. Woodley-Miller, C.; Chao, J.; Chao, L. Restriction fragment length polymorphisms mapped in spontaneously hypertensive rats using kallikrein probes. *J. Hypertens.* **1989**, *7*, 865–871. [CrossRef]

270. Madeddu, P.; Varoni, M.V.; Demontis, M.P.; Chao, J.; Simson, J.A.; Glorioso, N.; Anania, V. Kallikrein-kinin system and blood pressure sensitivity to salt. *Hypertension* **1997**, *29*, 471–477. [CrossRef]

271. Hua, H.; Zhou, S.; Liu, Y.; Wang, Z.; Wan, C.; Li, H.; Chen, C.; Li, G.; Zeng, C.; Chen, L.; et al. Relationship between the regulatory region polymorphism of human tissue kallikrein gene and essential hypertension. *J. Hum. Hypertens.* **2005**, *19*, 715–721. [CrossRef]

272. Alexander-Curtis, M.; Pauls, R.; Chao, J.; Volpi, J.J.; Bath, P.M.; Verdoorn, T.A. Human tissue kallikrein in the treatment of acute ischemic stroke. *Ther. Adv. Neurol. Disord.* **2019**, *12*. [CrossRef]

273. Anthony, T.G.; McDaniel, B.J.; Byerley, R.L.; McGrath, B.C.; Cavener, D.R.; McNurlan, M.A.; Wek, R.C. Preservation of liver protein synthesis during dietary leucine deprivation occurs at the expense of skeletal muscle mass in mice deleted for eIF2 kinase GCN2. *J. Biol. Chem.* **2004**, *279*, 36553–36561. [CrossRef]

274. Waypa, G.B.; Osborne, S.W.; Marks, J.D.; Berkelhamer, S.K.; Kondapalli, J.; Schumacker, P.T. Sirtuin 3 deficiency does not augment hypoxia-induced pulmonary hypertension. *Am. J. Respir. Cell Mol. Biol.* **2013**, *49*, 885–891. [CrossRef]

275. Kim, K.; Kim, I.-K.; Yang, J.M.; Lee, E.; Koh, B.I.; Song, S.; Park, J.; Lee, S.; Choi, C.; Kim, J.W.; et al. SoxF Transcription Factors Are Positive Feedback Regulators of VEGF Signaling. *Circ. Res.* **2016**, *119*, 839–852. [CrossRef]

276. Varshney, G.K.; Carrington, B.; Pei, W.; Bishop, K.; Chen, Z.; Fan, C.; Xu, L.; Jones, M.; LaFave, M.C.; Ledin, J.; et al. A high-throughput functional genomics workflow based on CRISPR/Cas9-mediated targeted mutagenesis in zebrafish. *Nat. Protoc.* **2016**, *11*, 2357–2375. [CrossRef]

277. Vickovic, S.; Eraslan, G.; Salmén, F.; Klughammer, J.; Stenbeck, L.; Schapiro, D.; Äijö, T.; Bonneau, R.; Bergenstråhle, L.; Navarro, J.F.; et al. High-definition spatial transcriptomics for in situ tissue profiling. *Nat. Methods* **2019**, *16*, 987–990. [CrossRef]

278. Grunewald, J.; Eklund, A. Sex-Specific Manifestations of Löfgren's Syndrome. *Am. J. Respir. Crit. Care Med.* **2007**, *175*, 40–44. [CrossRef]

279. Landini, S.; Mazzinghi, B.; Becherucci, F.; Allinovi, M.; Provenzano, A.; Palazzo, V.; Ravaglia, F.; Artuso, R.; Bosi, E.; Stagi, S.; et al. Reverse Phenotyping after Whole-Exome Sequencing in Steroid-Resistant Nephrotic Syndrome. *Clin. J. Am. Soc. Nephrol.* **2020**, *15*, 89–100. [CrossRef]

280. Nelson, M.R.; Tipney, H.; Painter, J.L.; Shen, J.; Nicoletti, P.; Shen, Y.; Floratos, A.; Sham, P.C.; Li, M.J.; Wang, J.; et al. The support of human genetic evidence for approved drug indications. *Nat. Genet.* **2015**, *47*, 856–860. [CrossRef]

281. Whicher, D.; Philbin, S.; Aronson, N. An overview of the impact of rare disease characteristics on research methodology. *Orphanet J. Rare Dis.* **2018**, *13*, 14. [CrossRef]

282. UK Biobank. Available online: https://www.ukbiobank.ac.uk (accessed on 17 October 2020).

283. Welcome to eMerge > Collaborate. Available online: https://emerge-network.org (accessed on 17 October 2020).

284. National Institutes of Health (NIH)—All of Us. Available online: https://www.nih.gov/precision-medicine-initiative-cohort-program (accessed on 17 October 2020).

285. Polubriaginof, F.C.G.; Vanguri, R.; Quinnies, K.; Belbin, G.M.; Yahi, A.; Salmasian, H.; Lorberbaum, T.; Nwankwo, V.; Li, L.; Shervey, M.M.; et al. Disease Heritability Inferred from Familial Relationships Reported in Medical Records. *Cell* **2018**, *173*, 1692–1704. [CrossRef]

286. Firth, H.V.; Richards, S.M.; Bevan, A.P.; Clayton, S.; Corpas, M.; Rajan, D.; Van Vooren, S.; Moreau, Y.; Pettett, R.M.; Carter, N.P. DECIPHER: Database of Chromosomal Imbalance and Phenotype in Humans Using Ensembl Resources. *Am. J. Hum. Genet.* **2009**, *84*, 524–533. [CrossRef]

287. Philippakis, A.A.; Azzariti, D.R.; Beltran, S.; Brookes, A.J.; Brownstein, C.A.; Brudno, M.; Brunner, H.G.; Buske, O.J.; Carey, K.; Doll, C.; et al. The Matchmaker Exchange: A platform for rare disease gene discovery. *Hum. Mutat.* **2015**, *36*, 915–921. [CrossRef]

288. Boycott, K.M.; Rath, A.; Chong, J.X.; Hartley, T.; Alkuraya, F.S.; Baynam, G.; Brookes, A.J.; Brudno, M.; Carracedo, A.; den Dunnen, J.T.; et al. International Cooperation to Enable the Diagnosis of All Rare Genetic Diseases. *Am. J. Hum. Genet.* **2017**, *100*, 695–705. [CrossRef]

289. Kopanos, C.; Tsiolkas, V.; Kouris, A.; Chapple, C.E.; Albarca Aguilera, M.; Meyer, R.; Massouras, A. VarSome: The human genomic variant search engine. *Bioinformatics* **2019**, *35*, 1978–1980. [CrossRef]

290. Freeman, P.J.; Hart, R.K.; Gretton, L.J.; Brookes, A.J.; Dalgleish, R. VariantValidator: Accurate validation, mapping, and formatting of sequence variation descriptions. *Hum. Mutat.* **2018**, *39*, 61–68. [CrossRef]

291. Corbin, L.J.; Tan, V.Y.; Hughes, D.A.; Wade, K.H.; Paul, D.S.; Tansey, K.E.; Butcher, F.; Dudbridge, F.; Howson, J.M.; Jallow, M.W.; et al. Formalising recall by genotype as an efficient approach to detailed phenotyping and causal inference. *Nat. Commun.* **2018**, *9*, 711. [CrossRef]

292. Koolen, D.A.; Kramer, J.M.; Neveling, K.; Nillesen, W.M.; Moore-Barton, H.L.; Elmslie, F.V.; Toutain, A.; Amiel, J.; Malan, V.; Tsai, A.C.-H.; et al. Mutations in the chromatin modifier gene KANSL1 cause the 17q21.31 microdeletion syndrome. *Nat. Genet.* **2012**, *44*, 639–641. [CrossRef]

293. Potocki, L.; Chen, K.-S.; Park, S.-S.; Osterholm, D.E.; Withers, M.A.; Kimonis, V.; Summers, A.M.; Meschino, W.S.; Anyane-Yeboa, K.; Kashork, C.D.; et al. Molecular mechanism for duplication 17p11.2—The homologous recombination reciprocal of the Smith-Magenis microdeletion. *Nat. Genet.* **2000**, *24*, 84–87. [CrossRef]

294. Potocki, L.; Bi, W.; Treadwell-Deering, D.; Carvalho, C.M.B.; Eifert, A.; Friedman, E.M.; Glaze, D.; Krull, K.; Lee, J.A.; Lewis, R.A.; et al. Characterization of Potocki-Lupski syndrome (dup(17)(p11.2p11.2)) and delineation of a dosage-sensitive critical interval that can convey an autism phenotype. *Am. J. Hum. Genet.* **2007**, *80*, 633–649. [CrossRef]

295. Ghigna, M.-R.; Guignabert, C.; Montani, D.; Girerd, B.; Jaïs, X.; Savale, L.; Hervé, P.; Thomas de Montpréville, V.; Mercier, O.; Sitbon, O.; et al. BMPR2 mutation status influences bronchial vascular changes in pulmonary arterial hypertension. *Eur. Respir. J.* **2016**, *48*, 1668–1681. [CrossRef] [PubMed]

296. Austin, E.D.; Phillips, J.A.; Cogan, J.D.; Hamid, R.; Yu, C.; Stanton, K.C.; Phillips, C.A.; Wheeler, L.A.; Robbins, I.M.; Newman, J.H.; et al. Truncating and missense BMPR2 mutations differentially affect the severity of heritable pulmonary arterial hypertension. *Respir. Res.* **2009**, *10*. [CrossRef] [PubMed]

297. Girerd, B.; Coulet, F.; Jaïs, X.; Eyries, M.; Van Der Bruggen, C.; De Man, F.; Houweling, A.; Dorfmüller, P.; Savale, L.; Sitbon, O.; et al. Characteristics of pulmonary arterial hypertension in affected carriers of a mutation located in the cytoplasmic tail of bone morphogenetic protein receptor type 2. *Chest* **2015**, *147*, 1385–1394. [CrossRef] [PubMed]

298. Ballif, B.C.; Theisen, A.; Rosenfeld, J.A.; Traylor, R.N.; Gastier-Foster, J.; Thrush, D.L.; Astbury, C.; Bartholomew, D.; McBride, K.L.; Pyatt, R.E.; et al. Identification of a recurrent microdeletion at 17q23.1q23.2 flanked by segmental duplications associated with heart defects and limb abnormalities. *Am. J. Hum. Genet.* **2010**, *86*, 454–461. [CrossRef] [PubMed]

299. Nimmakayalu, M.; Major, H.; Sheffield, V.; Solomon, D.H.; Smith, R.J.; Patil, S.R.; Shchelochkov, O.A. Microdeletion of 17q22q23.2 encompassing TBX2 and TBX4 in a patient with congenital microcephaly, thyroid duct cyst, sensorineural hearing loss, and pulmonary hypertension. *Am. J. Med Genet. Part A* **2011**, *155*, 418–423. [CrossRef]

300. German, K.; Deutsch, G.H.; Freed, A.S.; Dipple, K.M.; Chabra, S.; Bennett, J.T. Identification of a deletion containing TBX4 in a neonate with acinar dysplasia by rapid exome sequencing. *Am. J. Med Genet. Part A* **2019**, *179*, 842–845. [CrossRef]

301. Suhrie, K.; Pajor, N.M.; Ahlfeld, S.K.; Dawson, D.B.; Dufendach, K.R.; Kitzmiller, J.A.; Leino, D.; Lombardo, R.C.; Smolarek, T.A.; Rathbun, P.A.; et al. Neonatal Lung Disease Associated with TBX4 Mutations. *J. Pediatr.* **2019**, *206*, 286–292.e1. [CrossRef]

302. Iannuzzi, M.C.; Baughman, R.P. Reverse phenotyping in sarcoidosis. *Am. J. Respir. Crit. Care Med.* **2007**, *175*, 4–5. [CrossRef]

303. Stram, M.; Gigliotti, T.; Hartman, D.; Pitkus, A.; Huff, S.M.; Riben, M.; Henricks, W.H.; Farahani, N.; Pantanowitz, L. Logical Observation Identifiers Names and Codes for Laboratorians. *Arch. Pathol. Lab. Med.* **2020**, *144*, 229–239. [CrossRef]

304. Grubert, F.; Zaugg, J.B.; Kasowski, M.; Ursu, O.; Spacek, D.V.; Martin, A.R.; Greenside, P.; Srivas, R.; Phanstiel, D.H.; Pekowska, A.; et al. Genetic Control of Chromatin States in Humans Involves Local and Distal Chromosomal Interactions. *Cell* **2015**, *162*, 1051–1065. [CrossRef]

305. Perenthaler, E.; Yousefi, S.; Niggl, E.; Barakat, T.S. Beyond the Exome: The Non-coding Genome and Enhancers in Neurodevelopmental Disorders and Malformations of Cortical Development. *Front. Cell. Neurosci.* **2019**, *13*, 352. [CrossRef] [PubMed]

306. Reyes-Palomares, A.; Gu, M.; Grubert, F.; Berest, I.; Sa, S.; Kasowski, M.; Arnold, C.; Shuai, M.; Srivas, R.; Miao, S.; et al. Remodeling of active endothelial enhancers is associated with aberrant gene-regulatory networks in pulmonary arterial hypertension. *Nat. Commun.* **2020**, *11*, 1673. [CrossRef] [PubMed]

307. Machado, R.D.; James, V.; Southwood, M.; Harrison, R.E.; Atkinson, C.; Stewart, S.; Morrell, N.W.; Trembath, R.C.; Aldred, M.A. Investigation of second genetic hits at the BMPR2 locus as a modulator of disease progression in familial pulmonary arterial hypertension. *Circulation* **2005**, *111*, 607–613. [CrossRef]

308. Aldred, M.A.; Comhair, S.A.; Varella-Garcia, M.; Asosingh, K.; Xu, W.; Noon, G.P.; Thistlethwaite, P.A.; Tuder, R.M.; Erzurum, S.C.; Geraci, M.W.; et al. Somatic Chromosome Abnormalities in the Lungs of Patients with Pulmonary Arterial Hypertension. *Am. J. Respir. Crit. Care Med.* **2010**, *182*, 1153–1160. [CrossRef] [PubMed]

309. Drake, K.M.; Comhair, S.A.; Erzurum, S.C.; Tuder, R.M.; Aldred, M.A. Endothelial chromosome 13 deletion in congenital heart disease-associated pulmonary arterial hypertension dysregulates SMAD9 signaling. *Am. J. Respir. Crit. Care Med.* **2015**, *191*, 850–854. [CrossRef] [PubMed]

310. Best, D.H.; Vaughn, C.; McDonald, J.; Damjanovich, K.; Runo, J.R.; Chibuk, J.M.; Bayrak-Toydemir, P. Mosaic ACVRL1 and ENG mutations in hereditary haemorrhagic telangiectasia patients. *J. Med Genet.* **2011**, *48*, 358–360. [CrossRef] [PubMed]

311. Eyries, M.; Coulet, F.; Girerd, B.; Montani, D.; Humbert, M.; Lacombe, P.; Chinet, T.; Gouya, L.; Roume, J.; Axford, M.M.; et al. ACVRL1 germinal mosaic with two mutant alleles in hereditary hemorrhagic telangiectasia associated with pulmonary arterial hypertension. *Clin. Genet.* **2012**, *82*, 173–179. [CrossRef]

312. McDonald, J.; Gedge, F.; Burdette, A.; Carlisle, J.; Bukjiok, C.J.; Fox, M.; Bayrak-Toydemir, P. Multiple sequence variants in hereditary hemorrhagic telangiectasia cases: Illustration of complexity in molecular diagnostic interpretation. *J. Mol. Diagn.* **2009**, *11*, 569–575. [CrossRef]

313. Culley, M.K.; Chan, S.Y. Mitochondrial metabolism in pulmonary hypertension: Beyond mountains there are mountains. *J. Clin. Investig.* **2018**, *128*, 3704–3715. [CrossRef] [PubMed]

314. Rhodes, C.J.; Ghataorhe, P.; Wharton, J.; Rue-Albrecht, K.C.; Hadinnapola, C.; Watson, G.; Bleda, M.; Haimel, M.; Coghlan, G.; Corris, P.A.; et al. Plasma Metabolomics Implicates Modified Transfer RNAs and Altered Bioenergetics in the Outcomes of Pulmonary Arterial Hypertension. *Circulation* **2017**, *135*, 460–475. [CrossRef] [PubMed]

315. Farha, S.; Hu, B.; Comhair, S.; Zein, J.; Dweik, R.; Erzurum, S.C.; Aldred, M.A. Mitochondrial Haplogroups and Risk of Pulmonary Arterial Hypertension. *PLoS ONE* **2016**, *11*, e0156042. [CrossRef] [PubMed]

316. Bris, C.; Goudenege, D.; Desquiret-Dumas, V.; Charif, M.; Colin, E.; Bonneau, D.; Amati-Bonneau, P.; Lenaers, G.; Reynier, P.; Procaccio, V. Bioinformatics Tools and Databases to Assess the Pathogenicity of Mitochondrial DNA Variants in the Field of Next Generation Sequencing. *Front. Genet.* **2018**, *9*, 632. [CrossRef] [PubMed]

317. Bell, J.T.; Pai, A.A.; Pickrell, J.K.; Gaffney, D.J.; Pique-Regi, R.; Degner, J.F.; Gilad, Y.; Pritchard, J.K. DNA methylation patterns associate with genetic and gene expression variation in HapMap cell lines. *Genome Biol.* **2011**, *12*, R10. [CrossRef]

318. Cheung, W.A.; Shao, X.; Morin, A.; Siroux, V.; Kwan, T.; Ge, B.; Aïssi, D.; Chen, L.; Vasquez, L.; Allum, F.; et al. Correction to: Functional variation in allelic methylomes underscores a strong genetic contribution and reveals novel epigenetic alterations in the human epigenome. *Genome Biol.* **2019**, *20*. [CrossRef] [PubMed]

319. Archer, S.L.; Marsboom, G.; Kim, G.H.; Zhang, H.J.; Toth, P.T.; Svensson, E.C.; Dyck, J.R.B.; Gomberg-Maitland, M.; Thébaud, B.; Husain, A.N.; et al. Epigenetic Attenuation of Mitochondrial Superoxide Dismutase 2 in Pulmonary Arterial Hypertension. *Circulation* **2010**, *121*, 2661–2671. [CrossRef]

320. Bonnet, S.; Michelakis, E.D.; Porter, C.J.; Andrade-Navarro, M.A.; Thébaud, B.; Bonnet, S.; Haromy, A.; Harry, G.; Moudgil, R.; Sean McMurtry, M.; et al. An Abnormal Mitochondrial–Hypoxia Inducible Factor-1α–Kv Channel Pathway Disrupts Oxygen Sensing and Triggers Pulmonary Arterial Hypertension in Fawn Hooded Rats. *Circulation* **2006**, *113*, 2630–2641. [CrossRef]

321. Hyun, K.; Jeon, J.; Park, K.; Kim, J. Writing, erasing and reading histone lysine methylations. *Exp. Mol. Med.* **2017**, *49*, e324. [CrossRef]

322. Zhao, L.; Chen, C.-N.; Hajji, N.; Oliver, E.; Cotroneo, E.; Wharton, J.; Wang, D.; Li, M.; McKinsey, T.A.; Stenmark, K.R.; et al. Histone deacetylation inhibition in pulmonary hypertension: Therapeutic potential of valproic acid and suberoylanilide hydroxamic acid. *Circulation* **2012**, *126*, 455–467. [CrossRef]

323. Hon, C.-C.; Ramilowski, J.A.; Harshbarger, J.; Bertin, N.; Rackham, O.J.L.; Gough, J.; Denisenko, E.; Schmeier, S.; Poulsen, T.M.; Severin, J.; et al. An atlas of human long non-coding RNAs with accurate 5′ ends. *Nature* **2017**, *543*, 199–204. [CrossRef] [PubMed]

324. St Laurent, G.; Wahlestedt, C.; Kapranov, P. The Landscape of long noncoding RNA classification. *Trends Genet.* **2015**, *31*, 239–251. [CrossRef] [PubMed]

325. Su, H.; Xu, X.; Yan, C.; Shi, Y.; Hu, Y.; Dong, L.; Ying, S.; Ying, K.; Zhang, R. LncRNA H19 promotes the proliferation of pulmonary artery smooth muscle cells through AT1R via sponging let-7b in monocrotaline-induced pulmonary arterial hypertension. *Respir. Res.* **2018**, *19*. [CrossRef] [PubMed]

326. Jandl, K.; Thekkekara Puthenparampil, H.; Marsh, L.M.; Hoffmann, J.; Wilhelm, J.; Veith, C.; Sinn, K.; Klepetko, W.; Olschewski, H.; Olschewski, A.; et al. Long non-coding RNAs influence the transcriptome in pulmonary arterial hypertension: The role of PAXIP1-AS1. *J. Pathol.* **2019**, *247*, 357–370. [CrossRef]

327. Wang, D.; Xu, H.; Wu, B.; Jiang, S.; Pan, H.; Wang, R.; Chen, J. Long non-coding RNA MALAT1 sponges miR-124-3p.1/KLF5 to promote pulmonary vascular remodeling and cell cycle progression of pulmonary artery hypertension. *Int. J. Mol. Med.* **2019**, *44*, 871–884. [CrossRef]

328. Leung, A.; Trac, C.; Jin, W.; Lanting, L.; Akbany, A.; Sætrom, P.; Schones, D.E.; Natarajan, R. Novel long noncoding RNAs are regulated by angiotensin II in vascular smooth muscle cells. *Circ. Res.* **2013**, *113*, 266–278. [CrossRef]

329. Yang, L.; Liang, H.; Shen, L.; Guan, Z.; Meng, X. LncRNA Tug1 involves in the pulmonary vascular remodeling in mice with hypoxic pulmonary hypertension via the microRNA-374c-mediated Foxc1. *Life Sci.* **2019**, *237*, 116769. [CrossRef]

330. Zhang, H.; Liu, Y.; Yan, L.; Wang, S.; Zhang, M.; Ma, C.; Zheng, X.; Chen, H.; Zhu, D. Long noncoding RNA Hoxaas3 contributes to hypoxia-induced pulmonary artery smooth muscle cell proliferation. *Cardiovasc. Res.* **2019**, *115*, 647–657. [CrossRef]

331. Xing, Y.; Zheng, X.; Fu, Y.; Qi, J.; Li, M.; Ma, M.; Wang, S.; Li, S.; Zhu, D. Long Noncoding RNA-Maternally Expressed Gene 3 Contributes to Hypoxic Pulmonary Hypertension. *Mol. Ther.* **2019**, *27*, 2166–2181. [CrossRef]

332. Zhu, T.-T.; Sun, R.-L.; Yin, Y.-L.; Quan, J.-P.; Song, P.; Xu, J.; Zhang, M.-X.; Li, P. Long noncoding RNA UCA1 promotes the proliferation of hypoxic human pulmonary artery smooth muscle cells. *Pflugers Arch.* **2019**, *471*, 347–355. [CrossRef]

333. Sun, Z.; Nie, X.; Sun, S.; Dong, S.; Yuan, C.; Li, Y.; Xiao, B.; Jie, D.; Liu, Y. Long Non-Coding RNA MEG3 Downregulation Triggers Human Pulmonary Artery Smooth Muscle Cell Proliferation and Migration via the p53 Signaling Pathway. *Cell. Physiol. Biochem.* **2017**, *42*, 2569–2581. [CrossRef] [PubMed]

334. Zhu, B.; Gong, Y.; Yan, G.; Wang, D.; Qiao, Y.; Wang, Q.; Liu, B.; Hou, J.; Li, R.; Tang, C. Down-regulation of lncRNA MEG3 promotes hypoxia-induced human pulmonary artery smooth muscle cell proliferation and migration via repressing PTEN by sponging miR-21. *Biochem. Biophys. Res. Commun.* **2018**, *495*, 2125–2132. [CrossRef] [PubMed]

335. Gong, J.; Chen, Z.; Chen, Y.; Lv, H.; Lu, H.; Yan, F.; Li, L.; Zhang, W.; Shi, J. Long non-coding RNA CASC2 suppresses pulmonary artery smooth muscle cell proliferation and phenotypic switch in hypoxia-induced pulmonary hypertension. *Respir. Res.* **2019**, *20*, 53. [CrossRef] [PubMed]

336. Chen, J.; Guo, J.; Cui, X.; Dai, Y.; Tang, Z.; Qu, J.; Usha Raj, J.; Hu, Q.; Gou, D. The Long Noncoding RNA LnRPT Is Regulated by PDGF-BB and Modulates the Proliferation of Pulmonary Artery Smooth Muscle Cells. *Am. J. Respir. Cell Mol. Biol.* **2018**, *58*, 181–193. [CrossRef] [PubMed]

337. Xiang, Y.; Zhang, Y.; Tang, Y.; Li, Q. MALAT1 Modulates TGF-β1-Induced Endothelial-to-Mesenchymal Transition through Downregulation of miR-145. *Cell. Physiol. Biochem.* **2017**, *42*, 357–372. [CrossRef]

338. Neumann, P.; Jaé, N.; Knau, A.; Glaser, S.F.; Fouani, Y.; Rossbach, O.; Krüger, M.; John, D.; Bindereif, A.; Grote, P.; et al. The lncRNA GATA6-AS epigenetically regulates endothelial gene expression via interaction with LOXL2. *Nat. Commun.* **2018**, *9*, 237. [CrossRef] [PubMed]

339. Kang, H.; Louie, J.; Weisman, A.; Sheu-Gruttadauria, J.; Davis-Dusenbery, B.N.; Lagna, G.; Hata, A. Inhibition of microRNA-302 (miR-302) by bone morphogenetic protein 4 (BMP4) facilitates the BMP signaling pathway. *J. Biol. Chem.* **2012**, *287*, 38656–38664. [CrossRef]

340. Pullamsetti, S.S.; Doebele, C.; Fischer, A.; Savai, R.; Kojonazarov, B.; Dahal, B.K.; Ghofrani, H.A.; Weissmann, N.; Grimminger, F.; Bonauer, A.; et al. Inhibition of microRNA-17 improves lung and heart function in experimental pulmonary hypertension. *Am. J. Respir. Crit. Care Med.* **2012**, *185*, 409–419. [CrossRef]

341. Brock, M.; Samillan, V.J.; Trenkmann, M.; Schwarzwald, C.; Ulrich, S.; Gay, R.E.; Gassmann, M.; Ostergaard, L.; Gay, S.; Speich, R.; et al. AntagomiR directed against miR-20a restores functional BMPR2 signalling and prevents vascular remodelling in hypoxia-induced pulmonary hypertension. *Eur. Heart J.* **2014**, *35*, 3203–3211. [CrossRef]

342. Caruso, P.; MacLean, M.R.; Khanin, R.; McClure, J.; Soon, E.; Southgate, M.; MacDonald, R.A.; Greig, J.A.; Robertson, K.E.; Masson, R.; et al. Dynamic Changes in Lung MicroRNA Profiles During the Development of Pulmonary Hypertension due to Chronic Hypoxia and Monocrotaline. *Arterioscler. Thromb. Vasc. Biol.* **2010**, *30*, 716–723. [CrossRef]

343. Uchida, S.; Dimmeler, S. Long Noncoding RNAs in Cardiovascular Diseases. *Circ. Res.* **2015**, *116*, 737–750. [CrossRef]

344. Thum, T.; Condorelli, G. Long Noncoding RNAs and MicroRNAs in Cardiovascular Pathophysiology. *Circ. Res.* **2015**, *116*, 751–762. [CrossRef] [PubMed]

345. Chinnappan, M.; Mohan, A.; Agarwal, S.; Dalvi, P.; Dhillon, N.K. Network of MicroRNAs Mediate Translational Repression of Bone Morphogenetic Protein Receptor-2: Involvement in HIV-Associated Pulmonary Vascular Remodeling. *J. Am. Heart Assoc.* **2018**, *7*. [CrossRef] [PubMed]

346. Durrington, H.J.; Upton, P.D.; Hoer, S.; Boname, J.; Dunmore, B.J.; Yang, J.; Crilley, T.K.; Butler, L.M.; Blackbourn, D.J.; Nash, G.B.; et al. Identification of a lysosomal pathway regulating degradation of the bone morphogenetic protein receptor type II. *J. Biol. Chem.* **2010**, *285*, 37641–37649. [CrossRef] [PubMed]

347. Dalvi, P.; O'Brien-Ladner, A.; Dhillon, N.K. Downregulation of bone morphogenetic protein receptor axis during HIV-1 and cocaine-mediated pulmonary smooth muscle hyperplasia: Implications for HIV-related pulmonary arterial hypertension. *Arterioscler. Thromb. Vasc. Biol.* **2013**, *33*, 2585–2595. [CrossRef]

348. Mansoor, J.K.; Schelegle, E.S.; Davis, C.E.; Walby, W.F.; Zhao, W.; Aksenov, A.A.; Pasamontes, A.; Figueroa, J.; Allen, R. Analysis of volatile compounds in exhaled breath condensate in patients with severe pulmonary arterial hypertension. *PLoS ONE* **2014**, *9*, e95331. [CrossRef] [PubMed]

349. Hakim, M.; Broza, Y.Y.; Barash, O.; Peled, N.; Phillips, M.; Amann, A.; Haick, H. Volatile organic compounds of lung cancer and possible biochemical pathways. *Chem. Rev.* **2012**, *112*, 5949–5966. [CrossRef] [PubMed]

350. Montani, D.; Lau, E.M.; Descatha, A.; Jaïs, X.; Savale, L.; Andujar, P.; Bensefa-Colas, L.; Girerd, B.; Zendah, I.; Le Pavec, J.; et al. Occupational exposure to organic solvents: A risk factor for pulmonary veno-occlusive disease. *Eur. Respir. J.* **2015**, *46*, 1721–1731. [CrossRef]

351. Barker, D. Infant Mortality, Childhood Nutrition, and Ischaemic Heart Disease in England and Wales. *Lancet* **1986**, *327*, 1077–1081. [CrossRef]

352. Barker, D.J.P.; Osmond, C.; Winter, P.D.; Margetts, B.; Simmonds, S.J. Weight in Infancy and Death from Ischaemic Heart Disease. *Lancet* **1989**, *334*, 577–580. [CrossRef]

353. Hales, C.N.; Barker, D.J.; Clark, P.M.; Cox, L.J.; Fall, C.; Osmond, C.; Winter, P.D. Fetal and infant growth and impaired glucose tolerance at age 64. *BMJ* **1991**, *303*, 1019–1022. [CrossRef] [PubMed]

354. Stern, D.A.; Morgan, W.J.; Wright, A.L.; Guerra, S.; Martinez, F.D. Poor airway function in early infancy and lung function by age 22 years: A non-selective longitudinal cohort study. *Lancet* **2007**, *370*, 758–764. [CrossRef]

355. Berry, C.E.; Billheimer, D.; Jenkins, I.C.; Lu, Z.J.; Stern, D.A.; Gerald, L.B.; Carr, T.F.; Guerra, S.; Morgan, W.J.; Wright, A.L.; et al. A Distinct Low Lung Function Trajectory from Childhood to the Fourth Decade of Life. *Am. J. Respir. Crit. Care Med.* **2016**, *194*, 607–612. [CrossRef] [PubMed]

356. Mann, S.L.; Wadsworth, M.E.; Colley, J.R. Accumulation of factors influencing respiratory illness in members of a national birth cohort and their offspring. *J. Epidemiol. Community Health* **1992**, *46*, 286–292. [CrossRef]

357. Grigoriadis, S.; Vonderporten, E.H.; Mamisashvili, L.; Tomlinson, G.; Dennis, C.-L.; Koren, G.; Steiner, M.; Mousmanis, P.; Cheung, A.; Ross, L.E. Prenatal exposure to antidepressants and persistent pulmonary hypertension of the newborn: Systematic review and meta-analysis. *BMJ* **2014**, *348*, f6932. [CrossRef]

358. Marques, F.Z. Missing Heritability of Hypertension and Our Microbiome. *Circulation* **2018**, *138*, 1381–1383. [CrossRef]

359. Ranchoux, B.; Bigorgne, A.; Hautefort, A.; Girerd, B.; Sitbon, O.; Montani, D.; Humbert, M.; Tcherakian, C.; Perros, F. Gut-Lung Connection in Pulmonary Arterial Hypertension. *Am. J. Respir. Cell Mol. Biol.* **2017**, *56*, 402–405. [CrossRef]

360. Sofianopoulou, E.; Kaptoge, S.; Gräf, S.; Hadinnapola, C.; Treacy, C.M.; Church, C.; Coghlan, G.; Gibbs, J.S.R.; Haimel, M.; Howard, L.S.; et al. Traffic exposures, air pollution and outcomes in pulmonary arterial hypertension: A UK cohort study analysis. *Eur. Respir. J.* **2019**, *53*. [CrossRef]

361. Gazzo, A.; Raimondi, D.; Daneels, D.; Moreau, Y.; Smits, G.; Van Dooren, S.; Lenaerts, T. Understanding mutational effects in digenic diseases. *Nucleic Acids Res.* **2017**, *45*, e140. [CrossRef]

362. Eichstaedt, C.A.; Song, J.; Benjamin, N.; Harutyunova, S.; Fischer, C.; Grünig, E.; Hinderhofer, K. EIF2AK4 mutation as "second hit" in hereditary pulmonary arterial hypertension. *Respir. Res.* **2016**, *17*, 141. [CrossRef]

363. Pousada, G.; Baloira, A.; Valverde, D. Pulmonary arterial hypertension associated with several mutations. *Eur. Respir. J.* **2016**, *48*, PA2478.

Publisher's Note: MDPI stays neutral with regard to jurisdictional claims in published maps and institutional affiliations.

Article

BMPR2 Promoter Variants Effect Gene Expression in Pulmonary Arterial Hypertension Patients

Jie Song [1,2,3,4], Katrin Hinderhofer [1], Lilian T. Kaufmann [1], Nicola Benjamin [3,4], Christine Fischer [1], Ekkehard Grünig [3,4] and Christina A. Eichstaedt [1,3,4,*]

[1] Laboratory for Molecular Genetic Diagnostics, Institute of Human Genetics, Heidelberg University, Im Neuenheimer Feld 366, 69120 Heidelberg, Germany; jie.song@csu.edu.cn (J.S.); katrin.hinderhofer@med.uni-heidelberg.de (K.H.); lilian.kaufmann@med.uni-heidelberg.de (L.T.K.); christine.fischer@med.uni-heidelberg.de (C.F.)

[2] Department of Cardiovascular Medicine, The Second Xiangya Hospital of Central South University, Changsha 410011, China

[3] Centre for Pulmonary Hypertension, Thoraxklinik gGmbH Heidelberg at Heidelberg University Hospital, Röntgenstrasse 1, 69126 Heidelberg, Germany; nicola.benjamin@med.uni-heidelberg.de (N.B.); ekkehard.gruenig@med.uni-heidelberg.de (E.G.)

[4] Translational Lung Research Centre Heidelberg (TLRC), German Centre for Lung Research (DZL), 69120 Heidelberg, Germany

* Correspondence: christina.eichstaedt@med.uni-heidelberg.de; Tel.: +49-6221-396-1221; Fax: +49-6221-396-1222

Received: 4 September 2020; Accepted: 3 October 2020; Published: 6 October 2020

Abstract: Pathogenic variants have been identified in 85% of heritable pulmonary arterial hypertension (PAH) patients. These variants were mainly located in the bone morphogenetic protein receptor 2 (*BMPR2*) gene. However, the penetrance of *BMPR2* variants was reduced leading to a disease manifestation in only 30% of carriers. In these PAH patients, further modifiers such as additional pathogenic *BMPR2* promoter variants could contribute to disease manifestation. Therefore, the aim of this study was to identify *BMPR2* promoter variants in PAH patients and to analyze their transcriptional effect on gene expression and disease manifestation. *BMPR2* promoter variants were identified in PAH patients and cloned into plasmids. These were transfected into human pulmonary artery smooth muscle cells to determine their respective transcriptional activity. Nine different *BMPR2* promoter variants were identified in seven PAH families and three idiopathic PAH patients. Seven of the variants (c.-575A>T, c.-586dupT, c.-910C>T, c.-930_-928dupGGC, c.-933_-928dupGGCGGC, c.-930_-928delGGC and c.-1141C>T) led to a significantly decreased transcriptional activity. This study identified novel *BMPR2* promoter variants which may affect *BMPR2* gene expression in PAH patients. They could contribute to disease manifestations at least in some families. Further studies are needed to investigate the frequency of *BMPR2* promoter variants and their impact on penetrance and disease manifestation.

Keywords: *BMPR2* promoter; pathogenic variant; heritable pulmonary arterial hypertension

1. Introduction

Pulmonary arterial hypertension (PAH) is a rare disease characterized by remodeling of the small pulmonary vessels. This results in an increase of pulmonary artery pressure and resistance, eventually progressing to right heart failure [1]. In many forms of PAH, such as heritable (HPAH) or idiopathic PAH (IPAH), genetic defects have been identified [2]. In most cases of HPAH pathogenic variants (mutations) have been identified in the bone morphogenetic protein receptor 2 (*BMPR2*) gene leading to a loss of gene function [3]. The *BMPR2* gene encodes a cell membrane type II receptor of the transforming growth factor-β signaling pathway, which regulates expression of many target genes [4].

More than 600 pathogenic variants in the *BMPR2* gene have been identified so far, accounting for the disease development in about 85% of HPAH and 5–35% of IPAH patients [3]. These patients usually show an earlier onset of disease manifestation, more impaired hemodynamics [2] and reduced survival, compared to PAH patients without a genetic predisposition [5]. Apart from pathogenic variants in the *BMPR2* gene, genetic changes have also been described in recent years in the genes of its coreceptors *ACVRL1* and *ENG*, in the gene responsible for pulmonary venous occlusive disease *EIF2AK4*, and further BMPRII pathway genes [3]. So far, 17 genes have been classified as PAH-causing genes by the international task force for genetics and genomics in PAH [6].

While pathogenic variants in the *BMPR2* gene can lead to PAH, an incomplete penetrance of around 30% has been observed in family members with the same *BMPR2* variant [7]. In some families, up to 50% of *BMPR2* variant carriers can develop PAH [8]. It remains unclear why some variant carriers develop PAH while other carriers remain healthy for their entire life. It has been suggested that PAH patients could carry additional pathogenic variants in contrast to their healthy family members [9,10]. These so called "second hits" could serve as modifiers for disease penetrance leading to the disease manifestation together with the familial *BMPR2* variant [9]. Such second hits were also shown to be present in the noncoding or regulatory regions of the *BMPR2* gene [11]. However, in the normal routine genetic diagnostics setting, deep intronic regions and promoters are rarely investigated. Nevertheless, they can also contain pathogenic variants leading to PAH manifestation [12,13]. In previous studies, single substitution variants [14,15], a double substitution variant [13] and deleted or inserted tandem repeats [14,16] downstream of the transcription start site of the *BMPR2* gene were identified and led to reduced gene expression in a functional assessment. The contribution of *BMPR2* promoter variants to PAH manifestation, however, still remains unclear.

In this study we investigated whether *BMPR2* promoter variants in H/IPAH serve as second hits, by firstly identifying *BMPR2* promoter variants in H/IPAH patients and secondly, functionally characterizing the effect of the variants on *BMPR2* gene expression.

2. Materials and Methods

2.1. Study Population

PAH patients were clinically diagnosed by right heart catheterization following the guidelines of the European Society of Cardiology and the European Respiratory Society [1,17]. Further clinical examinations included medical history, physical examination, 12-lead electrocardiogram, chest radiograph, lung function testing and echocardiography at rest and during exercise. All patients of this study were either idiopathic or heritable PAH patients. In addition, healthy family members of H/IPAH patients performed echocardiography during exercise on a bicycle ergometer in the supine position. Pulmonary arterial systolic pressure (PASP) was measured with an increasing workload until maximum physical exhaustion was reached. A PASP value of >40 mmHg was used to define a hypertensive exercise response or a normal exercise response (PASP ≤ 40 mmHg) [18,19]. All subjects gave their informed consent for inclusion before they participated in the study. The study was conducted in accordance with the Declaration of Helsinki, and the protocol was approved by the Ethics Committee of the Medical Faculty of Heidelberg University (Project identification codes 065/2001 and S-426/2017).

2.2. Genetic Analysis

Genomic DNA was extracted from peripheral blood by an automated procedure (Autopure, Gentra Puregene Technology, Qiagen, Hilden, Germany). Variants in the *BMPR2* gene and the promoter region were identified either by direct Sanger sequencing (ABI 3130 genetic analyzer, ThermoFisher Scientific, Waltham, MA, USA) or by a PAH specific gene panel approach based on next-generation sequencing (Miseq, Illumina, San Diego, CA, USA) followed by Sanger sequencing confirmation [20]. Patients with no pathogenic variants in the *BMPR2* gene (RefSeq ID: NM_001204) were further evaluated by multiplex ligand-dependent probe amplification (MLPA) to search for large deletions or duplications (P093-C1, MRC-Holland, Amsterdam, the Netherlands). The PAH-specific gene panel of this study included 12 PAH genes (*ACVRL1, BMPR1B, BMPR2* including the *BMPR2* promoter region, *CAV1, EIF2AK4, ENG, KCNA5, KCNK3, SMAD1, SMAD4, SMAD9* and *TOPBP1*) and 17 further candidate genes as described previously [20]. Apart from coding exons, exon-intron boundaries reaching 20 bp into the introns were also analyzed. Each variant with a minor allele frequency in the database Ensembl < 1% was sought in the Human Gene Mutation Database and variants of *BMPR2, ENG, ACVRL1* and *SMAD4* also in the ARUP database. Putative transcription factor-binding sites containing promoter variants were assessed using the software MatInspector (version 3.9, Genomatix GmbH, München, Germany).

2.3. Plasmid Construction

Nine mutant 1520 base pair (bp) fragments plus the wild-type *BMPR2* promoter region (c.-1502 to c.18) were amplified from genomic DNA of PAH patients using a forward primer with a *Kpn*I restriction site at the 5′ end (5′-GAGGGTACCTCCCAAGCCATGCACATTTG-3′) and a reverse primer with a *Hind*III restriction site at the 3′ end (5′- CGCAAGCTTCTGCAGCGAGGAAGTCATC-3′) for cloning. The pGL4.10 (Promega GmbH, Walldorf, Germany) vector was cleaved by the *Kpn*I and *Hind*III restriction endonucleases and ligated with the ten amplified *BMPR2* promoter regions. The pGL4.10 vector included the firefly luciferase reporter gene sequence downstream of the inserted promoter region. The correct insertion of the recombinant gene constructs was confirmed by Sanger sequencing. The wild-type sequence contained no variants and corresponded to the human reference sequence (hg19). The pGL4.10 basic vector without any promoter element served as a negative control.

2.4. Cell Culture and Luciferase Assay

Commercial human pulmonary artery smooth muscle cells (PASMCs, Life Technologies GmbH, Darmstadt, Germany) were cultured in Medium 231 (Gibco, Thermo Fisher Scientific, Schwerte, Germany), supplemented with smooth muscle growth supplement (Gibco, Thermo Fisher Scientific, Schwerte, Germany) in a 37 °C, 5% carbon dioxide and 95% humidified cell culture incubator. Cells of passage 3 to 6 were plated onto 24-well plates at 5000 cells per well. After 24 h of incubation pGL4.10-*BMPR2*-variant constructs or pGL4.10-*BMPR2*-wild-type plasmids were added to the cells together with a transfection reagent (Lipofectamine 3000, Thermo Fisher Scientific, Schwerte, Germany). After 48 h of further incubation cells were lysed. Cleared cell lysates were aliquoted into 96-well plates. The activities of the firefly luciferase and the internal standard renilla luciferase were assayed with a Dual-Luciferase Reporter assay kit (Promega GmbH, Walldorf, Germany) using the auto-injector system Lucy2 Microplate Luminometer (Anthos Mikrosysteme GmbH, Krefeld, Germany). Each transfection was carried out five times and measured in triplicate by luciferase assay.

2.5. Statistical Analysis

Data were shown as mean ± standard deviation. Analysis of variance was performed by SPSS (version 21.0.0, IBM, Amonk, NY, USA), considering a p-value < 0.05 as statistically significant. Gene expression was normalized to the wild-type (corresponding to 1.0). Relative gene expression was calculated by: wild-type = value of every mutant plasmid/value of wild-type plasmid.

3. Results

3.1. Clinical Characteristics

Seven HPAH families with 53 family members and three IPAH patients were included in this study. The clinical characteristics of seven index patients were summarized in Table 1. The mean age of HPAH index patients was 34 years, 43% were female. The hemodynamic parameters measured by right heart catheterization revealed a mean pulmonary arterial pressure of 60.3 ± 9.4 mmHg, a pulmonary vascular resistance of 20.2 ± 5.7 Wood Units and a reduced cardiac index of 2.1 ± 0.9 mL/min/m^2. Of 53 sequenced family members, PASP during exercise was measured in 39 individuals.

Table 1. Clinical characteristics of the index patients from the seven HPAH families at initial diagnosis.

Characteristic *	Mean ± SD	Min	Max
Women [%]	43		
Age at diagnosis [years]	34.0 ± 14.2	13	56
Heart rate [min^{-1}]	88.6 ± 12.2	68	99
Oxygen saturation [%]	94.0 ± 5.0	87	98
Mean pulmonary artery pressure [mmHg]	60.3 ± 9.4	46	70
Pulmonary arterial wedge pressure [mmHg]	5.0 ± 1.6	3	7
Pulmonary vascular resistance [Wood Units]	20.2 ± 5.7	13	26
Cardiac index [ml/min/m^2]	2.1 ± 0.9	1.3	3.6

* Not all measurements were obtained from each patient. SD: standard deviation.

3.2. Genetic Analysis

To detect promoter variants that may have an impact on gene expression and disease penetrance, a segment of 1520 bp of the 5′ untranslated region (UTR) upstream region of the *BMPR2* gene was analyzed. All subjects were included regardless of whether they carried any other pathogenic variant or variant of uncertain significance (VUS) in the coding area of *BMPR2* or any other analyzed PAH gene. In total, nine variants were discovered in the promoter region (Figure 1) of which four had not been described before in PAH patients (Table 2). Four heterozygous nucleotide substitutions (c.-301G>A, c.-575A>T, c.-910C>T and c.-1141C>T) and one heterozygous duplication (c.-933_-928dupGGCGGC) were detected in four HPAH families (see Table 2). Additionally, the previously described c.-669G>A variant was identified in three HPAH families. Among these three families, only two index patients carried the c.-669G>A variant, while in the third family, this variant was only present in healthy family members.

In two further IPAH patients, a duplication or a deletion of a three base pair repeat out of a twelve repeat GGC region was identified (c.-930_-928dupGGC and c.-930_-928delGGC). This resulted in 11 and 13 GGC repeats in contrast to 12 GGC repeats in the wild-type sequence. The last variant in the *BMPR2* promoter region was a homozygous duplication (c.-586dupT) present in a third IPAH patient. The locations of the nine identified variants are depicted in Figure 1.

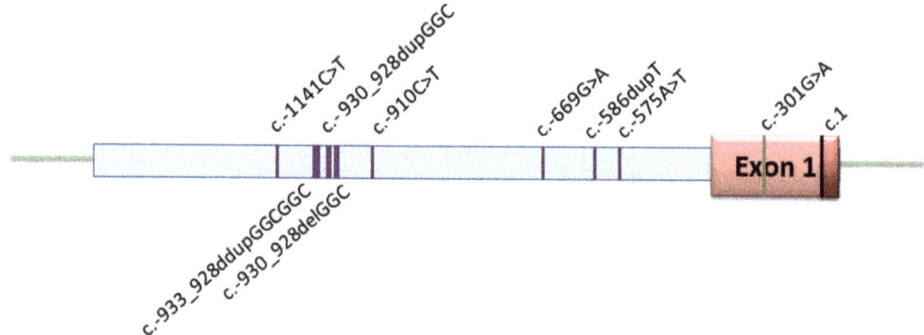

Figure 1. Distribution of variants in the promoter region of *BMPR2*. Nine variants were identified including five heterozygous substitutions, three heterozygous duplications/deletions at a single site, and one homozygous substitution (c.-586dupT). The 5′ untranslated region starts left of the nucleotide position c.1.

Table 2. Variants identified in the 5′UTR of the *BMPR2* in I/HPAH patients.

Nucleotide Change	Patients	Variant Carriers	GnomAD Frequency	rsID	Described in PAH Patients
c.-301G>A [1]	2 HPAH families	2 indices 6 family members [2]	0.69%	rs116154690	1 SSc-APAH [15]
c.-575A>T	1 HPAH family	1 index	0.04%	rs550462760	This study
c.-586dupT [3]	1 IPAH patient	1 index	0.17%	rs572725320	This study
c.-669G>A	3 HPAH families	2 indices 9 family members	0.95%	rs115604088	[11,14]
c.-910C>T [4]	1 HPAH family	1 index 5 family members	-	-	This study
c.-930_-928dupGGC (13 repeats)	1 IPAH patient	1 index	-	rs375624016	4 CHD-APAH and 10 controls; 1 HPAH [16,21]
c.-933_-928dupGGCGGC (14 repeats) [4]	1 HPAH family	1 index 2 family members	-	-	1 CHD-APAH [16]
c.-930_-928delGGC (11 repeats)	1 IPAH patient	1 index	-	rs886055459	1 CHD-APAH [16]
c.-1141C>T [1]	1 HPAH family	1 index 2 family members	-	-	This study

[1] c.-301G>A and c.-1141C>T were identified in the same HPAH family but in two different index patients of the same family. [2] All family members with any *BMPR2* promoter variant apart from one c.-301G>A carrier had no manifest PAH. [3] Homozygous variant, all other variants were heterozygous. [4] c.-910C>T and c.-933_-928dupGGCGGC were identified in the same HPAH family and index patient. CHD-APAH: congenital heart disease-associated pulmonary arterial hypertension; GnomAD: genome aggregation database; HPAH: heritable pulmonary arterial hypertension; IPAH: idiopathic pulmonary arterial hypertension; rsID: variant identifier; SSc-APAH: systemic sclerosis associated pulmonary arterial hypertension.

3.3. Effect of Promoter Variants on Gene Expression

The effect of the nine different *BMPR2* promoter variants on the transcriptional activity of *BMPR2* was analyzed using a luciferase reporter gene assay after transient transfection of human PASMCs. The transcriptional activity of the variants was compared to the *BMPR2* wild-type promoter. The results revealed that apart from the two variants c.-669G>A and c.-301G>A, all plasmids with *BMPR2* promoter variants led to a statistically significant decrease of transcriptional activity compared to the plasmid carrying *BMPR2* wild-type promoter (Figure 2).

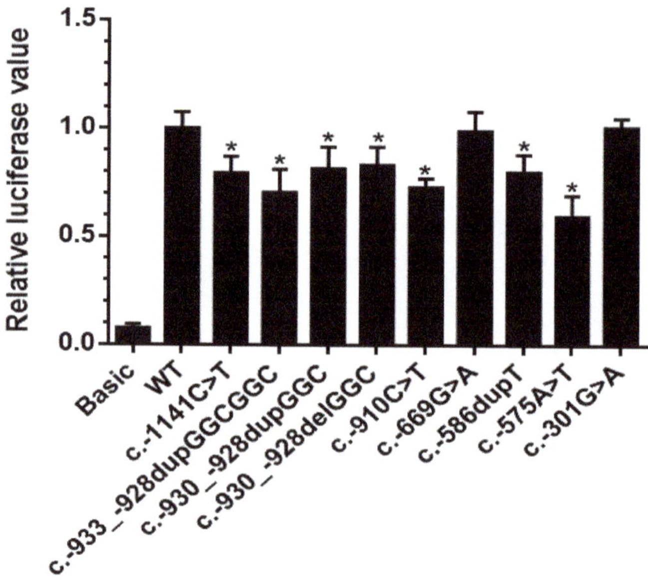

Figure 2. Transcriptional activity analysis of *BMPR2* promoter variants. The transcriptional activity of nine *BMPR2* promoter variants was compared with the wild-type *BMPR2* promoter by dual-luciferase assay. Seven of the nine variants showed significantly reduced gene expression in comparison to the wild-type. Basic: pGL4.10 plasmid (without promoter, negative control), WT: wild-type, c.-930_-928delGGC: 11GGC repeats, c.-930_-928dupGGC: 13GGC repeats, c.-933_-928dupGGCGGC: 14GGC repeats. Data are presented as mean ± standard error of the mean, normalized to the wild-type; *p*-value refers to the comparison between each mutant to wild-type; *: $p < 0.05$.

3.4. Association of Promoter Variants with an Abnormal Pulmonary Artery Pressure during Exercise or PAH Manifestation

One family (Family 1) was identified with two promoter variants in the index patient (Figure 3). The index patient (II: 2) was a 42-year-old woman, who was diagnosed with PAH 6 years prior to study inclusion. At diagnosis she presented with exertional dyspnea and limited exercise capacity. The echocardiogram revealed an enlarged right ventricle and right atrium. Right heart catheterization showed a mean pulmonary arterial pressure of 67 mmHg, pulmonary arterial wedge pressure of 5 mmHg, and pulmonary vascular resistance of 26 Wood Units. The patient carried the two variants c.-910C>T and c.-933_-928dupGGCGGC in the *BMPR2* promoter region. Each variant was inherited from one parent, thus being biallelic and compound heterozygous in the index patient (Figure 3). Both variants were shown to lead to a reduced gene expression in the functional analysis (Figure 2). Apart from the index patient, her brother (II: 5) also carried both promoter variants but did not develop PAH. The c.-910C>T variant was also found in her mother (I: 2) and in four other family members (II: 4, II: 5, III: 4, III: 5) without manifest PAH. In four of the five family members with the c.-910C>T variant an elevated PASP during exercise was measured by echocardiography, while only one variant carrier showed a normal pressure response (Figure 3, Family 1 in Table 3). This observation fits to the hypothesis of cosegregation of the variant with an elevated PASP assuming an autosomal dominant model of inheritance with reduced penetrance.

In contrast, the second *BMPR2* promoter variant (c.-933_-928dupGGCGGC) showed no cosegregation with abnormal PASP (Figure 3). Thus, while the first variant may be associated with a hypertensive exercise response the second variant does not seem to contribute to disease manifestation.

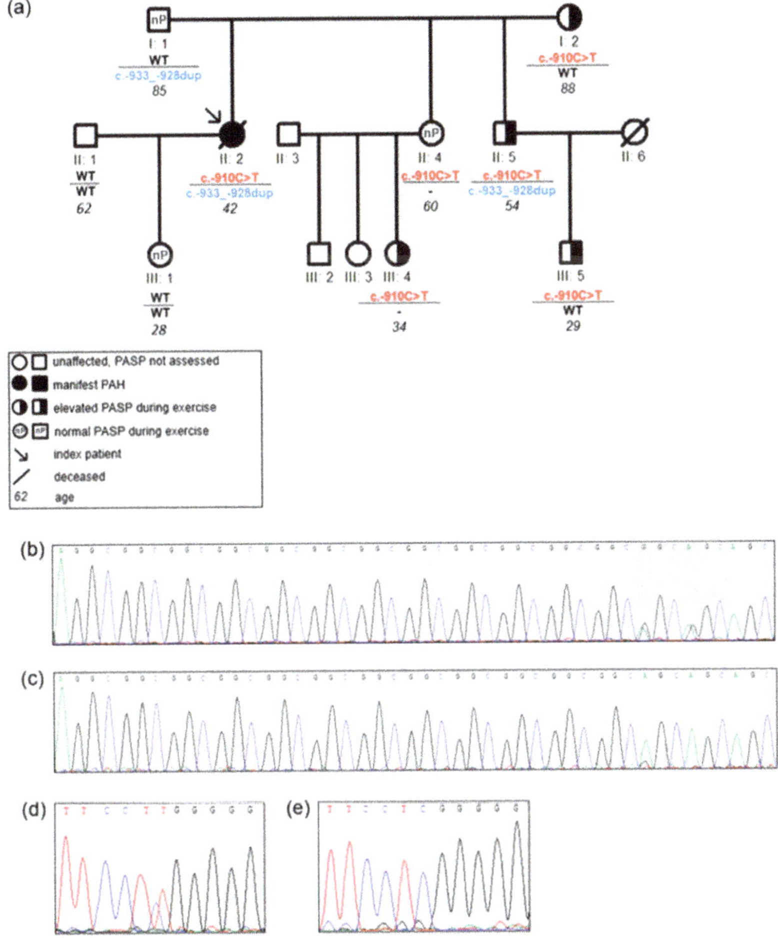

Figure 3. Pedigree of Family 1 with c.-910C>T and c.-933_-928dupGGCGGC in the 5′UTR of *BMPR2* gene: (**a**) the pedigree with *BMPR2* 5′UTR variants c.-910C>T and c.-933_-928dupGGCGGC. Only the c.-910C>T variant is present in all family members with an elevated pulmonary arterial systolic pressure (PASP) during exercise. The horizontal line separates the two loci in *BMPR2* 5′UTR c.-910C>T and c.-933_-928dupGGCGGC variants; c.-910C>T: *BMPR2* c.-910C>T; c.-933_-928dup: *BMPR2* c.-933_-928dupGGCGGC; WT: wild-type; -: not sequenced. Italic numbers represent the age of individuals; (**b**) sequencing analysis of the c.-933_-928dupGGCGGC shows the heterozygous G>A changes on the right side within the grey shaded area indicating a duplication of a 3 bp repeat; (**c**) c.-933_-928 wild-type sequence; (**d**) sequencing analysis of the heterozygous c.-910C>T variant within the grey shaded area; (**e**) c.-910 wild-type sequence.

In the other HPAH families no cosegregation of the respective promoter variant with an abnormal PASP was apparent (Table 3). In one family (Family 3, Table 3) the promoter variant (c.-575A>T) was not present in the second HPAH patient of the family. In another family (Family 2, Table 3) a single, different promoter variant was identified in the two HPAH patients of the family. However, one of the two variants (c.-301G>A) revealed no effect on gene expression in the functional analysis. Promoter variants, which reduced gene expression were identified in six of the ten I/HPAH index patients (Table 3). Out of these six patients only three patients had a known pathogenic variant in the coding region of a PAH gene. Thus, in these patients the promoter variant could be considered a possible second hit adding to a known pathogenic variant.

Table 3. *BMPR2* promoter and further gene variants in I/HPAH patients.

Family[-/-] Index	Other Pathogenic Variants/VUS	*BMPR2* Promoter Variants	Gene Expression Compared to Wild-Type	Variant in FM with ↑ Exercise PASP/all FM with ↑ Exercise PASP	Variant in FM with Normal Exercise PASP/All FM with Normal Exercise PASP
Family 1	none	c.-910C>T	0.73 x	4/4	1/2
		c.-933_-928 dupGGCGGC	0.70 x	1/4	1/2
Family 2	*EIF2AK4* c.641delA p.(Lys214Argfs*21)	c.-1141C>T	0.79 x	in index patient	NA
		c.-301G>A	1.01 x	in 2nd PAH patient	NA
Family 3	none	c.-575A>T	0.60 x	NA	not in 2nd PAH patient
Family 4	none	c.-301G>A	1.01 x	1/2	1/1
Family 5 [1]	*BMPR2* c.244C>T p.(Gln82*)	c.-669G>A	0.99 x	3/6	0/2
Family 6	*BMPR2* exon 2–3 deletion	c.-669G>A	0.99 x	4/6	0/2
Family 7	*ENG* c.1633G>A p.(Gly545Ser)	c.-669G>A	0.99 x	1/1	2/5
IPAH 1	none	c.-586dupT	0.80 x	NA	NA
IPAH 2	*BMPR2* c.1453G>A p.(Asp485Asn)	c.-930_-928delGGC	0.83 x	NA	NA
IPAH 3	*BMPR2* c.2695C>T p.(Arg899*)	c.-930_-928dupGGC	0.82 x	NA	NA

[1] in Family 5, *BMPR2* c.-669G>A was identified in healthy family members but not in the index patient. In other families, variants listed were identified in the index patients. Exercise PASP was measured in 39 of 53 sequenced family members.↑: hypertensive; FM: family member; NA: no PASP measured in family members during exercise; PASP: pulmonary arterial systolic pressure; VUS: variants of uncertain significance. Reference sequence IDs: *BMPR2*: NM_001204, *EIF2AK4*: NM_001013703, *ENG*: NM_001114753.

Two of the three IPAH patients, four HPAH patients and one HPAH family member were sequenced with a PAH-specific panel for 12 PAH genes and 17 candidate genes [20]. This identified a novel homozygous pathogenic *EIF2AK4* variant c.641delA p.(Lys214Argfs*21) in one family with two affected sisters clinically characterized as HPAH, albeit pulmonary veno-occlusive disease could not be excluded retrospectively. Further variants in the study cohort were reported previously (BMPR2 c.224C>T p.Gln82*; *BMPR2* c.1453G>A p.Asp485Asn; *BMPR2* c.2695C>T p.Arg899*; *BMPR2* exon 2–3 deletion; *ENG* c.1633G>A p.Gly545Ser) [2,11,20].

4. Discussion

This study identified four novel and five further variants in the *BMPR2* promoter of I/HPAH patients. The transcriptional impact of these variants revealed a reduced transcriptional activity for seven out of nine variants (78%).

4.1. Impact of Promoter Variants on Transcription Factor Binding Sites

The variant c.-910C>T was identified for the first time in this study. This patient harbored not only the c.-910C>T but additionally the c.-933_-928dupGGCGGC variant in the *BMPR2* promoter region. These two variants were both shown to decrease transcriptional activity of *BMPR2*. The c.-910C>T variant was predicted to create a new binding site for the transcription factor Gli-similar 3 (*GLIS3*). Glis3 is a Krüppel-like transcription factor as a member of Gli and Zinc finger families which shows a tissue specific expression mainly in the endocrinal system such as the thyroid gland and in low levels also in the lung [22]. So far, no study reported the relation of Glis3 and the *BMPR2* gene or any PAH pathway related gene. However, the new PAH gene Krüppel-like transcription factor 2 (*KLF2*) could indicate a possible relation to this Zinc finger family member and PAH pathogenesis [23,24]. The promoter variant c.-910C>T did not cosegregate with PAH in the family but with a hypertensive PASP during exercise. In contrast, the second variant of the index patient in Family 1 (c.-933_-928dupGGCGGC) showed no cosegregation with an abnormal PASP response in family members and was less likely to have contributed to PAH manifestation.

The variant c.-1141C>T of *BMPR2* promoter was also newly identified in a HPAH patient and a significant reduction in the *BMPR2* promoter transcriptional activity expression was shown. In addition, in silico a loss of a pleomorphic adenoma gene 1 (*PLAG1*) transcription factor binding site was predicted. It was reported that *PLAG1* could act as an activator for a gene promoter by upregulating the insulin-like growth factor gene *IGF1* [25]. However, no study has so far elucidated the relation between the transcription factor *PLAG1* and the *BMPR2* gene. If it served as an activator of the *BMPR2* promoter, the loss of the *PLAG1* binding site predicted in this study would be consistent with the measured reduced transcriptional activity. However, the variant showed no cosegregation with the disease or an abnormal PASP response in the family, possibly only attributing a weak functional impact on the *BMPR2* gene and PAH manifestation. Reduced transcriptional activity without cosegregation in the family was also identified for the heterozygous variant c.-575A>T.

Moreover, this study identified three different types of GGC repeats in the c.-963_-928 upstream sequence of *BMPR2*. All of these variants led to a reduced transcriptional activity. The wild-type sequence at position c.-963_-928 upstream of transcription start site is made up of 12 tandem GGC repeats. Limsuwan and colleagues first discovered an abnormal number of GGC repeats within these sequences in congenital heart disease-associated PAH children [16]. The transcriptional assessment of the altered number of GGC tandem repeats in our functional analysis (11, 13 and 14 GGC repeats) in Family 1 and two IPAH patients were in line with another study showing a decrease in gene expression with a promoter containing 13 or 10 GGC repeats [21] in comparison to the 12-GGC repeat. A further study identified a GC>AT change within this repeat sequence cosegregating with the disease in an HPAH family [13]. The authors could establish that the variant led to a preferred cryptic translational start site, followed by a premature stop codon and nonsense mediated decay. This variant could not be identified within the GGC-repeat region in our study.

The "GGCG" sequence is a possible early growth response-1 (*EGR1*) transcription factor binding site. EGR1 binds to the promoter region of many target genes, thereby participating in various pathways such as those relevant for cardiovascular homeostasis [26]. This factor can serve as an activator or repressor and it can be activated by hypoxia in vitro [26]. Gaddipati et al. demonstrated an increase of EGR1 can lead to a decreased expression of *BMPR2* [27]. However, in another study, knockdown of the *EGR1* gene by short interfering RNA treatment resulted in a reduction of *BMPR2* promoter transcription activity [21]. Thus, the complex molecular mechanisms between the EGR1 and *BMPR2* promoter remain to be further investigated.

4.2. Contradictions to Previous Studies

The two variants c.-301G>A and c.-669G>A neither indicated a cosegregation with the disease nor showed any influence on gene transcription. This suggests that both variants may be benign. This is supported by the presence of both alleles in non-PAH controls in the gnomAD database (c.-301G>A: 0.69% in 217 out of 31,336 individuals; c.-669G>A: 0.95% in 297 out of 31,142 individuals). The functional study results contradicted a previous study, which had shown a reduced transcriptional activity of the c.-669G>A variant [14]. The variant was present in only two index patients of three HPAH families in our study. In one family it co-occurred with a pathogenic *BMPR2* exon 2–3 deletion and in the second family with a VUS in the *ENG* gene. Therefore, both the pedigree and functional results suggest a benign nature of the variant, also contradicting our own previous assumptions, which suggested the c.-669G>A variant contributed to disease manifestation [11].

For c.-301G>A, the segregation data and functional analysis showed opposing results to Pousada and colleagues, who reported a decreased gene expression in plasmids harboring the c.-301G>A variant [15]. A predicted loss of an Msh Homeobox 2 (MSX2) transcription factor binding site may result in a reduced *BMPR2* expression. So far, no study reported any relation between the MSH2 and *BMPR2* expression. A previous study demonstrated in Msx2$^{-/-}$ mice an upregulation of *BMPR2* in smooth muscle cells in peripheral arteries [28]. Hence, the contradictory results above require further clarification to determine the effects of the promoter variants on *BMPR2* expression levels.

In contrast to the former studies, this study used a longer 5′UTR of the *BMPR2* gene as promoter sequence (1520 bp) in order to incorporate all detected variants and be able to compare them with each other in the same experimental set-up. In previous studies shorter 5′UTR fragments of 300 bp containing the region c.-770_c.-471 [14] and 539 bp including the base pairs c.-539_c.-1 [15] were used. A 5′UTR length-dependent up- or down-regulation of *BMPR2* has previously been described due to the inclusion of additional transcriptional modulators [27] and could have led to the different results. The origin of the cell culture cells used in functional analyses could also have influenced the results. In this study, PASMCs were used, which are at the heart of vascular remodeling in PAH, while in one of the two other studies the kidney cell line COS-1 was analyzed [15]. Expression data from different organs furthermore suggest the *BMPR2* gene expression in the lungs to be around three times higher than in the kidneys [29]. Thus, the different length of the promoter region encompassing different stretches and transcription factor binding sites of the *BMPR2* promotor, the different cell types and organ-specific gene expression may explain the different expression levels of the same variants between published works and this study. In future studies, expression levels should be compared in constructs with similar length of the inserted sequences and the same cell lines to have clearly comparable results to resolve the discrepancies.

4.3. Limitations and Future Directions

In this study, we only assessed one defined length of the promoter region in one specific cell line. Analyzing the effect of the promotor variants on *BMPR2*-gene expression using sequence fragments of different lengths in various cell lines side by side could help to elucidate the interaction of the transcription regulation domains within the promoter. Moreover, two promoter variants were identified in the same patient (Figure 3) but their effect was only analyzed separately. It would be interesting to evaluate their joint effect in the same plasmid and experimental setting to find out whether they are additive, lead to an even greater effect or cancel each other out.

In addition, the DNA-methylation of the promoter region should be measured in future studies to obtain a full picture of transcriptional activity. Equally, the evaluation of the subcellular localization of the respective BMPRII proteins could provide further functional evidence and valuable information regarding genetic variants in the *BMPR2* promoter region in future studies.

5. Conclusions

New variants in the *BMPR2* promoter region were discovered and investigated in this study. A down-regulated gene expression facilitated by variants in the *BMPR2* promoter suggested a possible impact on disease manifestation. However, the majority of variants showed no cosegregation with the disease or an elevated PASP during exercise in families, despite a statistically significant transcriptional impact. Thus, the promoter variants cannot fully explain the causality between the clinical manifestation and genotypes. There is still a lack of frequency data for promoter variants in a large PAH dataset. Predicted transcriptional factor binding sites remain to be validated and the complex mechanisms of altering these sites and the influence on gene transcription require further detailed investigations to determine the consequences on *BMPR2* expression levels. Thus, further studies are needed to investigate the frequency of *BMPR2* promoter variants and their impact on penetrance and disease manifestation.

Author Contributions: Conceptualization, J.S., K.H., E.G. and C.A.E.; methodology, J.S., L.T.K., N.B., C.F., E.G., C.A.E.; software, J.S., N.B. and C.F.; validation, J.S. and C.A.E.; formal analysis, J.S., N.B. and C.F.; investigation, J.S., L.T.K. and C.A.E.; resources, K.H., L.T.K. and E.G.; data curation, J.S. and C.A.E.; writing—original draft preparation, J.S. and C.A.E.; writing—review and editing, K.H., L.T.K., N.B., C.F. and E.G.; visualization, J.S. and C.A.E.; supervision, K.H., L.T.K., E.G. and C.A.E.; project administration, C.A.E.; funding acquisition, K.H. and E.G. All authors have read and agreed to the published version of the manuscript.

Funding: This research received no external funding.

Acknowledgments: The authors would like to thank Kirsten Linsmeier and Susanne Theiss from the Institute of Human Genetics at Heidelberg University for their kind technical support.

Conflicts of Interest: The authors declare no conflict of interest.

References

1. Galiè, N.; Humbert, M.; Vachiéry, J.-L.; Gibbs, S.; Lang, I.M.; Kaminski, K.A.; Simonneau, G.; Peacock, A.; Noordegraaf, A.V.; Beghetti, M.; et al. 2015 ESC/ERS Guidelines for the diagnosis and treatment of pulmonary hypertension. *Eur. Hear. J.* **2015**, *37*, 67–119. [CrossRef] [PubMed]

2. Pfarr, N.; Szamalek-Hoegel, J.; Fischer, C.; Hinderhofer, K.; Nagel, C.; Ehlken, N.; Tiede, H.; Olschewski, H.; Reichenberger, F.; Ghofrani, H.A.; et al. Hemodynamic and clinical onset in patients with hereditary pulmonary arterial hypertension and BMPR2 mutations. *Respir. Res.* **2011**, *12*, 99. [CrossRef] [PubMed]

3. Machado, R.D.; Southgate, L.; Eichstaedt, C.A.; Aldred, M.A.; Austin, E.D.; Best, D.H.; Chung, W.K.; Benjamin, N.; Elliott, C.G.; Eyries, M.; et al. Pulmonary Arterial Hypertension: A Current Perspective on Established and Emerging Molecular Genetic Defects. *Hum. Mutat.* **2015**, *36*, 1113–1127. [CrossRef] [PubMed]

4. Olschewski, A.; Berghausen, E.M.; Eichstaedt, C.A.; Fleischmann, B.K.; Grünig, E.; Grünig, G.; Hansmann, G.; Harbaum, L.; Hennigs, J.K.; Jonigk, D.; et al. Pathobiology, pathology and genetics of pulmonary hypertension: Update from the Cologne Consensus Conference 2018. *Int. J. Cardiol.* **2018**, *272*, 4–10. [CrossRef]

5. Evans, J.D.W.; Girerd, B.; Montani, D.; Wang, X.-J.; Galié, N.; Austin, E.D.; Elliott, G.; Asano, K.; Grunig, E.; Yan, Y.; et al. BMPR2 mutations and survival in pulmonary arterial hypertension: An individual participant data meta-analysis. *Lancet Respir. Med.* **2016**, *4*, 129–137. [CrossRef]

6. Morrell, N.W.; Aldred, M.A.; Chung, W.K.; Elliott, C.G.; Nichols, W.C.; Soubrier, F.; Trembath, R.C.; Loyd, J.E. Genetics and genomics of pulmonary arterial hypertension. *Eur. Respir. J.* **2019**, *53*, 1801899. [CrossRef]

7. Larkin, E.K.; Newman, J.H.; Austin, E.D.; Hemnes, A.R.; Wheeler, L.; Robbins, I.M.; West, J.; Phillips, J.A.; Hamid, R.; Loyd, J.E. Longitudinal Analysis Casts Doubt on the Presence of Genetic Anticipation in Heritable Pulmonary Arterial Hypertension. *Am. J. Respir. Crit. Care Med.* **2012**, *186*, 892–896. [CrossRef]

8. Frydman, N.A.; Steffann, J.; Girerd, B.; Frydman, R.; Munnich, A.; Simonneau, G.; Humbert, M. Pre-implantation genetic diagnosis in pulmonary arterial hypertension due to BMPR$_2$ mutation. *Eur. Respir. J.* **2012**, *39*, 1534–1535. [CrossRef]

9. Eichstaedt, C.A.; Song, J.; Benjamin, N.; Harutyunova, S.; Fischer, C.; Grünig, E.; Hinderhofer, K. EIF2AK4 mutation as "second hit" in hereditary pulmonary arterial hypertension. *Respir. Res.* **2016**, *17*, 1–8. [CrossRef]

10. Wang, G.; Knight, L.; Ji, R.; Lawrence, P.; Kanaan, U.; Li, L.; Das, A.; Cui, B.; Zou, W.; Penny, D.J.; et al. Early onset severe pulmonary arterial hypertension with 'two-hit' digenic mutations in both BMPR2 and KCNA5 genes. *Int. J. Cardiol.* **2014**, *177*, e167–e169. [CrossRef]

11. Viales, R.R.; Eichstaedt, C.A.; Ehlken, N.; Fischer, C.; Lichtblau, M.; Grünig, E.; Hinderhofer, K. Mutation in BMPR2 Promoter: A 'Second Hit' for Manifestation of Pulmonary Arterial Hypertension? *PLoS ONE* **2015**, *10*, e0133042. [CrossRef] [PubMed]

12. Hinderhofer, K.; Fischer, C.; Pfarr, N.; Szamalek-Hoegel, J.; Lichtblau, M.; Nagel, C.; Egenlauf, B.; Ehlken, N.; Grünig, E. Identification of a New Intronic BMPR2-Mutation and Early Diagnosis of Heritable Pulmonary Arterial Hypertension in a Large Family with Mean Clinical Follow-up of 12 Years. *PLoS ONE* **2014**, *9*, e91374. [CrossRef] [PubMed]

13. Aldred, M.; Machado, R.D.; James, V.; Morrell, N.W.; Trembath, R. Characterization of theBMPR25′-Untranslated Region and a Novel Mutation in Pulmonary Hypertension. *Am. J. Respir. Crit. Care Med.* **2007**, *176*, 819–824. [CrossRef] [PubMed]

14. Wang, H.; Li, W.; Zhang, W.; Sun, K.; Song, X.; Gao, S.; Zhang, C.; Hui, R.; Hu, H. Novel promoter and exon mutations of the BMPR2 gene in Chinese patients with pulmonary arterial hypertension. *Eur. J. Hum. Genet.* **2009**, *17*, 1063–1069. [CrossRef]

15. Pousada, G.; Lupo, V.; Cástro-Sánchez, S.; Álvarez-Satta, M.; Sánchez-Monteagudo, A.; Baloira, A.; Espinós, C.; Valverde, D. Molecular and functional characterization of the BMPR2 gene in Pulmonary Arterial Hypertension. *Sci. Rep.* **2017**, *7*, 1923. [CrossRef]

16. Limsuwan, A.; Choubtum, L.; Wattanasirichaigoon, D. 5′UTR Repeat Polymorphisms of the BMPR2 gene in Children with Pulmonary Hypertension associated with Congenital Heart Disease. *Hear. Lung Circ.* **2013**, *22*, 204–210. [CrossRef]

17. Galiè, N.; Hoeper, M.M.; Humbert, M.; Torbicki, A.; Vachiery, J.-L.; Barbera, J.A.; Beghetti, M.; Corris, P.; Gaine, S.; Gibbs, J.S.; et al. Guidelines for the diagnosis and treatment of pulmonary hypertension: The Task Force for the Diagnosis and Treatment of Pulmonary Hypertension of the European Society of Cardiology (ESC) and the European Respiratory Society (ERS), endorsed by the International Society of Heart and Lung Transplantation (ISHLT). *Eur. Hear. J.* **2009**, *30*, 2493–2537. [CrossRef]

18. Nagel, C.; Henn, P.; Ehlken, N.; D'Andrea, A.; Blank, P.D.N.; Bossone, E.; Böttger, A.; Fiehn, C.; Fischer, C.; Lorenz, H.-M.; et al. Stress Doppler echocardiography for early detection of systemic sclerosis-associated pulmonary arterial hypertension. *Arthritis Res.* **2015**, *17*, 165. [CrossRef]

19. Grünig, E.; Janssen, B.; Mereles, D.; Barth, U.; Borst, M.M.; Vogt, I.R.; Fischer, C.; Olschewski, H.; Kuecherer, H.F.; Kübler, W. Abnormal Pulmonary Artery Pressure Response in Asymptomatic Carriers of Primary Pulmonary Hypertension Gene. *Circulation* **2000**, *102*, 1145–1150. [CrossRef]

20. Song, J.; Eichstaedt, C.A.; Viales, R.R.; Benjamin, N.; Harutyunova, S.; Fischer, C.; Grünig, E.; Hinderhofer, K. Identification of genetic defects in pulmonary arterial hypertension by a new gene panel diagnostic tool. *Clin. Sci.* **2016**, *130*, 2043–2052. [CrossRef]

21. Wang, J.; Zhang, C.; Liu, C.; Wang, W.; Zhang, N.; Hadadi, C.; Huang, J.; Zhong, N.; Lu, W. Functional mutations in 5′UTR of the BMPR2 gene identified in Chinese families with pulmonary arterial hypertension. *Pulm. Circ.* **2016**, *6*, 103–108. [CrossRef] [PubMed]

22. Kim, Y.-S.; Nakanishi, G.; Lewandoski, M.; Jetten, A.M. GLIS3, a novel member of the GLIS subfamily of Kruppel-like zinc finger proteins with repressor and activation functions. *Nucleic Acids Res.* **2003**, *31*, 5513–5525. [CrossRef] [PubMed]

23. Eichstaedt, C.A.; Song, J.; Viales, R.R.; Pan, Z.; Benjamin, N.; Fischer, C.; Hoeper, M.; Ulrich, S.; Hinderhofer, K.; Grünig, E. First identification of Krüppel-like factor 2 mutation in heritable pulmonary arterial hypertension. *Clin. Sci.* **2017**, *131*, 689–698. [CrossRef] [PubMed]

24. Sindi, H.A.; Russomanno, G.; Satta, S.; Abdul-Salam, V.B.; Jo, K.B.; Qazi-Chaudhry, B.; Ainscough, A.J.; Szulcek, R.; Bogaard, H.J.; Morgan, C.C.; et al. Author Correction: Therapeutic potential of KLF2-induced exosomal microRNAs in pulmonary hypertension. *Nat. Commun.* **2020**, *11*, 1. [CrossRef]

25. Voz, M.L.; Agten, N.S.; Van De Ven, W.J.; Kas, K. PLAG1, the main translocation target in pleomorphic adenoma of the salivary glands, is a positive regulator of IGF-II. *Cancer Res.* **2000**, *60*, 106–113.

26. Nishi, H.; Nishi, K.H.; Johnson, A.C. Early Growth Response-1 gene mediates up-regulation of epidermal growth factor receptor expression during hypoxia. *Cancer Res.* **2002**, *62*, 827–834.

27. Gaddipati, R.; West, J.; Loyd, J.E.; Blackwell, T.; Lane, K.A. EGR1 is essential for transcriptional regulation of BMPR2. *Am. J. Mol. Biol.* **2011**, *1*, 131–139. [CrossRef]

28. Lopes, M.; Goupille, O.; Cloment, C.S.; Lallemand, Y.; Cumano, A.; Robert, B. Msx genes define a population of mural cell precursors required for head blood vessel maturation. *Development* **2011**, *138*, 3055–3066. [CrossRef]

29. GTExPortal Database Release V8. Available online: www.gtexportal.org (accessed on 8 March 2020).

Article

Expression Quantitative Trait Locus Mapping in Pulmonary Arterial Hypertension

Anna Ulrich [1], Pablo Otero-Núñez [1], John Wharton [1], Emilia M. Swietlik [2,3], Stefan Gräf [2,4,5], Nicholas W. Morrell [2,4], Dennis Wang [6,7], Allan Lawrie [8], Martin R. Wilkins [1], Inga Prokopenko [9,10], Christopher J. Rhodes [1,*], on behalf of The NIHR BioResource—Rare Diseases Consortium [†] and UK PAH Cohort Study Consortium [†]

[1] National Heart and Lung Institute, Hammersmith Campus, Imperial College London, London SW7 2BU, UK; anna.ulrich15@imperial.ac.uk (A.U.); pablo.otero-nunez17@imperial.ac.uk (P.O.-N.); j.wharton@imperial.ac.uk (J.W.); m.wilkins@imperial.ac.uk (M.R.W.)
[2] Department of Medicine, University of Cambridge, Cambridge CB2 3AX, UK; es740@cam.ac.uk (E.M.S.); sg550@cam.ac.uk (S.G.); nwm23@cam.ac.uk (N.W.M.)
[3] Pulmonary Vascular Disease Unit, Royal Papworth Hospital NHS Foundation Trust, Cambridge CB2 0AY, UK
[4] NIHR BioResource-Rare Diseases, Cambridge, CB2 0QQ, UK
[5] Department of Haematology, University of Cambridge, Cambridge CB2 3AX, UK
[6] Sheffield Institute for Translational Neuroscience, University of Sheffield, Sheffield S10 2TN, UK; dennis.wang@sheffield.ac.uk
[7] Sheffield Bioinformatics Core, University of Sheffield, Sheffield S10 2TN, UK
[8] Department of Infection, Immunity & Cardiovascular Disease, University of Sheffield, Sheffield S10 2TN, UK; a.lawrie@sheffield.ac.uk
[9] Department of Clinical and Experimental Medicine, University of Surrey, Guildford GU2 7XH, UK; i.prokopenko@surrey.ac.uk
[10] Department of Metabolism, Digestion and Reproduction, Imperial College London, London SW7 2BU, UK
* Correspondence: crhodes@imperial.ac.uk
† Members listed in the Supplementary Materials.

Received: 10 September 2020; Accepted: 21 October 2020; Published: 22 October 2020

Abstract: Expression quantitative trait loci (eQTL) can provide a link between disease susceptibility variants discovered by genetic association studies and biology. To date, eQTL mapping studies have been primarily conducted in healthy individuals from population-based cohorts. Genetic effects have been known to be context-specific and vary with changing environmental stimuli. We conducted a transcriptome- and genome-wide eQTL mapping study in a cohort of patients with idiopathic or heritable pulmonary arterial hypertension (PAH) using RNA sequencing (RNAseq) data from whole blood. We sought confirmation from three published population-based eQTL studies, including the GTEx Project, and followed up potentially novel eQTL not observed in the general population. In total, we identified 2314 eQTL of which 90% were cis-acting and 75% were confirmed by at least one of the published studies. While we observed a higher GWAS trait colocalization rate among confirmed eQTL, colocalisation rate of novel eQTL reported for lung-related phenotypes was twice as high as that of confirmed eQTL. Functional enrichment analysis of genes with novel eQTL in PAH highlighted immune-related processes, a suspected contributor to PAH. These potentially novel eQTL specific to or active in PAH could be useful in understanding genetic risk factors for other diseases that share common mechanisms with PAH.

Keywords: expression quantitative trait locus; eQTL; pulmonary arterial hypertension; blood; genetics

1. Introduction

The relationship between genomic variability, disease risk and endophenotypes has been a focus of many research groups with access to suitable disease cohorts. This has been aided by the increasing affordability of genotyping and sequencing methodologies. Interpretation of the mechanisms behind the effects of genetic variants discovered to associate with phenotypes poses a challenge as many are located in the non-coding space of the genome [1]. Non-coding variants exert no direct effect on protein structure, making the biological link to the disease or phenotype more difficult to discern. Variants in the non-coding space may instead exert their effect by influencing gene expression [1]. Characterising the genomic control of gene expression can offer a handle on understanding the role of disease-associated variants by linking them to disrupted pathways [2]. Variants associated with the expression of a gene, commonly called expression quantitative trait loci (eQTL), have been described, primarily in population-based studies [3–5].

Differential gene expression analyses comparing cases and controls may be useful for diagnostic and/or prognostic purposes or in identifying genes, which may have a causal role in developing the disease [6–8]. In the case of the latter, distinguishing between differentially expressed genes that are secondary to the disease and those that contribute to disease development is a significant challenge. If the frequency of one or several eQTL for a differentially expressed gene is also different between cases and controls, a causal association may be established, given the assumptions of causal inference [9] are met. In brief, for a causal relationship to be established between a gene product proxied by an eQTL and a disease, the eQTL has to be independent of the confounders of the association between the gene and the disease, and it has to have no direct effect on the disease via a pathway that does not involve the gene being instrumented.

Currently, one of the main limiting factors for causal inference analysis using genomic data is the number of known eQTL. Certain eQTL effects, which go undetected in healthy cohorts commonly used for eQTL mapping, may be unmasked by disease and development. The diagnostic value of gene expression profiles, i.e., that they can be used to distinguish affected individuals from non-affected ones, underlines the uniqueness of the set of genes expressed in a disease state [6,7]. Differential gene expression in diverse states (carcinogenesis, inflammation, etc.) or at diverse stages (such as stages of embryonic development) becomes evident when comparing the gene correlation matrices of a biological system under multiple different conditions. Such techniques are called 'differential co-expression networks' [10] and have been used successfully to identify genes and gene sets that are important in the state or stage under investigation in a network of genes [11,12]. By principle, functionally involved nodes or modules rewire more frequently than uninvolved ones in states they create and/or maintain and, therefore, can be identified [13].

With the exception of housekeeping genes, the majority of genes in the human transcriptome are tissue-specific. Therefore, tissue selection for gene expression studies of disease is of importance. The lung is the most relevant tissue in the pathogenesis of pulmonary arterial hypertension (PAH), a disease in which pulmonary vascular remodelling drives right ventricular failure. Whole blood also has relevance to PAH and might capture more than just systemic effects given the immune component of this disease. The role of inflammation in PAH has been extensively studied, based on the observation that inflammatory cells infiltrate the remodeled vascular wall [14]. Since PAH can be a complication of many inflammatory diseases, including connective tissue disease, thyroiditis, scleroderma, systemic lupus erythematosus, human immunodeficiency virus infection and schistosomiasis, it might be reasonable to suspect that inflammatory processes, even in the absence of a diagnosed comorbidity, contribute to disease development and maintenance. Another advantage of using whole blood instead of lung tissue is the non-invasiveness of sampling and availability. Lung samples are often only obtainable from the explanted organ after lung transplantation or from post-mortem tissues. Appropriate control samples for differential gene expression analyses are equally hard to come by.

This study aimed to characterise the genetic variability of transcriptome-wide gene expression in whole blood from 276 consecutively sampled patients with PAH and defined potentially novel eQTL active in this disease state. These potentially disease-specific eQTL could be useful in elucidating known genotype-phenotype associations or causal analyses.

2. Materials and Methods

2.1. Study Participants and Sample Processing

A total of 276 patients with idiopathic, heritable or drug-induced PAH diagnosed following international guidelines [15] were recruited from expert centres across the UK with whole-genome sequence conducted as part of the UK National Institute for Health Research BioResource (NIHRBR) study [16,17]. Transcriptome profiling through RNAseq was completed as part of the PAH cohort study [18]. Demographic characteristics and white blood cell counts of the individuals in this study are shown in Supplementary Table S1.

2.2. Gene Expression Data

RNA sequencing and transcript abundance estimation procedures are described in the Supplementary Material. White blood cell composition was quantified using quanTIseq, a novel deconvolution algorithm [19]. Predicted white blood cell fractions from quanTIseq correlated with clinical white cell fractions available in a subset of PAH patients (Spearman correlation = 0.44–0.73). Well-detected transcripts with a minimum of two reads in at least 95% of samples ($n = 26{,}050$) were taken forward to the eQTL analyses.

2.3. Genetic Data

Whole-genome sequence Hg19-aligned data were available from the NIHRBR study described in detail elsewhere [16]. Genetic variants with a minimum minor allele frequency of 5% were extracted from the NIHRBR variant call format (VCF) files and further filtered for variants called in at least 95% of the samples and had a *p*-value greater than 10^{-5} for Hardy Weinberg equilibrium, leaving a total of 7,362,566 variants for eQTL mapping. Software tools used in the eQTL mapping pipeline included PLINKv1.90 [20] for data extraction and computing multidimensional scaling components, QCTOOLv2.0 [21], for variant filtering, and QUICKTESTv1.1 [22] for association testing.

2.4. eQTL Effect Estimation

Raw gene count data were first variance stabilised to achieve homoskedasticity and then quantile normalised to the median distribution. The sample means were then centered to zero, and the sample variance was linearly scaled, such that each sample had a standard deviation of one. A total of 7,362,566 markers were tested for their effects on the expression of 26,050 well-detected genes. Variants within 1 megabase (Mb) on either side of the gene's transcription start site (TSS) were regarded cis-acting or *cis*-eQTL while all other markers farther on the same chromosome or on different chromosomes were regarded *trans*-eQTL. Each well-detected transcript was modelled as a function of genotype (coded 0, 1 or 2 reflecting the number of alternate alleles) in a linear regression framework while adjusting for the following covariates: Sex, first four components from multidimensional scaling and white blood cell fractions (Supplementary Table S1). The genome-wide results were subsequently pruned using the 'clump' command in PLINK 1.90 (parameters: —clump-p1 1.9×10^{-12}; —clump-r2 0.01; —clump-kb 1000) to pick the lead variant with the lowest *p*-value in a block of variants in linkage disequilibrium, and thus obtain independent effects. Genetic variants that reached the Bonferroni-adjusted significance threshold of 1.9×10^{-12} given by dividing the genome-wide significance threshold ($p \leq 5 \times 10^{-8}$) by the number of transcripts tested ($n = 26{,}050$) were considered eQTL in this study.

2.5. Confirmation Rate

The confirmation rate of eQTL in this study was calculated in the 2 largest published eQTL studies to date and the Genotype-Tissue Expression (GTEx) Project [3,4,23] (Table 1) to assess the extent to which eQTL in PAH overlapped with eQTL described in healthy populations. These studies were selected for having measured gene expression in the same tissue as our study, in addition to being the largest eQTL mapping studies to date. No eQTL study other than the GTEx Project with available genome-wide results applied RNAseq in whole blood at the time this study was conducted. Confirmation rate was defined as the number of PAH eQTL confirmed by the published study, divided by the total number of PAH eQTL tested by the published study and multiplied by 100 to give a percent value. The matching of gene transcripts and genetic variants between our study and the published studies as well as similarities and differences between the designs of this study and the 3 published studies are described in Table 1 and in the Supplementary Appendix.

Table 1. Characteristics of the PAH Cohort eQTL mapping study and published eQTL studies used for confirmation.

	Westra et al.	Joehanes et al.	GTEx	This Study
Gene expression array	Illumina HumanHT-12 v4.0	Affymetrix HuEx 1.0 ST	RNAseq	RNAseq
Genotyping panel	HapMap2	1000-Genomes	WGS	WGS
MAF threshold	≥5%	≥1%	≥1%	≥5%
n variants in analysis	not reported	8,510,936	10,008,325	7,362,566
cis-eQTL	≤250 kb from PMP	≤1 Mb from the TSS	≤1 Mb from the TSS	≤1 Mb from the TSS
trans-eQTL	≥5 Mb from PMP	>1 Mb from the TSS	>1 Mb from the TSS	>1 Mb from the TSS

Comparison of gene expression- and genetic data and expression quantitative trait locus (eQTL) definitions. MAF = minor allele frequency; PMP = probe midpoint; TSS = transcription start site; RNAseq = RNA sequencing; WGS = whole-genome sequencing.

2.6. Overlap with Variants from the GWAS Catalog

An important aim of this study was to identify eQTL in PAH with effects that had not been observed in population-based studies and to assess the relevance of these PAH-specific eQTL to related phenotypes and diseases. The NHGRI-EBI GWAS Catalog of published genome-wide association studies (GWAS) [24] provided a curated database of genetic markers associated with a wide range of traits and diseases. All variant-phenotype pairs below an association p-value threshold of 9×10^{-6} were downloaded from the GWAS Catalog website on 5 March 2020. The GWAS Catalog at the time contained 113,510 unique variants (including variants in linkage disequilibrium [LD]) reported as lead variants for 4314 phenotypes. The list of GWAS variants were intersected with each PAH eQTL (including the lead eQTL and variants in LD with the lead eQTL [$r^2 \geq 40\%$]). The difference between the proportions of novel eQTL and previously reported eQTL overlapping lung-related phenotypes and diseases curated from the full GWAS Catalog phenotype list was tested using the 2-sample equality of proportions test.

2.7. Functional Enrichment Analysis of Genes with Novel eQTL

Functional enrichment analysis of genes with novel eQTL in PAH using the WEB-based GEne SeT AnaLysis Toolkit (WebGestalt) [25] was conducted to determine if genes with novel eQTL were predominantly from pathways with relevance to the pathobiology of PAH. A form of pathway analysis called over-representation analysis (ORA) [26] was run using the Kyoto Encyclopedia of Genes and Genomes (KEGG) pathways [27] database to group these genes into a smaller number of gene-sets relating to biological processes. We queried the list of genes with novel eQTL whilst all the tested genes also present on the expression array of at least one of the published eQTL studies used for the confirmation process constituted the list of background genes. Pathways below a false-discovery rate (FDR)-corrected significance threshold of 0.05 were considered significant.

3. Results

We observed 2354 eQTL in total with the majority (90%) acting in cis (Figure 1A), accounting for 9% of all genes tested in this transcriptome- and genome-wide eQTL mapping analysis in PAH whole blood samples. Of these eQTL, 146 were associated with unmapped transcripts not yet annotated in the Ensembl database (Figure 1B). Lead eQTL ranked by the percent of variance explained (R^2) in gene expression are presented in the Supplementary Table S2. The proportion of variation explained (R^2) in gene expression levels was generally lower for *trans*-eQTL than for *cis*-eQTL (median$_{cis\text{-eQTL}}$ = 23.1%, IQR$_{cis\text{-eQTL}}$ = 12.5%; median$_{trans\text{-eQTL}}$ = 21.5%, IQR$_{trans\text{-eQTL}}$ = 11.5%; $t_{(df=1332)}$ = 4.5, *p*-value = 8.3×10^{-6}).

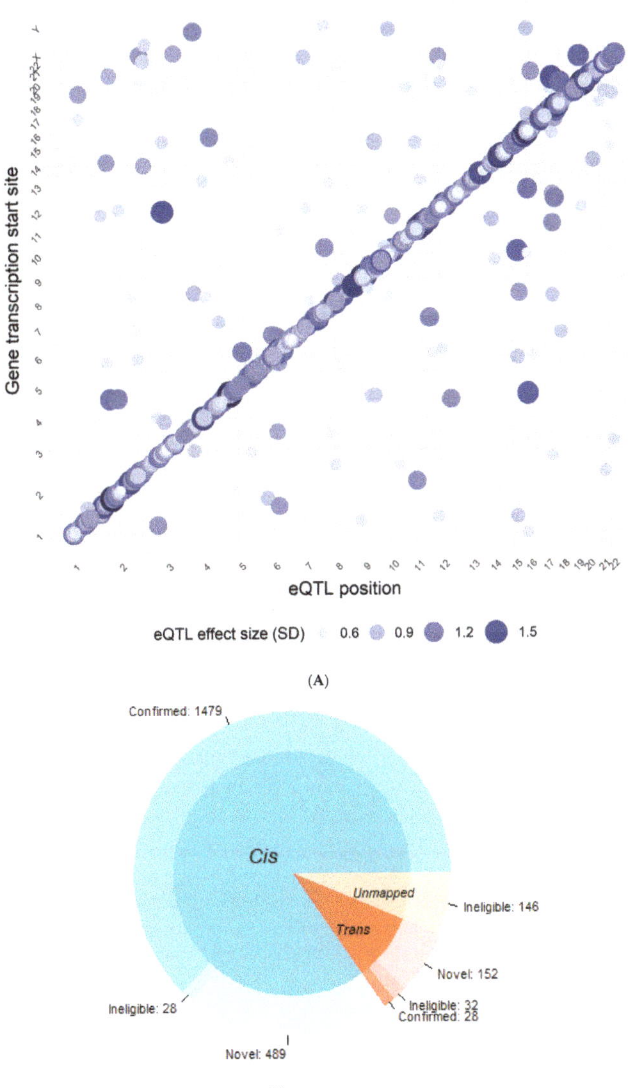

Figure 1. (**A**) Results of the pulmonary arterial hypertension (PAH) Cohort eQTL mapping. Genomic eQTL location (*x*-axis) plotted against the transcription start site of the gene associated with the eQTL. Bubble size and color are proportional to the effect size of the eQTL-gene association. (**B**) Pie chart showing the proportions of *cis*-, *trans*- and unmapped eQTL identified in the PAH Cohort eQTL mapping. Counts of eQTL within each category are shown next to the category name. An eQTL was considered 'novel' if it was eligible to be confirmed in at least one of the previously published studies

and was not reported as an eQTL previously. Those eQTL that reached the study-specific significance threshold in at least one of the published studies were considered 'confirmed'. Unmapped transcripts not yet annotated in the Ensembl database could not be confirmed. Ineligible eQTL were not tested by any of the three studies used for confirmation.

3.1. Confirmation Rate

Out of the 1986 unique genes with *cis*-eQTL, 1509 (76%) could be mapped to the Illumina HT12v3 array used by Westra et al. and 2134 (95%) to the Affymetrix Human Exon chip used by Joehanes et al., respectively. Results from the GTEx Portal were obtained for 73% of eQTL in this study. Twenty-eight percent of *cis*-eQTL were confirmed by Westra et al., 31% by Joehanes et al. and 90% of tested *cis*-eQTL confirmed in GTEx. Overall, 75% of *cis*-eQTL were confirmed in at least one of the published studies (Figure 1). The overall confirmation rate for *trans*-eQTL reached 16%, with all but one being confirmed by GTEx alone. Joehanes et al. confirmed only one *trans*-eQTL for *JAM3* (chr1:248039451 [C/T]).

3.2. Overlap with Variants from the GWAS Catalog

In order to assess the relevance of eQTL identified in this study to a wide range of phenotypes and diseases, the 2173 unique eQTL were intersected with the database of published genotype-phenotype associations from the GWAS Catalog. In total, 929 eQTL were reported for at least one trait previously (median$_{N\ GWAS\ traits/eQTL}$ = 2; IQR$_{N\ GWAS\ traits/eQTL}$ = 3). Ninety-eight (11%) of the 929 eQTL had at least 10 or more unique GWAS phenotypes associated with them. Figure 2 shows the number of overlapping GWAS phenotypes per eQTL in the confirmed and novel eQTL categories. The proportion of *cis*-acting eQTL overlapping with at least one published GWAS association in the confirmed subset was 47%, while in the novel subset, it was 37%. Among *trans*-eQTL, the proportion of loci overlapping at least one GWAS trait was 50% in the confirmed and 27% in the novel group. The higher proportion of overlapping loci in the confirmed eQTL groups was significant for both *cis*- and *trans*-acting eQTL (two-sample equality of proportions test *cis*-eQTL: $\chi^2_{(df=1)}$ = 13.9; 95% CI = 0.05–0.15; *p*-value = 0.0002. *trans*-eQTL: $\chi^2_{(df=1)}$ = 4.2; 95% CI = −0.003–0.45; *p*-value = 0.04). Interestingly, the proportion of novel eQTL previously reported for lung-related phenotypes (11%) was twice as high as the proportion of confirmed eQTL associated with lung-related phenotypes (5.5%) ($\chi^2_{(df=1)}$ = 6.9; 95% CI = 0.006–0.1; *p*-value = 0.009). Lung-related phenotypes reported for the novel eQTL included chronic obstructive pulmonary disease (COPD), lung function (forced vital capacity, forced expiratory volume), low versus high forced expiratory volume, interstitial lung disease, emphysema, lung cancer and lung adenocarcinoma. The full list of 63 QTL overlapping lung-related phenotypes in the GWAS Catalog can be found in Supplementary Table S3.

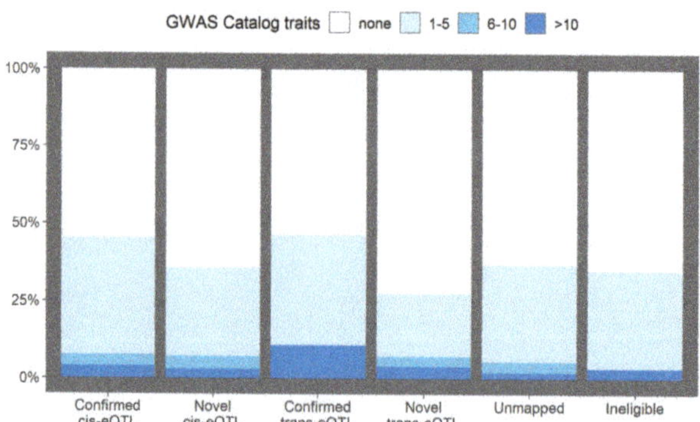

Figure 2. GWAS trait-associations reported for eQTL in the PAH Cohort. Percentage of binned

numbers of overlapping GWAS traits (*y*-axis) per eQTL in each eQTL category (*x*-axis). Reported genotype-trait associations were downloaded from the NHGRI-EBI Catalog of published genome-wide association studies.

3.3. Functional Enrichment Analysis of Genes with Novel eQTL

We assessed the list of 606 genes with novel eQTL in the PAH Cohort for their involvement in certain biological processes or pathways in the KEGG knowledgebase. Five pathways were found to be significantly (FDR <0.05) enriched, with the pathways taste transduction, graft-versus-host disease and autoimmune thyroid disease being the three most significant (Table 2). Supplementary Table S4 lists the overlapping genes driving the association of these five pathways and our results with the corresponding eQTL in PAH, including several human leukocyte antigen (HLA) genes. Furthermore, all genes overlapping with type I diabetes and allograft rejection pathways come from the HLA class. Genes encoded by the more than 200 HLA genes in the highly polymorphic HLA region play an important role in antigen processing and presentation. In total, 20 genes with eQTL in this study came from the HLA classes I and II (*n* = 18) and from the HLA class IB (*n* = 2) genes. Half of these eQTL-HLA gene associations were found to be novel. Out of the 5 *trans*- and 17 *cis*-eQTL, 3 *trans*- and 8 *cis*-eQTL were confirmed by published studies [3,4,23].

Table 2. Pathways enriched for genes with novel eQTL in the PAH Cohort.

Gene Set	Description	Size	Overlap	Expectation	Enrichment Ratio	*p*-Value	FDR
hsa04742	Taste transduction	48	8	0.87	9.24	1.78×10^{-6}	5.72×10^{-4}
hsa05332	Graft-versus-host disease	32	6	0.58	10.40	1.85×10^{-5}	2.86×10^{-3}
hsa05320	Autoimmune thyroid disease	34	6	0.61	9.79	2.66×10^{-5}	2.86×10^{-3}
hsa05330	Allograft rejection	33	5	0.59	8.40	2.75×10^{-4}	0.02
hsa04940	Type I diabetes mellitus	36	5	0.65	7.70	4.19×10^{-4}	0.03

Results of the over-representation analysis using the Kyoto Encyclopedia of Genes and (KEGG) pathway database. Column names: Gene set = searchable identifier of the gene set in KEGG; Size = number of genes participating in the gene set and overlap gives the number of genes in the gene set that overlap with the queried gene list; Expectation = number of genes that are expected to be shared between the queried gene list and the gene set if the queried list was a random sample of the background gene list; Enrichment ratio = fold enrichment of the gene set for the queried gene list; *p*-value = enrichment significance of the Fisher's exact test; FDR = false discovery rate adjusted significance.

4. Discussion

Gene expression serves as an intermediate phenotype between genetic variants and associated phenotypes, such as clinical diagnoses, accepted biomarkers and anthropometric and behavioural traits. Previous eQTL studies provided support for the idea that genetic effects on gene expression have phenotypic consequences and that eQTL aid in the biological interpretation of associated genetic markers in disease [3–5]. However, eQTL effects are not only dependent on the tissue or cell type under investigation [23,28] but also on the environment or context [5,29–31] in which gene expression is measured. This implies that a more comprehensive eQTL map could be constructed by extending mapping efforts beyond population-based studies. We ran a transcriptome-wide eQTL analysis in a cohort of 276 PAH patients to characterise the genetic control of gene expression variability in this condition, uncovering potentially novel eQTL not detected in healthy populations. The resulting 2354 significant eQTL were intersected with the results of two previously published population-based eQTL studies and the results of the GTEx Project. In this study, 25% of *cis*-eQTL and 84% of *trans*-eQTL were not found in any of the three studies used for confirmation. Novel and confirmed eQTL were investigated for their relevance to a wide range of diseases, and results focused on lung-related phenotypes from the NHGRI-EBI Catalog of published genome-wide association studies (GWAS Catalog).

Nearly half (43%) of all eQTL identified in this study colocalised with at least one trait or disease in the GWAS Catalog. Even though the proportion of colocalising eQTL was higher for confirmed *cis*- and *trans*-eQTL than for novel eQTL, the proportion of eQTL associated with lung-related phenotypes was twice as high among novel eQTL than confirmed eQTL previously detected in population-based studies. This indicates that these novel eQTL identified in blood samples of PAH patients might be highly informative for pulmonary diseases such as COPD, interstitial lung disease and emphysema. The higher proportion of GWAS-trait associated eQTL in the confirmed subset might be explained by the lack of replication of novel eQTL by an independent PAH cohort; therefore, it is expected that a proportion of these novel association signals is spurious.

As an example for the overlap between lung-related GWAS traits and eQTL in PAH, the novel *GTF2IRD2B cis*-eQTL rs13238996 (Supplementary Table S3) overlaps with multiple phenotypes including COPD [32], lung function [33], cardiovascular disease [33] and diastolic blood pressure [34]. The Deletion of *GTF2IRD2B* leads to a rare congenital disease called Williams-Beuren syndrome, which frequently presents with supravalvular aortic stenosis (SVAS; OMIM 185500), a congenital heart defect characterised by the narrowing of the aorta, pulmonary and coronary arteries and other blood vessels [35].

Pathway analysis of genes associated with novel eQTL identified five biological processes, including four immune-related phenotypes enriched for this list of genes from the KEGG database. Six genes from the graft-versus-host disease and allograft rejection pathways overlap with the list of genes with novel eQTL, five of which belong to the Human Leukocyte Antigen (HLA) class. The immune-related pathways enriched for genes with novel eQTL in this study demonstrate that novel eQTL could be identified in disease populations since their gene expression profiles differ from the profiles of healthy individuals. The complex pathophysiological mechanisms behind PAH involve multiple driving factors of which immune dysfunction and inflammation are suspected to be among the major contributors (Rabinovitch et al., 2014). The importance of antigen-presenting and recognition in PAH is underlined by the most significant genetic variant discovered in the PAH GWAS in the *HLA-DPA1/DPB1* locus encoding class II major histocompatibility complex (MHC) molecules [16], which associated with three *HLA-DPB1* alleles, all containing a glutamic acid at amino acid residue 69 (Glu^{69}). The *HLA-DPA1/DPB1* PAH susceptibility locus emphasises the role of immune dysregulation in PAH development [36,37] and warrants further investigation.

Overall, the confirmation rate of eQTL in this study was comparable to that seen in published studies [3,4]. A third of *cis*-eQTL confirmed in the two population-based studies by Westra et al. and Joehanes et al., while 90% of *cis*-eQTL were confirmed by the GTEx Project. Fifteen percent of *trans*-eQTL confirmed in either the GTEx project or the study of Joehanes et al. *Trans*-eQTL validation rates are generally much lower (under 10%) than validation rates of *cis*-eQTL [3,4], reflecting the higher average effect size of *cis*-eQTL and a stronger tissue-specificity of *trans*-eQTL effects compared to *cis*-eQTL, which may render *trans*-eQTL more susceptible to confounding by differing experimental conditions and environmental factors. However, the confirmation rate reported by this study and other studies possibly underestimate the true extent of eQTL overlap between studies since the list of variants and genes that passed quality control and were tested by any one study are usually not made available. We observed a higher confirmation rate with the GTEx Project that also used RNAseq for assaying gene expression. Similarly, Joehanes et al. reported higher validation rates between two array-based studies than between array-based and RNAseq-based studies. This may be related to platform differences, for example, hybridisation in arrays is less sensitive than high-depth sequencing and potentially affected by differing background signals in various tissues. Differences between the study populations being compared can also give rise to imperfect validation and also to novel discoveries, as all genetic effects depend on both the genetic (epistasis) and environmental context of the population they were estimated in and, therefore, do not necessarily apply to another population or the same population at a different time. A more accurate way of identifying novel eQTL in this study would have been to contrast PAH eQTL effects with eQTL effects estimated in a control population assayed on the same platforms and in one batch with the PAH samples to reduce the variability due to technical factors. In addition, novel eQTL are yet to be replicated in an independent population of PAH patients to identify true positives.

The effects of eQTL can vary by the tissue and even cell type under investigation [28,38] and may be modified by external and environmental factors [29,30]. Previous studies have successfully identified 'response' eQTL that are associated with gene expression levels in cells under one of two experimental conditions, but not both. For example, in the study of Barreiro, nearly 200 eQTL were identified in primary dendritic cells from 65 individuals with effects specific to either the Mycobacterium tuberculosis-infected cells or the uninfected ones [29]. These response eQTL were argued to constitute natural regulatory variation that likely affect host-Mycobacterium tuberculosis interaction and account for interindividual variation in the immune response and susceptibility of tuberculosis. The authors of the study found when integrating their data with genome-wide genetic susceptibility of pulmonary tuberculosis that these response eQTL were more likely to be associated with the disease, uncovering potential susceptibility genes in pulmonary tuberculosis. Another study by Fairfax et al. [30] investigated the effect of innate immune stimuli on eQTL effects by exposing primary CD14+ cells from 432 European volunteers to the inflammatory cytokine interferon-γ or the endotoxin lipopolysaccharide. Inflammatory stimulation revealed hundreds of eQTL specific to either stimulus, which were enriched for disease-risk loci. In this study, the proportion of eQTL also associated with lung diseases or lung function was twice as high in the novel eQTL subset than in the confirmed subset, highlighting the importance of genotype-environment interaction in understanding the genetic variation of disease susceptibility characterised by genome-wide association studies.

To date, large-scale eQTL-mapping has been done in healthy individuals from population-based studies (Joehanes et al., 2017, Westra et al., 2013, Zhernakova et al., 2017), providing a valuable knowledge base for understanding associations between genetic variation and various phenotypes. It has been shown that by recapitulating the environmental context relevant to disease, it is possible to decipher the genetic variation of disease susceptibility more extensively (Fairfax et al., 2014, Barreiro et al., 2012, Çalışkan et al., 2015). Genome- and transcriptome-wide eQTL-mapping in this cohort of idiopathic and heritable PAH patients identified hundreds of potentially novel eQTL with twice the proportion of lung disease associated with genetic variants than eQTL confirmed by population-based studies. Apart from pulmonary conditions, these novel eQTL specific to, or active in, PAH could be useful in understanding genetic risk factors for other diseases that share common mechanisms with PAH, such as those with immune dysregulation.

Supplementary Materials: The following are available online at http://www.mdpi.com/2073-4425/11/11/1247/s1, Table S1: Patient characteristics and white blood cell fractions in the PAH Cohort study, Table S2: Lead variants for novel eQTL in the PAH Cohort ordered by variance explained in gene expression, Table S3: Phenotypes from the NHGRI-EBI GWAS Catalog of published genome-wide association studies reported for novel and confirmed eQTL, Table S4: Significant KEGG pathways from the enrichment analysis of genes with novel eQTL in the PAH Cohort.

Author Contributions: Conceptualization, C.J.R., M.R.W., D.W., A.L., N.W.M. and I.P.; methodology, A.U.; formal analysis, A.U., P.O.-N.; data curation, J.W., E.M.S., S.G.; writing—original draft preparation, A.U.; writing—review and editing, C.J.R., M.R.W. and I.P.; visualization, A.U.; supervision, C.J.R., M.R.W. and I.P.; funding acquisition, M.R.W., C.J.R., A.L. and N.W.M. All authors have read and agreed to the published version of the manuscript.

Funding: The work cited here is supported by funding from the NIHR BR-RD, the British Heart Foundation (SP/12/12/29836), the BHF Cambridge and Imperial Centres of Cardiovascular Research Excellence (RE/18/4/34215), UK Medical Research Council (MR/K020919/1), the Dinosaur Trust and BHF Programme grants to RCT (RG/08/006/25302), NWM (RG/13/4/30107) and MRW (RG/10/16/28575). Funding for the PAH Biobank is provided by NIH/NHLBI HL105333. The authors acknowledge use of BRC Core Facilities provided by the financial support from the Department of Health via the National Institute for Health Research (NIHR) comprehensive Biomedical Research Centre award to Imperial College NHS Trust, Cambridge University Hospitals and Guy's and St Thomas' NHS Foundation Trust in partnership with King's College London and King's College Hospital NHS Foundation Trust and by NIHR funding to the Imperial NIHR Clinical Research Facility. CJR is supported by a British Heart Foundation Intermediate Basic Science Research Fellowship (FS/15/59/31839) and Academy of Medical Sciences Springboard fellowship (SBF004\1095). DW is supported by the NIHR Sheffield Biomedical Research Centre, and the Academy of Medical Sciences Springboard (ref: SBF004/1052). I.P. is in part funded by the World Cancer Research Fund (WCRF UK) and World Cancer Research Fund International (2017/1641), the Wellcome Trust (WT205915/Z/17/Z), the European Union's Horizon 2020 research, and innovation programme (LONGITOOLS, H2020-SC1-2019-874739), and the Royal Society (IEC\R2\181075). NWM is a British Heart Foundation Professor and National Institute of Health Research (NIHR) Senior Investigator.

Acknowledgments: We gratefully acknowledge the participation of patients recruited to the UK National Institute of Health Research BioResource—Rare Diseases (NIHR BR-RD) Study, the UK National Cohort of Idiopathic and Heritable PAH, and the National Institutes of Health/National Heart, Lung, and Blood Institute—sponsored National Biological Sample and Data Repository for PAH (aka PAH Biobank). We thank the physicians, research nurses and coordinators in the UK, Europe and at the 38 pulmonary hypertension centers across the US involved in the PAH Biobank (www.pahbiobank.org).

Conflicts of Interest: The authors declare no conflict of interest.

References

1. Maurano, M.T.; Humbert, R.; Rynes, E.; Thurman, R.E.; Haugen, E.; Wang, H.; Reynolds, A.P.; Sandstrom, R.; Qu, H.; Brody, J.; et al. Systematic localization of common disease-associated variation in regulatory DNA. *Science* **2012**, *337*, 1190–1195. [CrossRef] [PubMed]

2. Gamazon, E.R.; Segrè, A.V.; van de Bunt, M.; Wen, X.; Xi, H.S.; Hormozdiari, F.; Ongen, H.; Konkashbaev, A.; Derks, E.M.; Aguet, F.; et al. Using an atlas of gene regulation across 44 human tissues to inform complex disease- and trait-associated variation. *Nat. Genet.* **2018**, *50*, 956–967. [CrossRef] [PubMed]

3. Westra, H.J.; Peters, M.J.; Esko, T.; Yaghootkar, H.; Schurmann, C.; Kettunen, J.; Christiansen, M.W.; Fairfax, B.P.; Schramm, K.; Powell, J.E.; et al. Systematic identification of trans eQTLs as putative drivers of known disease associations. *Nat. Genet.* **2013**, *45*, 1238–1243. [CrossRef] [PubMed]

4. Joehanes, R.; Zhang, X.; Huan, T.; Yao, C.; Ying, S.X.; Nguyen, Q.T.; Demirkale, C.Y.; Feolo, M.L.; Sharopova, N.R.; Sturcke, A.; et al. Integrated genome-wide analysis of expression quantitative trait loci aids interpretation of genomic association studies. *Genome Biol.* **2017**, *18*, 6. [CrossRef] [PubMed]

5. Zhernakova, D.V.; Deelen, P.; Vermaat, M.; van Iterson, M.; van Galen, M.; Arindrarto, W.; van 't Hof, P.; Mei, H.; van Dijk, F.; Westra, H.J.; et al. Identification of context-dependent expression quantitative trait loci in whole blood. *Nat. Genet.* **2017**, *49*, 139–145. [CrossRef]

6. Gonorazky, H.D.; Naumenko, S.; Ramani, A.K.; Nelakuditi, V.; Mashouri, P.; Wang, P.; Kao, D.; Ohri, K.; Viththiyapaskaran, S.; Tarnopolsky, M.A.; et al. Expanding the Boundaries of RNA Sequencing as a Diagnostic Tool for Rare Mendelian Disease. *Am. J. Hum. Genet.* **2019**, *104*, 466–483. [CrossRef]

7. Dancik, G.M. An online tool for evaluating diagnostic and prognostic gene expression biomarkers in bladder cancer. *BMC Urol.* **2015**, *15*, 59. [CrossRef]

8. Parker, J.S.; Mullins, M.; Cheang, M.C.U.; Leung, S.; Voduc, D.; Vickery, T.; Davies, S.; Fauron, C.; He, X.; Hu, Z.; et al. Supervised risk predictor of breast cancer based on intrinsic subtypes. *J. Clin. Oncol. Off. J. Am. Soc. Clin. Oncol.* **2009**, *27*, 1160–1167. [CrossRef]

9. VanderWeele, T.J.; Tchetgen Tchetgen, E.J.; Cornelis, M.; Kraft, P. Methodological challenges in mendelian randomization. *Epidemiology* **2014**, *25*, 427–435. [CrossRef]

10. Singh, A.J.; Ramsey, S.A.; Filtz, T.M.; Kioussi, C. Differential gene regulatory networks in development and disease. *Cell. Mol. Life Sci.* **2018**, *75*, 1013–1025. [CrossRef]

11. Gov, E.; Arga, K.Y. Differential co-expression analysis reveals a novel prognostic gene module in ovarian cancer. *Sci. Rep.* **2017**, *7*, 4996. [CrossRef]

12. Zhu, L.; Ding, Y.; Chen, C.-Y.; Wang, L.; Huo, Z.; Kim, S.; Sotiriou, C.; Oesterreich, S.; Tseng, G.C. MetaDCN: Meta-analysis framework for differential co-expression network detection with an application in breast cancer. *Bioinformatics* **2016**, *33*, 1121–1129. [CrossRef] [PubMed]

13. Hou, L.; Chen, M.; Zhang, C.K.; Cho, J.; Zhao, H. Guilt by rewiring: Gene prioritization through network rewiring in genome wide association studies. *Hum. Mol. Genet.* **2014**, *23*, 2780–2790. [CrossRef] [PubMed]

14. Tuder, R.M.; Groves, B.; Badesch, D.B.; Voelkel, N.F. Exuberant endothelial cell growth and elements of inflammation are present in plexiform lesions of pulmonary hypertension. *Am. J. Pathol.* **1994**, *144*, 275–285. [PubMed]

15. Galie, N.; Humbert, M.; Vachiery, J.L.; Gibbs, S.; Lang, I.; Torbicki, A.; Simonneau, G.; Peacock, A.; Vonk Noordegraaf, A.; Beghetti, M.; et al. 2015 ESC/ERS Guidelines for the diagnosis and treatment of pulmonary hypertension: The Joint Task Force for the Diagnosis and Treatment of Pulmonary Hypertension of the European Society of Cardiology (ESC) and the European Respiratory Society (ERS): Endorsed by: Association for European Paediatric and Congenital Cardiology (AEPC), International Society for Heart and Lung Transplantation (ISHLT). *Eur. Respir. J.* **2015**, *46*, 903–975. [PubMed]

16. Rhodes, C.J.; Batai, K.; Bleda, M.; Haimel, M.; Southgate, L.; Germain, M.; Pauciulo, M.W.; Hadinnapola, C.; Aman, J.; Girerd, B.; et al. Genetic determinants of risk in pulmonary arterial hypertension: International genome-wide association studies and meta-analysis. *Lancet Respir. Med.* **2019**, *7*, 227–238. [CrossRef]

17. Graf, S.; Haimel, M.; Bleda, M.; Hadinnapola, C.; Southgate, L.; Li, W.; Hodgson, J.; Liu, B.; Salmon, R.M.; Southwood, M.; et al. Identification of rare sequence variation underlying heritable pulmonary arterial hypertension. *Nat. Commun.* **2018**, *9*, 1416. [CrossRef]

18. Rhodes, C.J.; Otero-Nunez, P.; Wharton, J.; Swietlik, E.M.; Kariotis, S.; Harbaum, L.; Dunning, M.J.; Elinoff, J.M.; Errington, N.; Thompson, A.A.R.; et al. Whole-Blood RNA Profiles Associated with Pulmonary Arterial Hypertension and Clinical Outcome. *Am. J. Respir. Crit. Care Med.* **2020**, *202*, 586–594. [CrossRef]

19. Finotello, F.; Mayer, C.; Plattner, C.; Laschober, G.; Rieder, D.; Hackl, H.; Krogsdam, A.; Loncova, Z.; Posch, W.; Wilflingseder, D.; et al. Molecular and pharmacological modulators of the tumor immune contexture revealed by deconvolution of RNA-seq data. *Genome Med.* **2019**, *11*, 34. [CrossRef]

20. Chang, C.C.; Chow, C.C.; Tellier, L.C.A.M.; Vattikuti, S.; Purcell, S.M.; Lee, J.J. Second-generation PLINK: Rising to the challenge of larger and richer datasets. *GigaScience* **2015**, *25*. [CrossRef]

21. Band, G.; Marchini, J. Qctool. Available online: https://www.well.ox.ac.uk/~{}gav/qctool/index.html (accessed on 10 September 2020).

22. QUICKTEST. Available online: https://wp.unil.ch/sgg/program/quicktest/ (accessed on 10 September 2020).

23. Aguet, F.; Barbeira, A.N.; Bonazzola, R.; Brown, A.; Castel, S.E.; Jo, B.; Kasela, S.; Kim-Hellmuth, S.; Liang, Y.; Oliva, M.; et al. The GTEx Consortium atlas of genetic regulatory effects across human tissues. *bioRxiv* **2019**, 787903v1. [CrossRef]

24. Buniello, A.; MacArthur, J.A.L.; Cerezo, M.; Harris, L.W.; Hayhurst, J.; Malangone, C.; McMahon, A.; Morales, J.; Mountjoy, E.; Sollis, E.; et al. The NHGRI-EBI GWAS Catalog of published genome-wide association studies, targeted arrays and summary statistics 2019. *Nucleic Acids Res.* **2019**, *8*, D1005–D1012. [CrossRef] [PubMed]

25. Wang, J.; Vasaikar, S.; Shi, Z.; Greer, M.; Zhang, B. WebGestalt 2017: A more comprehensive, powerful, flexible and interactive gene set enrichment analysis toolkit. *Nucleic Acids Res.* **2017**, *45*, W130–W137. [CrossRef] [PubMed]

26. Khatri, P.; Sirota, M.; Butte, A.J. Ten Years of Pathway Analysis: Current Approaches and Outstanding Challenges. *PLoS Comput. Biol.* **2012**, *8*, e1002375. [CrossRef]

27. Kanehisa, M.; Araki, M.; Goto, S.; Hattori, M.; Hirakawa, M.; Itoh, M.; Katayama, T.; Kawashima, S.; Okuda, S.; Tokimatsu, T.; et al. KEGG for linking genomes to life and the environment. *Nucleic Acids Res.* **2007**, *36*, D480–D484. [CrossRef]

28. Powell, J.E.; Henders, A.K.; McRae, A.F.; Wright, M.J.; Martin, N.G.; Dermitzakis, E.T.; Montgomery, G.W.; Visscher, P.M. Genetic control of gene expression in whole blood and lymphoblastoid cell lines is largely independent. *Genome Res.* **2012**, *22*, 456–466. [CrossRef]

29. Barreiro, L.B.; Tailleux, L.; Pai, A.A.; Gicquel, B.; Marioni, J.C.; Gilad, Y. Deciphering the genetic architecture of variation in the immune response to Mycobacterium tuberculosis infection. *Proc. Natl. Acad. Sci. USA* **2012**, *109*, 1204. [CrossRef]

30. Fairfax, B.P.; Humburg, P.; Makino, S.; Naranbhai, V.; Wong, D.; Lau, E.; Jostins, L.; Plant, K.; Andrews, R.; .McGee, C.; et al. Innate Immune Activity Conditions the Effect of Regulatory Variants upon Monocyte Gene Expression. *Science* **2014**, *343*, 1246949. [CrossRef]

31. Çalışkan, M.; Baker, S.W.; Gilad, Y.; Ober, C. Host Genetic Variation Influences Gene Expression Response to Rhinovirus Infection. *PLoS Genet.* **2015**, *11*, e1005111. [CrossRef]

32. Zhu, Z.; Wang, X.; Li, X.; Lin, Y.; Shen, S.; Liu, C.L.; Hobbs, B.D.; Hasegawa, K.; Liang, L.; Boezen, H.M.; et al. Genetic overlap of chronic obstructive pulmonary disease and cardiovascular disease-related traits: A large-scale genome-wide cross-trait analysis. *Respir. Res.* **2019**, *20*, 64. [CrossRef]

33. Kichaev, G.; Bhatia, G.; Loh, P.-R.; Gazal, S.; Burch, K.; Freund, M.K.; Schoech, A.; Pasaniuc, B.; Price, A.L. Leveraging Polygenic Functional Enrichment to Improve GWAS Power. *Am. J. Hum. Genet.* **2019**, *104*, 65–75. [CrossRef] [PubMed]

34. Evangelou, E.; Warren, H.R.; Mosen-Ansorena, D.; Mifsud, B.; Pazoki, R.; Gao, H.; Ntritsos, G.; Dimou, N.; Cabrera, C.P.; Karaman, I.; et al. Genetic analysis of over 1 million people identifies 535 new loci associated with blood pressure traits. *Nat. Genet.* **2018**, *50*, 1412–1425. [CrossRef] [PubMed]

35. Merla, G.; Brunetti-Pierri, N.; Piccolo, P.; Micale, L.; Loviglio Maria, N. Supravalvular Aortic Stenosis. *Circul. Cardiovasc. Genet.* **2012**, *5*, 692–696. [CrossRef]

36. Rabinovitch, M.; Guignabert, C.; Humbert, M.; Nicolls, M.R. Inflammation and Immunity in the Pathogenesis of Pulmonary Arterial Hypertension. *Circul. Res.* **2014**, *115*, 165–175. [CrossRef] [PubMed]

37. Kumar, R.; Graham, B. How does inflammation contribute to pulmonary hypertension? *Eur. Respir. J.* **2018**, *51*, 1702403. [CrossRef]

38. Andiappan, A.K.; Melchiotti, R.; Poh, T.Y.; Nah, M.; Puan, K.J.; Vigano, E.; Haase, D.; Yusof, N.; San Luis, B.; Lum, J.; et al. Genome-wide analysis of the genetic regulation of gene expression in human neutrophils. *Nat. Commun.* **2015**, *6*, 7971. [CrossRef]

Publisher's Note: MDPI stays neutral with regard to jurisdictional claims in published maps and institutional affiliations.

Review

DNA Damage and Repair in Pulmonary Arterial Hypertension

Samantha Sharma and Micheala A. Aldred *

Division of Pulmonary, Critical Care, Sleep & Occupational Medicine, Department of Medicine,
Indiana University School of Medicine, Indianapolis, IN 46202, USA; samshar@iu.edu
* Correspondence: maaldred@iu.edu; Tel.: +1-317-278-7716

Received: 19 September 2020; Accepted: 13 October 2020; Published: 19 October 2020

Abstract: Pulmonary arterial hypertension (PAH) is a complex multifactorial disease with both genetic and environmental dynamics contributing to disease progression. Over the last decade, several studies have demonstrated the presence of genomic instability and increased levels of DNA damage in PAH lung vascular cells, which contribute to their pathogenic apoptosis-resistant and proliferating characteristics. In addition, the dysregulated DNA damage response pathways have been indicated as causal factors for the presence of persistent DNA damage. To understand the significant implications of DNA damage and repair in PAH pathogenesis, the current review summarizes the recent advances made in this field. This includes an overview of the observed DNA damage in the nuclear and mitochondrial genome of PAH patients. Next, the irregularities observed in various DNA damage response pathways and their role in accumulating DNA damage, escaping apoptosis, and proliferation under a DNA damaging environment are discussed. Although the current literature establishes the pertinence of DNA damage in PAH, additional studies are required to understand the temporal sequence of the above-mentioned events. Further, an exploration of different types of DNA damage in conjunction with associated impaired DNA damage response in PAH will potentially stimulate early diagnosis of the disease and development of novel therapeutic strategies.

Keywords: pulmonary arterial hypertension; endothelial cells; smooth muscle cells; DNA damage; DNA repair

---

1. Introduction

Pulmonary arterial hypertension (PAH) is a potentially fatal vascular disease of enigmatic origin [1]. Following recent revision at the 6th World Symposium on Pulmonary Hypertension, it is now defined as a mean pulmonary artery pressure greater than 20 mmHg at rest, pulmonary artery wedge pressure of ≤15 mmHg, and pulmonary vascular resistance ≥3 Wood Units [2]. It is sub-divided into the following categories: heritable PAH (HPAH) with a known genetic mutation and/or a family history; idiopathic (IPAH) with no known cause; associated PAH (APAH) that occurs in concert with a predisposing condition, such as connective tissue disorder (CTD), congenital heart disease (CHD), or ingestion of certain drugs and toxins; and PAH with overt features of venous/capillary involvement (PVOD/PCH) [2]. PAH is characterized by remodeling of the pre-capillary arterioles that is directed by endothelial cell (EC) dysfunction, abnormal smooth muscle cell (SMC) proliferation, and proinflammatory cytokines [2–6]. These vascular abnormalities lead to progressive obliteration of the proximal pulmonary arterioles and pruning of the distal microvessels, thus increasing resistance in the pulmonary artery (PA) and ultimately leading to right heart failure [1,4,5]. Current therapies provide symptomatic relief; however, the 5-year survival rate of PAH remains below 60%, with lung transplantation being the only curative option [4,7–9]. A better understanding of the early molecular events in PAH pathogenesis would potentially aid earlier diagnosis and the development of targeted therapies.

PAH is a complex multifactorial disease with both genetic and environmental factors contributing to the pathogenesis. HPAH is inherited as an autosomal dominant trait with reduced penetrance. Mutations in the bone morphogenetic protein receptor II gene (*BMPR2*) are by far the predominant cause [10–12], but recent whole-exome and genome sequencing efforts have identified numerous other genes, including several other members of the BMP pathway [12–17]. However, the low penetrance of these mutations suggests that additional factors are required to trigger the disease. Such events may include other aberrations within the genome [18,19], and/or environmental stressors such as hypoxia or inflammation [5,20–25].

Over the last two decades, several reports have highlighted a role for DNA damage in PAH pathogenesis. The primary cause of this DNA damage remains uncertain. Depending on the type of DNA damage, the cell manifests a specific DNA damage response (DDR) cascade that is responsible for the recognition, signaling, and repair process [26]. DDR is accompanied by increased nucleotide synthesis, transcriptional and epigenetic changes, and metabolic amendments via higher consumption of glucose [26,27]. These alterations disrupt the cellular homeostasis and prompt the cell's signaling network to determine the cell's survival fate. In this review, we discuss the hallmark discoveries surrounding DNA damage and associated repair pathways to understand their potential roles in PAH vascular remodeling.

2. Background: DNA Damage and Response Pathways

Continuous exposure of genomic DNA to cellular metabolites and exogenous agents can damage the structural integrity of DNA. These alterations can range from a single base to complex structural changes. Based on the type of DNA damage, relevant DDR pathways are activated [27]. These responses work towards (a) restoration of the DNA duplex, (b) activation of DNA damage checkpoint kinase 1 and 2 (CHK1 and CHK2) to prevent transmission of damaged DNA, (c) transcriptional alterations to maintain cellular health, and (d) apoptosis signaling if the damage is unrepairable (Figure 1a) [26].

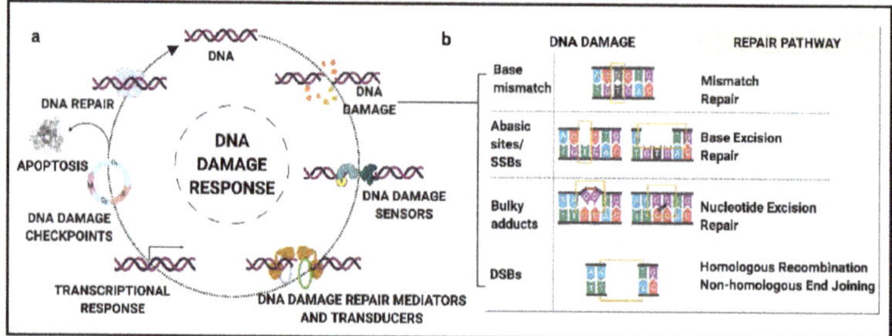

Figure 1. Pictorial representation of (**a**) DNA damage response and (**b**) types of DNA damage and their repair pathways. (**a**) DNA damage response includes the recognition of DNA damage by DNA damage sensors, recruitment of DNA damage repair mediators and transducers, modulation of transcriptional response, activation of DNA damage checkpoint kinases, and lastly, restoration of DNA duplex or apoptosis signaling if the damage is unrepairable. (**b**) Types of DNA damage include base mismatches repaired via mismatch repair pathway, abasic site or SSBs repaired via base excision repair, bulky adducts repaired via nucleotide excision repair, and DSBs repaired via homologous recombination or non-homologous end joining. DNA, deoxyribonucleic acid; SSBs, single-stranded DNA breaks; DSBs, double-stranded DNA breaks. Created with BioRender.com.

DNA damage includes single-stranded DNA breaks (SSBs), abasic sites, modified bases, double-stranded DNA breaks (DSBs), and inter- and intra-strand crosslinks (Figure 1b) [26]. Abasic sites, SSBs, and DSBs fall into the category of DNA backbone damage, which is the most frequent type of DNA damage [26]. Abasic or AP sites (apurinic/apyrimidinic site) are characterized by the absence of a single base from the DNA backbone. SSBs are the type of DNA damage that affect only one strand of the duplex DNA, characterized by gaps in the range of up to 30 nucleotides. DSBs, or damage to both the strands of DNA, are the most lethal to DNA integrity, with the capability to generate large chromosomal aberrations. Certain chemical agents are known to generate intra- and inter-strand DNA-DNA crosslinks that can halt transcription or replication machinery [27].

DDR initiates DNA repair pathways that replace potentially damaged sites with newly synthesized DNA via base excision or recombination mechanism [26]. In humans, based on the type of DNA damage, the DNA repair mechanism can be classified into five major types (Figure 1b):

(a) Mismatch Repair (MMR): MMR is responsible for the recognition and repair of base mismatches. Base mismatches can arise as a result of covalent or non-covalent structural changes, or due to insertion/deletions resulting from replication errors or recombination [28]. For example, methylated guanine base, O6MeGua, has a high frequency of pairing to thymine (T), activating MMR to excise the mismatched T residue. Loss of MMR can lead to a significant increase in spontaneous mutations. Major known genes in the MMR pathway include *MGMT*, *MSH6*, and *MLH3*.

(b) Base Excision Repair (BER): This repair process is governed by DNA glycosylases along with endonucleases that recognize and eliminate the modified or damaged bases, such as oxidized, reduced, alkylated or deaminated bases, to generate an abasic site [29]. For example, in humans, 8-oxoguanine glycosylase-1 (OGG1) recognizes and removes the oxidatively modified guanine base, 8-oxoGuanine (8-oxoG) via incision of the 3′-phosphodiester bond. Following this step, the apurinic/apyrimidinic endodeoxyribonuclease 1 (APEX1) cleaves the 5′-bond generating a 1-nt abasic site [30]. Major genes of the BER pathway include *MBD4*, *OGG1*, *MUTYH*, and *NEIL1*.

(c) Nucleotide Excision Repair (NER): Unlike BER, NER involves a complex of enzymes that work in coordination to recognize SSBs and remove bulky lesions [31]. Briefly, the steps include recognition of the damaged site, a dual incision at extreme ends of the lesion, elimination of damaged oligomer, and new base synthesis followed by ligation [32]. Major known NER genes include *XPC*, *XPA*, and *ERCC1-5*.

(d) Homologous Recombination (HR): As compared to the excision repair pathways, HR is a far more complex phenomenon. HR involves multiple-step processing of DSBs by several different proteins with specific functions [33]. The key characteristic of HR is that it uses a homologous duplex template to retrieve the lost information. It is a complex phenomenon, with the potential for incorrect template usage that can lead to gene conversion. Major genes involved in the HR pathway include *RAD51*, *BRCA1*, *BRCA2*, and the Mre11/Rad50/NBS1 complex [34].

(e) Non-Homologous End Joining (NHEJ): Similar to HR, NHEJ involves multiple-step repair processing of DSBs. In this mechanism, the two ends of DSBs are stabilized by DNA-protein kinases and ligated together [35]. It is believed to be the main repair pathway for DSBs induced by ionizing radiation. Major proteins implicated in NHEJ include KU70/80 heterodimer and XRCC4 [36,37]. A lack of specific recognition criteria for the ligated ends can lead to erroneous joining of non-contiguous DNA sequences, giving rise to structural rearrangements.

After the initiation of the DNA repair pathway, the cell deploys DNA repair checkpoint kinases, CHK1 and CHK2, that delay or inhibit the DDR-associated cell cycle progression. Checkpoint kinases arrest cell cycle until the repair process is complete, avoiding transmission of damaged DNA to the daughter cells [38]. This step is a significant player in DDR as it halts the replication of damaged DNA and ensures the transmission of intact healthy DNA. If the DNA damage cannot be repaired, checkpoint kinases initiate apoptosis signaling, leading to cell death. Owing to their pertinence in DDR, checkpoint-specific damage sensors, ataxia telangiectasia mutated (ATM) and Ataxia Telangiectasia and Rad3-Related Protein (ATR), have received the highest recognition. Checkpoint kinases also regulate the biochemical pathways that guard different steps of the cell cycle and hence play a vital role in DDR.

3. DNA Damage and Genetic Instability in PAH

In 1998, it was first discovered that pulmonary artery ECs (PAECs) within the majority of plexiform lesions microdissected from IPAH lung tissues were monoclonal, suggesting that each lesion arose from the proliferation of a single EC [39,40]. Similar results were obtained in lungs from patients with appetite suppressant-associated PH, whereas lesions from the lungs of patients with CHD-PAH or CTD-PAH were polyclonal [39,40]. Thereafter, subsequent studies focused on exploring the mechanisms that confer such ECs with a unique selective growth advantage. The similarity of PAECs monoclonal expansion with neoplasia anticipated the role of an unstable genome in PAECs that favors the disruption of the apoptotic signals [41]. Soon after, the first paper to report microsatellite instability within the PAECs from plexiform lesions of PAH patients affirmed this hypothesis [42]. PAH patients reportedly had microsatellite instabilities within the transforming growth factor-β receptor II (*TGFBR2*) and BCL-2 associated X, apoptosis regulator (*BAX*) genes, known to regulate cell proliferation and apoptosis, respectively. Further, mutations within these microsatellite sites resulted in truncated proteins, thereby producing lower levels of the functional proteins. Microsatellite instability, or a condition of genetic hypermutability, results from the loss of functional DNA mismatch repair process. Hence, these quasi-neoplastic PAECs bearing unstable microsatellite mutations suggested the relevance of DNA damage and DNA repair regulation in the pathophysiology of PAH.

Subsequent to the identification of *BMPR2* mutations in heritable PAH, Machado et al. tested the hypothesis that this gene might follow a classical two-hit tumor suppressor model, with somatic loss of the wildtype allele in lung vascular lesions triggering disease onset [43]. However, careful microdissection and genetic analysis disproved this hypothesis, suggesting that if somatic mutations existed, they lie elsewhere in the genome. Our lab then performed a genome-wide microarray copy number analysis in cultured PAECs and PASMCs isolated from the lungs of patients with idiopathic, heritable, and associated forms of PAH [44]. Chromosomal deletions were identified in PAECs from five of the nine cases studied and validated in the uncultured lung tissue. These aberrations were not detectable in paired DNA samples from blood or other lung cell types, confirming that they were somatic events. In the same study, a second-hit mechanism became evident when a patient with a germline *BMPR2* mutation was found to harbor somatic loss of one copy of chromosome-13, which deleted one of the BMP signaling transducer genes, SMAD family member 9 (*SMAD9*) [44]. Subsequently, a detailed analysis of an interstitial chromosome-13 deletion in PAEC from an APAH patient revealed dysregulated BMP signaling, similar to that seen in HPAH cells bearing a germline *SMAD9* mutation [45]. In contrast, PASMC from the same patient did not carry the deletion and showed normal BMP signaling. Validation of chromosomal abnormalities in a larger sample size confirmed a significant excess of copy number changes in PAH-PAECs (30.9%) as compared to the control-PAECs (5.3%), whereas the frequency in PASMCs did not differ between patients and controls [46].

These studies identified a surprisingly high level of DNA damage in the lung tissues of PAH patients. However, its relevance to PAH pathophysiology was yet to be understood. Is increased DNA damage an intrinsic property of a cell acting as a disease driver, or simply an end-stage consequence in PAH pathogenesis? To answer this question, we directly measured DNA damage in PAECs and blood cells using micronucleus assay and immunocytochemistry for phosphorylated Histone H2a Family Member X (γH2AX), a marker of double-strand break repair. PAH cells showed higher levels of DNA damage than controls, both in PAEC and in blood cells [46]. Reactive oxygen species (ROS) levels were also elevated in PAH cells. Treatment with antioxidants reduced the level of DNA damage to a similar baseline as control cells, suggesting that excess DNA damage in PAH cells may be due, at least in part, to oxidative stress. Notably, similar results were shown in blood cells from healthy first-degree relatives of PAH patients, suggesting there may be an intrinsic genetic or epigenetic basis [46].

Overall, these studies suggested ROS and DNA damage as biomarkers for PAH susceptibility across multiple PAH sub-groups, including IPAH, HPAH, and APAH. This is supported by evidence of increased DNA damage in PAECs from amphetamine-PAH lungs [47], as well as pulmonary microvascular ECs (PMVECs) from IPAH patients [48]. Furthermore, although PASMCs did not show significant evidence of chromosomal abnormalities [46], they do also exhibit higher levels of DNA damage than control cells [49].

4. Role of Mutagens and Environmental Modifiers

With the cellular environment as a key factor, several studies established low oxygen levels, or hypoxia, as a powerful stimulant for pathological conditions [21,24]. With relevance to DNA damage, hypoxia is reported to downregulate a cell's DNA repair machinery, leading to an increased prevalence of genomic damage [50–53]. Amphetamines, a potent synthetic neuro-stimulant, reportedly have vasoconstrictive and mutagenic properties in vascular cells [54]. Further, oxidant injury (caused by a hypoxic environment) enhances their neurotoxic effects, proposing amphetamines as a potent trigger for vasculopathy [55]. It has been reported that people exposed to amphetamine have a 3-fold higher risk factor for the development of PAH [54]. To understand the mechanism of amphetamine-associated PAH, Chen and colleagues performed an elaborate translational study in PAECs from drug and toxin-induced PAH patients [47]. In addition to elevated levels of baseline DNA damage, PAECs derived from amphetamine-PAH lungs were also more susceptible to genotoxins as compared to the controls. Further, the genotoxic effect of doxorubicin was exaggerated under hypoxic conditions and persisted even after recovery under normoxic conditions. In vivo, animals treated with amphetamines demonstrated increased DNA damage, but no significant change in hemodynamics was observed, supporting the involvement of additional factors causing the pathogenic vulnerability in the vascular cells [47].

Other groups have also reported an augmented susceptibility to mutagens in PAH cells, including etoposide, bleomycin, and hydroxyurea [46,48]. Topoisomerase-II binding protein 1 (TOPBP1) plays a role in the rescue of stalled replication forks and checkpoint control, binding both dsDNA and ssDNA breaks. It is downregulated in PMVECs from IPAH patients, which showed evidence of increased DNA damage and apoptosis [48]. Common single-nucleotide variants in the *TOPBP1* gene modified the susceptibility of normal PMVECs to hydroxyurea, and it was proposed as a novel gene in IPAH [48]. However, this has not been validated in a recent large genome-wide association study [56].

The most direct evidence that mutagens can precipitate the development of pulmonary hypertension comes from pulmonary veno-occlusive disease (PVOD), a rare form of PH with an especially poor prognosis. Perros and colleagues reported that cancer patients undergoing treatment with mitomycin-C (MMC) had a significantly higher annual incidence of PVOD (3.9 out of 1000) as compared to the general population (0.5 per million) [57]. In vivo treatment of rats with MMC induced severe pulmonary vascular resistance and right ventricular (RV) hypertrophy accompanied by vascular remodeling and EC proliferation in the capillary bed. At the expression level, MMC induced depletion of GCN2 in rats. Interestingly, GCN2 loss has been reported to promote oxidative stress and inflammatory-mediated DNA damage [58], a proposed pathogenic setup for PAH development. GCN2, encoded by the gene eukaryotic translation initiation factor 2α kinase 4 (*EIF2AK4*), is biallelically inactivated in familial PVOD/PCH, which is inherited as an autosomal recessive trait [59,60]. Based on these observations, it can be hypothesized that MMC, an alkylating agent, can directly trigger pulmonary vascular dysfunction via DNA damage signaling, and that a similar mechanism may underlie familial PVOD/PCH with biallelic inactivation of *EIF2AK4* (Figure 2).

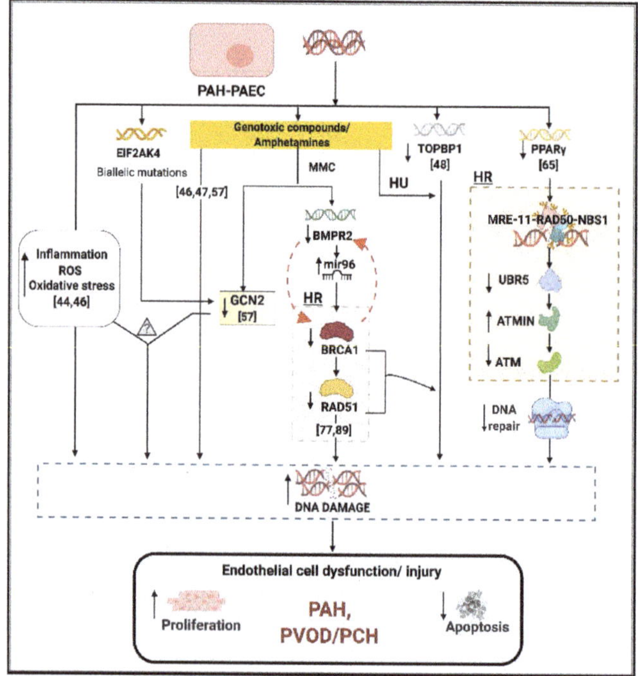

Figure 2. Dysregulated DNA damage response genes involved in pro-proliferative and apoptosis resistance characteristics of PAH-PAECs. Increased inflammation, ROS, oxidative stress, and susceptibility to genotoxic compounds are associated with increased DNA damage in PAH-PAECs. Reduced expression of PPARγ and BMPR2 in PAH-PAECs downregulates the downstream HR DNA repair genes, ATM and RAD51, respectively. The BMPR2-BRCA1-RAD51 pathway is also dysregulated in presence of MMC, a genotoxin. MMC is also proposed to trigger pulmonary vascular dysfunction associated with DNA damage signaling via reduced GCN2 expression, leading to PVOD. A similar mechanism may contribute to familial PVOD/PCH, where GCN2 is inactivated by biallelic *EIF2AK4* mutations. Similarly, reduced expression of TOPBP1, a DNA repair gene, in PAH-PAECs is reported to associate with PAH pathogenesis. These factors cause increased DNA damage, leading to endothelial cell dysfunction/injury in PAH and PVOD/PCH. The question mark symbol represents that the exact mechanism of this pathway in inducing the disease is still uncertain. DNA, Deoxyribonucleic acid; ROS, Reactive oxygen species; EIF2AK4, Eukaryotic translation initiation factor 2α kinase 4; GCN2, eIF-2α kinase GCN2; MMC, Mitomycin-C; BMPR2, bone morphogenetic protein receptor type 2; miR96, microRNA-96; HR, Homologous recombination; BRCA1, Breast and ovarian cancer susceptibility protein 1; RAD51, RAD51 recombinase; HU, Hydroxyurea; TopBP1, DNA Topoisomerase II Binding Protein 1; PPARγ, Peroxisome proliferator-activated receptor γ; MRE11, Double-strand break repair protein MRE11; RAD50, RAD50 double-strand break repair protein; NBS1, Nibrin; UBR5, Ubiquitin protein ligase e3 component n-recognin 5; ATMIN, ATM interacting protein; ATM, Ataxia telangiectasia mutated; PAH, Pulmonary arterial hypertension; PVOD, Pulmonary veno-occlusive disease; PCH, Pulmonary capillary hemangiomatosis. The numbers represented in square brackets cite the reference for that study. Created with BioRender.com.

5. DNA Repair Pathways and Cell Cycle Checkpoints in PAH

Despite the relatively high frequency of chromosomal abnormalities in PAH PAECs, clonal analysis of endothelial colony-forming cells from these cultures showed that the genome remained grossly stable for up to 15 passages [61]. This suggests there is no major defect in DNA repair, and indeed it may even be enhanced in PAH cells. As we review below, most studies of DNA repair to date have been performed in PASMCs, with relatively little information known about PAECs. However, the consensus thus far suggests opposite effects in these two cell types, with evidence of decreased repair in PAH-PAECs, but increased repair in PASMCs.

Peroxisome proliferator-activated receptor γ (PPAR-γ) is a nuclear receptor known to regulate fatty acid storage and glucose metabolism. It has been implicated in multiple diseases, including cancer, obesity, diabetes, and cardiovascular diseases, particularly in ECs [62]. Mice with EC-specific deletion of PPAR-γ develop PH under hypoxia, which is persistent after re-oxygenation [63]. In PAECs and PASMCs, PPAR-γ promotes cell survival and suppresses proliferation via interaction with Apelin [64]. Using an unbiased proteomics approach, Li and colleagues showed that PPAR-γ interacts with MRN (MRE-11-RAD50-NBS1), promoting ATM signaling, and is also essential for UBR5 activity targeting ATM interactor (ATMIN) [65]. Hence, upon DNA damage, ATMIN is degraded by UBR5, leading to ATM activation. However, this axis is dysfunctional in PAH-PAECs, with reduced PPAR-γ-UBR5 interaction, elevated ATMIN which leads to progressive DNA damage, and impaired repair in these cells (Figure 2). Consistent with the in vitro pathway findings, the results were validated in PAH-PAECs and PAH-lung tissues [65]. Furthermore, the reduction of ATMIN in PAH-PAECs reduced synthetic DNA damage comparable to the control cells. Overall, the study established a non-canonical pathway for DDR and DNA repair via PPAR-γ suggesting the importance of PPAR-γ in EC homeostasis and maintenance of genome integrity (Figure 2).

One of the first responder proteins to detect DNA damage is Poly [ADP-ribose] polymerase 1 (PARP1). Meloche and colleagues first reported decreased microRNA miR-223, increased PARP-1 expression and associated proliferation/apoptosis imbalance in PAH [66] (Figure 3). Expectedly, treatment with PARP-1 inhibitor, ABT-888, induced more DNA damage in PASMCs; however, it also initiated anti-proliferative and pro-apoptotic signaling via reversal of miR-204-dependent NFAT and Hif1-α levels (Figure 3) [49]. These observations were recapitulated in the in vivo experimental PAH setup where treatment with ABT-888 reversed the effects of monocrotaline (MCT) and Sugen-induced PAH, as represented by reduced PA pressure and RV hypertrophy. Overall, the study was the first to report the augmented DNA damage response pathway in PAH via PARP-1 activation. Based on the in vivo results, PARP-1 inactivation was proposed as a potential therapeutic marker. However, the myriad effects of PARP-1 in the regulation of stress-regulated cell signaling along with other potential DNA damage inducers like tumor necrosis factor-α (TNF-α) and interleukin-6 (IL-6) cannot be ruled out. The immediate effect of PARP-1 inhibitor on increased DNA damage may route cells towards a far more unfavorable cellular environment, encouraging a broader exploration of multifaceted PARP-1 in PAH pathogenesis.

In another study from the same group, Lampron and colleagues explored the role of Moloney murine leukemia provirus integration site (PIM1) [67], a regulator of NHEJ repair pathway gene in PAH [68]. Increased PIM1 expression was observed in PAH lungs and PASMCs as compared to the controls [67]. PIM1 inhibition itself did not increase DNA damage; however, it reduced the expression of its downstream target, KU70, encoded by the gene X-Ray Repair Cross Complementing 6 (*XRCC6*) that stabilizes the ends of DSBs. Overall, PIM1 reduction significantly reduced the DNA repair capacity of the cells by inhibiting the primary events involved in DNA repair. Treatment with PIM1 inhibitors SGI-1776 and TP-3654 affected proliferation and apoptosis in PAH-PASMCs in an anti-pathological manner (Figure 3) [67]. These observations were recapitulated in PH animal models, where pharmacological intervention in two experimental rat models with PIM1 inhibitors improved the hemodynamic characteristics, reduced vascular remodeling, reduced proliferation, and enhanced apoptosis [67]. As opposed to the previous pharmacological inhibitor of the DNA repair pathway studied by this group [66], PIM1 inhibition does not expose vascular cells to additional genetic insults.

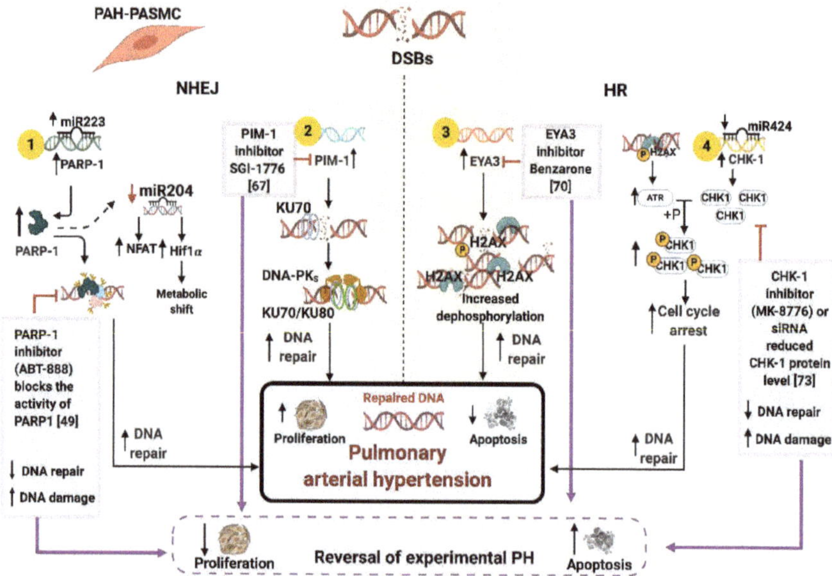

Figure 3. Dysregulated DNA damage response genes involved in pro-proliferative and apoptosis resistance characteristics of PAH-PASMCs. The figure represents the DDR genes and their downstream pathways implicated in PAH pathogenesis (1) PARP-1, (2) PIM-1, (3) EYA3, and (4) CHK1. These genes are upregulated in PAH-PASMCs leading to increased DNA repair and proliferation and reduced apoptosis. Studies have shown specific pharmacological inhibitors like ABT-888 (PARP1 inhibitor), SGI-1776 (PIM1 inhibitor), Benzarone (EYA3 inhibitor), and MK-8776 (CHK1 inhibitor) downregulate their target genes, leading to reduced proliferation and increased apoptosis, suggesting their therapeutic potential in PAH. DSBs, Double-stranded breaks; NHEJ, Non-Homologous End Joining; HR, Homologous Recombination; DSBs, Double-strand breaks; PARP1, Poly (ADP-Ribose) Polymerase 1; PIM1, Proto-Oncogene Serine/Threonine Kinase; EYA-3, Eyes Absent Homolog 3; CHK-1, Checkpoint Kinase 1; KU70, X-Ray Repair Cross Complementing 6; KU80, X-Ray Repair Cross Complementing 5; NFAT, Nuclear Factor of Activated T-Cells; HIF1α, Hypoxia-Inducible Factor 1-α; H2AX, H2a Histone Family Member X; P, phosphoryl group; ATR, Ataxia Telangiectasia and Rad3-Related Protein; miR223, microRNA 223; miR204, microRNA 204; miR424, microRNA 223; DNA-PKs, DNA-Dependent Protein Kinases; PAH, Pulmonary Arterial Hypertension. The numbers represented in square brackets cite the reference for that study. Created with BioRender.com.

The majority of the studies focused on H2AX to quantify the repair of DSBs. H2AX is the central player that binds to the DSBs, and the status of phosphorylation and dephosphorylation of H2AX governs the assembly of DDR complexes [69]. The phosphorylation of H2AX at Tyrosine-142 (Y142) residue is constitutive. Under a DNA-damaging environment, H2AX is phosphorylated at Serine-139 (S139) residue, allowing cells to initiate the DNA damage response. To initiate the repair machinery, EYA3, a tyrosine phosphatase, dephosphorylates the Y142 residue, permitting the assembly of repair complexes at the site of DNA damage. If Y142 is not de-phosphorylated, the cell initiates apoptosis [69]. In this context, Wang et al., reported increased levels of EYA3 protein in PAH-PASMCs and distal pulmonary arteries, suggesting an elevated repair mechanism in PAH pathogenesis (Figure 3) [70]. They also showed that in the presence of hydrogen peroxide, EYA3 provides a protective mechanism aiding in PASMC survival which was completely reversed in a setting with attenuated EYA3 expression. The treatment of cells with the small molecule inhibitor, benzarone, established the relationship between EYA3 tyrosine phosphatase activity and cell survival under a DNA damaging environment

(Figure 3) [70]. Inhibition of EYA3 tyrosine phosphatase activity in Sugen-hypoxia rats improved pulmonary hemodynamics and vascular remodeling [70]. Using genetically modified mice with an inactive EYA3 (Eya3^{D262N}), the authors showed that under hypoxic conditions, Eya3^{D262N} mice are protected from developing PH as opposed to the control mice under similar conditions [70].

In cancerous cells, checkpoint kinase-1 (CHK-1) acts as a nexus for cell survival under DNA damaging conditions by halting cell progression and triggering DNA repair [71,72]. Bourgeois and colleagues reported increased levels of checkpoint kinase-1 (CHK-1) in PAH-PASMCs and distal pulmonary arteries but not in PAH-PAECs (Figure 3) [73]. This expression correlated with the increased DNA damage markers, γH2AX and RPA32. Further, in PAH-PASMCs, the direct upstream activator of CHK-1, phospho-ATK [74,75], was upregulated, and mir-424, which targets CHK-1 [76], was also found to be downregulated. Upon treatment with DNA-damage-inducing compounds, an increase in CHK-1 levels was observed with anti-proliferative and pro-apoptosis characteristics (Figure 3) [73]. To understand the pharmacological effects, treatment of PAH-PASMCs with the CHK1 kinase inhibitor, MK-8776, exacerbated the DNA damage while controlling the proliferation and enhancing apoptosis. The results also validated with small interfering RNA (siRNA) revealed enhanced expression of CHK-1 upregulates RAD-51 [73]. This contrasts with the findings in PAECs in a separate study, where BMPR2-deficient PAECs showed a reduced level of RAD51 and lung tissue from IPAH patients had attenuated RAD51 levels [77]. In vivo studies supported the therapeutic potential of the CHK-1 inhibitor by reducing the hemodynamic parameters associated with increased DNA damage in fawn-hooded rats with already developed PAH; however, no reduction in RV hypertrophy was observed [73]. In a separate experiment on MCT rats, pharmacological inhibition of CHK-1 using MK-8776 revealed marginal improvement in hemodynamic parameters and enhanced reduction in vascular remodeling [73]. Overall, the study suggests that ubiquitous expression of CHK-1 allows the vascular cells to thrive under excessive DNA damaging conditions by providing a survival advantage [78].

6. DNA Damage and Mitochondria

Mitochondria serve an indispensable role in cellular health under various stresses. They have their own genome, a circular DNA molecule (mtDNA) of approximately 16,500 base pairs. Despite the higher levels of ROS, mtDNA repair mechanisms are reduced compared with the nucleus, increasing the likelihood of mtDNA damage. Several studies have reported abnormalities of mitochondrial function in PAH pathogenesis, including increased glycolytic metabolism, decreased mitochondrial copy number, and enhanced fission [79–83]. PARP1 plays a role in regulating mitochondrial energy metabolism and may therefore contribute to the alterations seen in PAH cells [84,85]. However, mitochondrial function is a very broad topic that has been extensively reviewed by others [83,86], and thus here we focus on the studies that directly relate to mitochondrial DNA damage.

Diebold and colleagues studied the role of BMPR2 signaling in mitochondrial DNA (mtDNA) damage and metabolism in apoptosis of PAECs in PAH [78]. Using human PAECs with siRNA-downregulated BMPR2 and PAH-PAECs with inherent *BMPR2* mutations, it was revealed that BMPR2 dysregulation can promote mtDNA damage when exposed to reoxygenation. Hypoxia-reoxygenation leads to reduced expression of TFAM and mitofusin 1 and 2 proteins. While TFAM is involved in mtDNA maintenance and repair, mitofusin 1 and 2 regulate mitochondrial fission. Consistent with their functions, enhanced accumulation of a 4977-bp mtDNA deletion and increased mitochondrial fission was observed in BMPR2-depleted PAECs [78] (Figure 4). This suggested that hypoxia-reoxygenation was a severe pathogenic trigger linked to the mitochondrial genetic abnormality. Overall, these studies imply that the genomic instability in PAH vasculature is not only confined to the nuclear genome but also impacts the mtDNA, which is far more susceptible to damage as compared to the nuclear genome [87].

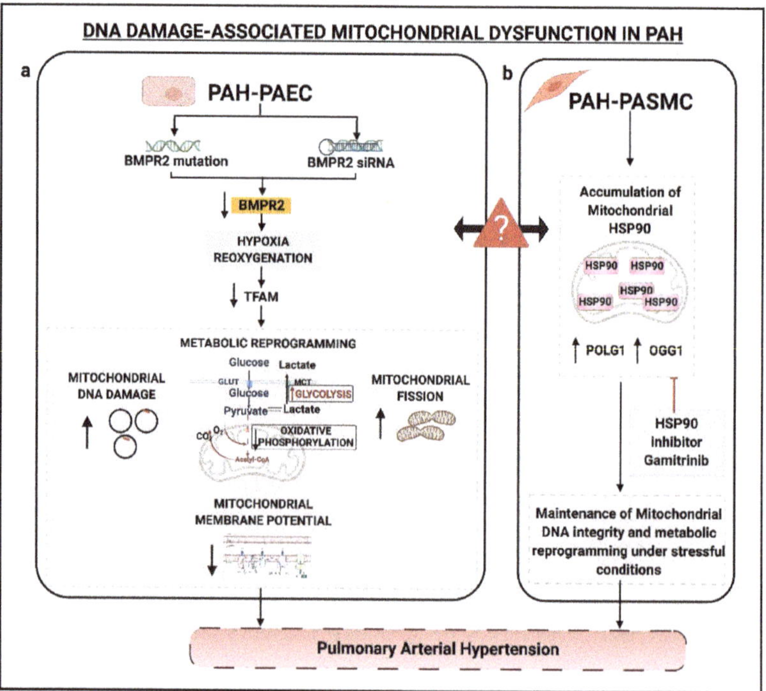

Figure 4. DNA damage-associated mitochondrial dysfunction in (**a**) PAH-PAEC and (**b**) PAH-PASMC. (**a**) In PAH-PAECs, reduced expression of BMPR2 accompanied by hypoxia-reoxygenation is associated with reduced expression of the mitochondrial DNA maintenance gene, TFAM, increased mitochondrial DNA damage, and decreased mitochondrial membrane potential. These factors, associated with reduced oxidative phosphorylation and increased glycolysis, trigger endothelial dysfunction in PAH-PAECs [78]. (**b**) In PAH-PASMCs, increased mitochondrial HSP90 accumulation along with an increased expression of POLG1 and OGG1 is proposed as a regulatory mechanism in the maintenance of mitochondrial DNA and metabolic reprogramming under stressful conditions [88]. The question mark symbol between the two cell types indicates a lack of studies showing crosstalk between the two cell types. DNA, Deoxyribonucleic acid; TFAM, Transcription Factor A Mitochondrial; GLUT, Glucose transporter; MCT, monocarboxylate transporter; O_2, Oxygen; CO_2, Carbon dioxide; HSP90, Heat shock protein 90; POLG1, Mitochondrial DNA polymerase γ; OGG1, 8-Oxoguanine glycosylase. Created with BioRender.com.

Boucherat et al. showed an important role of stress responder chaperone HSP90 in the maintenance of mitochondrial function [88]. Under a pathogenic PAH environment, HSP90 favorably localized within the mitochondria of PAH-PASMCs to preserve the mtDNA and its bioenergetic functions (Figure 4) [88]. Contrary to that, the cytosolic HSP90 displayed no conclusive role in PAH pathogenesis. On the same lines, in vitro and in vivo studies established a potential therapeutic role of Gamitrinib, a selective inhibitor of mtHSP90, in repressing PAH-PASMCs proliferation and initiating apoptosis, and reversing experimental PAH in MCT treated rats [88]. Within PAH-PASMCs, Gamitrinib reduced the endogenous overexpression of mtDNA maintenance genes, DNA polymerase γ (*POLG1*) and *OGG1* (Figure 4) [88]. OGG1, involved in BER, is pertinent to the elimination of oxidative DNA damage. Along with its role in mitochondrial maintenance, overexpression of OGG1 might implicate higher oxidative stress-induced damage in mtDNA, although this needs further validation.

7. BMPR2 and DNA Damage

Using a meta-analysis approach, Li and colleagues analyzed the expression profiles of IPAH (PASMCs and whole lung) and HPAH patients with *BMPR2* genetic variants (PAECs) to reveal 586 up-regulated and 372 down-regulated genes in PAH [89]. In silico analyses revealed significant enrichment of chromatin organization genes, predominantly regulated by the transcription factors, SP1 and NKX3. This also included over 35 genes involved in DNA repair. More importantly, both SP1 and NKX3 are well-known regulators of DSB repair. This observation suggests that the multifactorial channeling of cell signaling in PAH vasculopathy could be initiated by genomic instability. Similar to human expression profiles, an enrichment of DNA repair genes was also reported in rats in the setting of PAH. The authors went on to understand the relationship between two known hallmarks of PAH pathogenesis: altered BMP signaling and DNA damage. Using discrete in vitro methods, the authors established that MMC-induced DNA damage significantly downregulates BMPR2 expression and that BMPR2 is critical for DNA damage control in ECs [89]. However, these observations were confined to ECs; PASMCs showed no such correlation. Also, a transcription binding site for BRCA1, an important gene in cancer pathogenesis, was revealed using the ChIP assay and established a regulatory feed-back loop mechanism for *BMPR2* and *BRCA1*. Over the years, several theories and relevant studies have suggested a cancer-like pathogenesis in PAH. Establishing the interaction between BMPR2 and BRCA1 strengthened the link between cancer and PAH; however, the findings need to be corroborated with a larger dataset.

Consistent with the downregulation of BMPR2 under a DNA-damaging environment, Vattulainen-Collanus and colleagues uncovered a previously unknown role of BMP signaling in the protection of cells from genomic insult [77]. They showed that MMC treatment reduced the expression of BMPR2, BRCA1, and RAD51. Also, siRNA against BMPR2 reduced BRCA1 and RAD51 transcript and protein levels by over 50% and also lead to a 61% higher level of fragmented DNA. Further, activation of BMPR2 signaling by BMP9 partially rescues RAD51 and reduces sensitivity to DNA damage agents. At the epigenetic level, miR-96 is known to target the coding region of RAD51 and downregulate its expression [90]. Vattulainen-Collanus and colleagues hypothesized that repressing miR-96 via BMP9–BMPR2 signaling was critical for RAD51 expression and maintained pulmonary microvascular ECs homeostasis via managing DNA repair (Figure 2). To validate the in vitro results, the role of RAD-51 was studied in $BMPR2^{R899X/+}$ mice, which develop mild PH beyond 6 months [91]. It was observed that RAD51 expression was repressed in PMVECs and PASMCs extracted from $BMPR2^{R899X/+}$ mice as compared to control mice, suggesting a regulatory link between BMPR2 and RAD51 [77]. Similarly, lung sections from IPAH patients had a significant reduction of RAD51 in the endothelium of pulmonary arteries compared to control samples.

8. Conclusions and Future Directions

Over the last two decades, considerable progress has been made in understanding the pathobiology of PAH, with both genetic and environmental factors playing crucial roles in disease pathogenesis. It was very recently that the DNA damage and DDR malfunction was recognized. Several studies have reported persistent DNA damage (nuclear and mitochondrial) in the vascular cells across different types of PAH. Intriguingly, the increased DNA damage in blood cells of PAH patients and their first-degree relatives suggests that it may be a genetically-determined susceptibility factor that pre-dates disease initiation [46]. The increased incidence of PVOD following chemotherapy with MMC also suggests DNA damage as an early driver of pulmonary vascular remodeling [57]. Importantly, however, the types of DNA damage most prevalent in PAH and their cause(s) remain to be fully determined. Some studies suggest that the downregulation of BMPR2 may play a pivotal role in DNA damage [77,89], yet the incidence of DNA damage measured by micronucleus assay and γH2AX staining trended lower in HPAH cells compared with APAH-CHD [46], emphasizing that gaps in knowledge remain.

Beyond DNA damage, various studies focused on the status of DDR in PAH have ascertained a dysfunctional repair mechanism leading to an augmented apoptosis-resistant and proliferative phenotype [49,67,73,88]. Taken together, it is likely that the pathways underlying DDR play a key role in PAH. DDR involves a complex network of proteins that recognize different types of DNA damage and orchestrate the repair process. At present, there is not enough evidence to draw conclusions on the status of different DNA repair pathways in PAH. There are reports suggesting amplified DDR in PAH-PASMCs [49,67], while a few stand in contradiction (Table 1: OGG1, RAD51), reporting reduced DDR in PAH-PAECs [47,65,89] under different DNA damaging environments (Table 1). One of the major reasons behind this gap is a lack of paired cell types from the patients in the same study, as well as replication of the findings across different types of PAH. The absence of information on the underlying genetic mutations and polymorphisms that may act as genetic modifiers is an additional limitation, but this can be addressed fairly readily by more extensive use of whole-genome sequencing. PAH is a complex disease, and in this framework, it is important to get a comprehensive view of each variable of the disease to make significant progress.

Table 1. Genes associated with the DNA damage and response pathways in pulmonary arterial hypertension.

DNA Damage and Response Genes	PAH-PAECs	PAH-PASMCs
Base Excision Repair		
OGG1 (8-Oxoguanine DNA Glycosylase)	Not known	Increased expression [88] Reduced expression ** [67]
Homologous recombination		
RAD51 (RAD51 Recombinase)	Reduced expression [77]	Increased expression [73]
BRCA1 (Breast and Ovarian Cancer Susceptibility Protein 1)	Reduced expression [#] [77]	Reduced expression [89]
NBS1 (Nibrin)	Not known	Reduced expression ***
Non-homologous end-joining		
XRCC6 (Ku70) (X-Ray Repair Cross Complementing 6)	Not known	Reduced expression **
PARP-1 (Poly (ADP-Ribose) Polymerase 1)	Not known	Increased expression [49]
Other genes involved in regulation of DNA damage		
BMPR2	Reduced expression [#] [89]	No change in expression * [89]
TFAM	Reduced expression [78]	Not known
TOPBP1 (DNA Topoisomerase II Binding Protein 1)	Reduced expression [48]	Not known
PPARG-UBR5	No change in expression but reduced interaction observed [65]	Not known
ATMIN	Increased expression [65]	Not known
PIM1	Not known	Increased expression [67]
EYA3	Not known	Increased expression [70]
CHK1 (Check point Kinase-1)	No association [73]	Increased expression [73]

PAH, Pulmonary arterial hypertension; PAECs, Pulmonary artery endothelial cells; PASMCs, Pulmonary artery smooth muscle cells; [#], Experiment performed in control PAECs with MMC treatment; *, Experiment performed in control PASMCs with MMC treatment; **, Experiment performed in control PASMCs with *PIM-1* inhibition; *** Experiment performed in control PAH-PASMCs with siRNA for *FOXM1*.

The similarity of PAH to a quasi-neoplastic phenotype has triggered the study of DDR inhibitors as a potential therapeutic approach. As outlined in Figure 3, specific pharmacological inhibitors like ABT-888 (PARP1 inhibitor), SGI-1776 (PIM1 inhibitor), Benzarone (EYA3 inhibitor), and MK-8776 (CHK1 inhibitor) downregulate their target genes, leading to reduced proliferation and increased apoptosis, suggesting their therapeutic potential in PAH [49,67,70,73]. However, the cancer-like mechanism has a broad spectrum of phenotypes delineated by different stages of the disease, organs, and cell types. DDR inhibitors have shown some success, but the additional damage and associated detrimental effects on health across different vascular compartments cannot be ignored [49,67,70,73].

It is important to note that, by necessity, the majority of these studies are performed on PAH-vascular cells, PAECs and PASMCs, derived from end-stage tissues, since such cells cannot be obtained at earlier stages of the disease [39,40,42,44–49,61,66,67,70,73,77,78,88,89]. The dysfunctional DDR or augmented DNA damage could be the result of surrounding variables like inflammation, hypoxia etc. Efforts need to be made to understand the cellular microenvironment and the transition of cells from an adaptive to maladaptive DDR. A prospective therapeutic can only be developed once a delineation between the adaptive and maladaptive switch has been established.

PAH goes undiagnosed until the later stages of the disease. This imposes a major limitation to define the early diagnosis markers. The lung tissue samples are only available once the patient undergoes a transplant. In such a scenario, it is difficult to find biomarkers for early disease prediction. Additional reports on the first-degree relatives and in blood cells from PAH patients may assist in establishing novel prognosis markers along with validation of DNA damage and ROS levels. Different cell types in PAH have different pathogenic presentations, and the lack of data on more than one cell type emerges as another major limitation across most of these studies. Further, the majority of studies have explored DNA damage and repair pathways only in Group 1 PAH. Studying these mechanisms in other groups of PH will be important to understand the similarities and differences in molecular mechanisms. Thought should be given to innovative methods to get access to patient cells and tissues at different developmental stages. If not the lung, are there other tissues that can mirror the lung phenotype? In this context, the use of iPSCs and 3D printing of organs with known hits for disease pathogenesis can be explored. To produce significant improvements in patient outcomes, priority should be given to the development of comprehensive functional-based predictive biomarker assays. With current advancements, techniques like single-cell transcriptional and translational studies will give a better understanding of the individual pathogenic (or pro-pathogenic) events in a cell in context to its microenvironment. These studies, in combination with high-throughput sequencing to underscore genetic and epigenetic changes, will supplement the exploration of the pathological mechanism at the systemic and lung-specific level.

Author Contributions: Writing—original draft preparation, review and editing, S.S. and M.A.A.; funding acquisition, M.A.A. Both authors have read and agreed to the published version of the manuscript.

Funding: The authors are supported by the National Heart, Lung, and Blood Institute of the National Institutes of Health under Award Number R35HL140019. The content of this work is solely the responsibility of the authors and does not necessarily represent the official views of the National Institutes of Health.

Conflicts of Interest: The authors declare no conflict of interest.

References

1. Hoeper, M.M.; Humbert, M.; Souza, R.; Idrees, M.; Kawut, S.M.; Sliwa-Hahnle, K.; Jing, Z.C.; Gibbs, J.S. A global view of pulmonary hypertension. *Lancet Respir. Med.* **2016**, *4*, 306–322. [CrossRef]

2. Simonneau, G.; Montani, D.; Celermajer, D.S.; Denton, C.P.; Gatzoulis, M.A.; Krowka, M.; Williams, P.G.; Souza, R. Haemodynamic definitions and updated clinical classification of pulmonary hypertension. *Eur. Respir. J.* **2019**, *53*. [CrossRef] [PubMed]

3. Humbert, M.; Montani, D.; Perros, F.; Dorfmuller, P.; Adnot, S.; Eddahibi, S. Endothelial cell dysfunction and cross talk between endothelium and smooth muscle cells in pulmonary arterial hypertension. *Vascul. Pharmacol.* **2008**, *49*, 113–118. [CrossRef] [PubMed]

4. Rabinovitch, M. Molecular pathogenesis of pulmonary arterial hypertension. *J. Clin. Investig.* **2012**, *122*, 4306–4313. [CrossRef]

5. Tuder, R.M.; Archer, S.L.; Dorfmuller, P.; Erzurum, S.C.; Guignabert, C.; Michelakis, E.; Rabinovitch, M.; Schermuly, R.; Stenmark, K.R.; Morrell, N.W. Relevant issues in the pathology and pathobiology of pulmonary hypertension. *J. Am. Coll. Cardiol.* **2013**, *62*, D4–D12. [CrossRef]

6. Potus, F.; Graydon, C.; Provencher, S.; Bonnet, S. Vascular remodeling process in pulmonary arterial hypertension, with focus on miR-204 and miR-126 (2013 Grover Conference series). *Pulm. Circ.* **2014**, *4*, 175–184. [CrossRef]

7. Benza, R.L.; Miller, D.P.; Barst, R.J.; Badesch, D.B.; Frost, A.E.; McGoon, M.D. An evaluation of long-term survival from time of diagnosis in pulmonary arterial hypertension from the REVEAL Registry. *Chest* **2012**, *142*, 448–456. [CrossRef]

8. Hoeper, M.M.; McLaughlin, V.V.; Dalaan, A.M.; Satoh, T.; Galie, N. Treatment of pulmonary hypertension. *Lancet Respir. Med.* **2016**, *4*, 323–336. [CrossRef]

9. Lau, E.M.T.; Giannoulatou, E.; Celermajer, D.S.; Humbert, M. Epidemiology and treatment of pulmonary arterial hypertension. *Nat. Rev. Cardiol.* **2017**, *14*, 603–614. [CrossRef]

10. Deng, Z.; Morse, J.H.; Slager, S.L.; Cuervo, N.; Moore, K.J.; Venetos, G.; Kalachikov, S.; Cayanis, E.; Fischer, S.G.; Barst, R.J.; et al. Familial primary pulmonary hypertension (gene PPH1) is caused by mutations in the bone morphogenetic protein receptor-II gene. *Am. J. Hum. Genet.* **2000**, *67*, 737–744. [CrossRef]

11. International, P.P.H.C.; Lane, K.B.; Machado, R.D.; Pauciulo, M.W.; Thomson, J.R.; Phillips, J.A., 3rd; Loyd, J.E.; Nichols, W.C.; Trembath, R.C. Heterozygous germline mutations in BMPR2, encoding a TGF-β receptor, cause familial primary pulmonary hypertension. *Nat. Genet.* **2000**, *26*, 81–84. [CrossRef]

12. Morrell, N.W.; Aldred, M.A.; Chung, W.K.; Elliott, C.G.; Nichols, W.C.; Soubrier, F.; Trembath, R.C.; Loyd, J.E. Genetics and genomics of pulmonary arterial hypertension. *Eur. Respir. J.* **2019**, *53*. [CrossRef] [PubMed]

13. Shintani, M.; Yagi, H.; Nakayama, T.; Saji, T.; Matsuoka, R. A new nonsense mutation of SMAD8 associated with pulmonary arterial hypertension. *J. Med. Genet.* **2009**, *46*, 331–337. [CrossRef]

14. Austin, E.D.; Ma, L.; LeDuc, C.; Berman Rosenzweig, E.; Borczuk, A.; Phillips, J.A., 3rd; Palomero, T.; Sumazin, P.; Kim, H.R.; Talati, M.H.; et al. Whole exome sequencing to identify a novel gene (caveolin-1) associated with human pulmonary arterial hypertension. *Circ. Cardiovasc. Genet.* **2012**, *5*, 336–343. [CrossRef] [PubMed]

15. Ma, L.; Roman-Campos, D.; Austin, E.D.; Eyries, M.; Sampson, K.S.; Soubrier, F.; Germain, M.; Tregouet, D.A.; Borczuk, A.; Rosenzweig, E.B.; et al. A novel channelopathy in pulmonary arterial hypertension. *N. Engl. J. Med.* **2013**, *369*, 351–361. [CrossRef]

16. Graf, S.; Haimel, M.; Bleda, M.; Hadinnapola, C.; Southgate, L.; Li, W.; Hodgson, J.; Liu, B.; Salmon, R.M.; Southwood, M.; et al. Identification of rare sequence variation underlying heritable pulmonary arterial hypertension. *Nat. Commun.* **2018**, *9*, 1416. [CrossRef]

17. Zhu, N.; Pauciulo, M.W.; Welch, C.L.; Lutz, K.A.; Coleman, A.W.; Gonzaga-Jauregui, C.; Wang, J.; Grimes, J.M.; Martin, L.J.; He, H.; et al. Novel risk genes and mechanisms implicated by exome sequencing of 2572 individuals with pulmonary arterial hypertension. *Genome Med.* **2019**, *11*, 69. [CrossRef]

18. McLaughlin, V.V.; Archer, S.L.; Badesch, D.B.; Barst, R.J.; Farber, H.W.; Lindner, J.R.; Mathier, M.A.; McGoon, M.D.; Park, M.H.; Rosenson, R.S.; et al. ACCF/AHA 2009 expert consensus document on pulmonary hypertension a report of the American College of Cardiology Foundation Task Force on Expert Consensus Documents and the American Heart Association developed in collaboration with the American College of Chest Physicians; American Thoracic Society, Inc.; and the Pulmonary Hypertension Association. *J. Am. Coll. Cardiol.* **2009**, *53*, 1573–1619. [CrossRef]

19. Loyd, J.E. Pulmonary arterial hypertension: Insights from genetic studies. *Proc. Am. Thorac. Soc.* **2011**, *8*, 154–157. [CrossRef]

20. Soon, E.; Holmes, A.M.; Treacy, C.M.; Doughty, N.J.; Southgate, L.; Machado, R.D.; Trembath, R.C.; Jennings, S.; Barker, L.; Nicklin, P.; et al. Elevated levels of inflammatory cytokines predict survival in idiopathic and familial pulmonary arterial hypertension. *Circulation* **2010**, *122*, 920–927. [CrossRef]

21. Demarco, V.G.; Whaley-Connell, A.T.; Sowers, J.R.; Habibi, J.; Dellsperger, K.C. Contribution of oxidative stress to pulmonary arterial hypertension. *World J. Cardiol.* **2010**, *2*, 316–324. [CrossRef] [PubMed]

22. Kherbeck, N.; Tamby, M.C.; Bussone, G.; Dib, H.; Perros, F.; Humbert, M.; Mouthon, L. The role of inflammation and autoimmunity in the pathophysiology of pulmonary arterial hypertension. *Clin. Rev. Allergy Immunol.* **2013**, *44*, 31–38. [CrossRef] [PubMed]

23. Cohen-Kaminsky, S.; Hautefort, A.; Price, L.; Humbert, M.; Perros, F. Inflammation in pulmonary hypertension: What we know and what we could logically and safely target first. *Drug Discov. Today* **2014**, *19*, 1251–1256. [CrossRef] [PubMed]

24. Van Houten, B. Pulmonary Arterial Hypertension Is Associated with Oxidative Stress-induced Genome Instability. *Am. J. Respir. Crit. Care Med.* **2015**, *192*, 129–130. [CrossRef]

25. Boucherat, O.; Vitry, G.; Trinh, I.; Paulin, R.; Provencher, S.; Bonnet, S. The cancer theory of pulmonary arterial hypertension. *Pulm. Circ.* **2017**, *7*, 285–299. [CrossRef]

26. Sancar, A.; Lindsey-Boltz, L.A.; Unsal-Kacmaz, K.; Linn, S. Molecular mechanisms of mammalian DNA repair and the DNA damage checkpoints. *Annu. Rev. Biochem.* **2004**, *73*, 39–85. [CrossRef]
27. Harper, J.W.; Elledge, S.J. The DNA damage response: Ten years after. *Mol. Cell* **2007**, *28*, 739–745. [CrossRef]
28. Li, G.M. Mechanisms and functions of DNA mismatch repair. *Cell Res.* **2008**, *18*, 85–98. [CrossRef]
29. Izumi, T.; Wiederhold, L.R.; Roy, G.; Roy, R.; Jaiswal, A.; Bhakat, K.K.; Mitra, S.; Hazra, T.K. Mammalian DNA base excision repair proteins: Their interactions and role in repair of oxidative DNA damage. *Toxicology* **2003**, *193*, 43–65. [CrossRef]
30. Wilson, D.M., 3rd; Barsky, D. The major human abasic endonuclease: Formation, consequences and repair of abasic lesions in DNA. *Mutat. Res.* **2001**, *485*, 283–307. [CrossRef]
31. Sancar, A. DNA excision repair. *Annu. Rev. Biochem.* **1996**, *65*, 43–81. [CrossRef] [PubMed]
32. Wood, R.D. DNA damage recognition during nucleotide excision repair in mammalian cells. *Biochimie* **1999**, *81*, 39–44. [CrossRef]
33. Modesti, M.; Kanaar, R. Homologous recombination: From model organisms to human disease. *Genome Biol.* **2001**. [CrossRef] [PubMed]
34. Carney, J.P.; Maser, R.S.; Olivares, H.; Davis, E.M.; Le Beau, M.; Yates, J.R., 3rd; Hays, L.; Morgan, W.F.; Petrini, J.H. The hMre11/hRad50 protein complex and Nijmegen breakage syndrome: Linkage of double-strand break repair to the cellular DNA damage response. *Cell* **1998**, *93*, 477–486. [CrossRef]
35. Gottlieb, T.M.; Jackson, S.P. The DNA-dependent protein kinase: Requirement for DNA ends and association with Ku antigen. *Cell* **1993**, *72*, 131–142. [CrossRef]
36. Ramsden, D.A.; Gellert, M. Ku protein stimulates DNA end joining by mammalian DNA ligases: A direct role for Ku in repair of DNA double-strand breaks. *EMBO J.* **1998**, *17*, 609–614. [CrossRef]
37. Nick McElhinny, S.A.; Snowden, C.M.; McCarville, J.; Ramsden, D.A. Ku recruits the XRCC4-ligase IV complex to DNA ends. *Mol. Cell. Biol.* **2000**, *20*, 2996–3003. [CrossRef]
38. Harrison, J.C.; Haber, J.E. Surviving the breakup: The DNA damage checkpoint. *Annu. Rev. Genet.* **2006**, *40*, 209–235. [CrossRef]
39. Tuder, R.M.; Radisavljevic, Z.; Shroyer, K.R.; Polak, J.M.; Voelkel, N.F. Monoclonal endothelial cells in appetite suppressant-associated pulmonary hypertension. *Am. J. Respir. Crit. Care Med.* **1998**, *158*, 1999–2001. [CrossRef]
40. Lee, S.D.; Shroyer, K.R.; Markham, N.E.; Cool, C.D.; Voelkel, N.F.; Tuder, R.M. Monoclonal endothelial cell proliferation is present in primary but not secondary pulmonary hypertension. *J. Clin. Investig.* **1998**, *101*, 927–934. [CrossRef]
41. Voelkel, N.F.; Cool, C.; Lee, S.D.; Wright, L.; Geraci, M.W.; Tuder, R.M. Primary pulmonary hypertension between inflammation and cancer. *Chest* **1998**, *114*, 225S–230S. [CrossRef] [PubMed]
42. Yeager, M.E.; Halley, G.R.; Golpon, H.A.; Voelkel, N.F.; Tuder, R.M. Microsatellite instability of endothelial cell growth and apoptosis genes within plexiform lesions in primary pulmonary hypertension. *Circ. Res.* **2001**, *88*, E2–E11. [CrossRef] [PubMed]
43. Machado, R.D.; James, V.; Southwood, M.; Harrison, R.E.; Atkinson, C.; Stewart, S.; Morrell, N.W.; Trembath, R.C.; Aldred, M.A. Investigation of second genetic hits at the BMPR2 locus as a modulator of disease progression in familial pulmonary arterial hypertension. *Circulation* **2005**, *111*, 607–613. [CrossRef]
44. Aldred, M.A.; Comhair, S.A.; Varella-Garcia, M.; Asosingh, K.; Xu, W.; Noon, G.P.; Thistlethwaite, P.A.; Tuder, R.M.; Erzurum, S.C.; Geraci, M.W.; et al. Somatic chromosome abnormalities in the lungs of patients with pulmonary arterial hypertension. *Am. J. Respir. Crit. Care Med.* **2010**, *182*, 1153–1160. [CrossRef] [PubMed]
45. Drake, K.M.; Comhair, S.A.; Erzurum, S.C.; Tuder, R.M.; Aldred, M.A. Endothelial chromosome 13 deletion in congenital heart disease-associated pulmonary arterial hypertension dysregulates SMAD9 signaling. *Am. J. Respir. Crit. Care Med.* **2015**, *191*, 850–854. [CrossRef]
46. Federici, C.; Drake, K.M.; Rigelsky, C.M.; McNelly, L.N.; Meade, S.L.; Comhair, S.A.; Erzurum, S.C.; Aldred, M.A. Increased Mutagen Sensitivity and DNA Damage in Pulmonary Arterial Hypertension. *Am. J. Respir. Crit. Care Med.* **2015**, *192*, 219–228. [CrossRef]
47. Chen, P.I.; Cao, A.; Miyagawa, K.; Tojais, N.F.; Hennigs, J.K.; Li, C.G.; Sweeney, N.M.; Inglis, A.S.; Wang, L.; Li, D.; et al. Amphetamines promote mitochondrial dysfunction and DNA damage in pulmonary hypertension. *JCI Insight* **2017**, *2*, e90427. [CrossRef] [PubMed]

48. de Jesus Perez, V.A.; Yuan, K.; Lyuksyutova, M.A.; Dewey, F.; Orcholski, M.E.; Shuffle, E.M.; Mathur, M.; Yancy, L., Jr.; Rojas, V.; Li, C.G.; et al. Whole-exome sequencing reveals TopBP1 as a novel gene in idiopathic pulmonary arterial hypertension. *Am. J. Respir. Crit. Care Med.* **2014**, *189*, 1260–1272. [CrossRef]

49. Meloche, J.; Pflieger, A.; Vaillancourt, M.; Paulin, R.; Potus, F.; Zervopoulos, S.; Graydon, C.; Courboulin, A.; Breuils-Bonnet, S.; Tremblay, E.; et al. Role for DNA damage signaling in pulmonary arterial hypertension. *Circulation* **2014**, *129*, 786–797. [CrossRef]

50. Chan, N.; Ali, M.; McCallum, G.P.; Kumareswaran, R.; Koritzinsky, M.; Wouters, B.G.; Wells, P.G.; Gallinger, S.; Bristow, R.G. Hypoxia provokes base excision repair changes and a repair-deficient, mutator phenotype in colorectal cancer cells. *Mol. Cancer Res.* **2014**, *12*, 1407–1415. [CrossRef]

51. Hammond, E.M.; Denko, N.C.; Dorie, M.J.; Abraham, R.T.; Giaccia, A.J. Hypoxia links ATR and p53 through replication arrest. *Mol. Cell. Biol.* **2002**, *22*, 1834–1843. [CrossRef]

52. Hammond, E.M.; Dorie, M.J.; Giaccia, A.J. ATR/ATM targets are phosphorylated by ATR in response to hypoxia and ATM in response to reoxygenation. *J. Biol. Chem.* **2003**, *278*, 12207–12213. [CrossRef]

53. Scanlon, S.E.; Glazer, P.M. Multifaceted control of DNA repair pathways by the hypoxic tumor microenvironment. *DNA Repair (Amst.)* **2015**, *32*, 180–189. [CrossRef] [PubMed]

54. de Jesus Perez, V.; Kudelko, K.; Snook, S.; Zamanian, R.T. Drugs and toxins-associated pulmonary arterial hypertension: Lessons learned and challenges ahead. *Int. J. Clin. Pract. Suppl.* **2011**, 8–10. [CrossRef]

55. Yamamoto, B.K.; Raudensky, J. The role of oxidative stress, metabolic compromise, and inflammation in neuronal injury produced by amphetamine-related drugs of abuse. *J. Neuroimmune Pharmacol.* **2008**, *3*, 203–217. [CrossRef] [PubMed]

56. Rhodes, C.J.; Batai, K.; Bleda, M.; Haimel, M.; Southgate, L.; Germain, M.; Pauciulo, M.W.; Hadinnapola, C.; Aman, J.; Girerd, B.; et al. Genetic determinants of risk in pulmonary arterial hypertension: International genome-wide association studies and meta-analysis. *Lancet Respir. Med.* **2019**, *7*, 227–238. [CrossRef]

57. Perros, F.; Gunther, S.; Ranchoux, B.; Godinas, L.; Antigny, F.; Chaumais, M.C.; Dorfmuller, P.; Hautefort, A.; Raymond, N.; Savale, L.; et al. Mitomycin-Induced Pulmonary Veno-Occlusive Disease: Evidence From Human Disease and Animal Models. *Circulation* **2015**, *132*, 834–847. [CrossRef] [PubMed]

58. Wilson, G.J.; Bunpo, P.; Cundiff, J.K.; Wek, R.C.; Anthony, T.G. The eukaryotic initiation factor 2 kinase GCN2 protects against hepatotoxicity during asparaginase treatment. *Am. J. Physiol. Endocrinol. Metab.* **2013**, *305*, E1124–E1133. [CrossRef]

59. Eyries, M.; Montani, D.; Girerd, B.; Perret, C.; Leroy, A.; Lonjou, C.; Chelghoum, N.; Coulet, F.; Bonnet, D.; Dorfmuller, P.; et al. EIF2AK4 mutations cause pulmonary veno-occlusive disease, a recessive form of pulmonary hypertension. *Nat. Genet.* **2014**, *46*, 65–69. [CrossRef]

60. Best, D.H.; Sumner, K.L.; Austin, E.D.; Chung, W.K.; Brown, L.M.; Borczuk, A.C.; Rosenzweig, E.B.; Bayrak-Toydemir, P.; Mao, R.; Cahill, B.C.; et al. EIF2AK4 mutations in pulmonary capillary hemangiomatosis. *Chest* **2014**, *145*, 231–236. [CrossRef]

61. Drake, K.M.; Federici, C.; Duong, H.T.; Comhair, S.A.; Erzurum, S.C.; Asosingh, K.; Aldred, M.A. Genomic stability of pulmonary artery endothelial colony-forming cells in culture. *Pulm. Circ.* **2017**, *7*, 421–427. [CrossRef]

62. Ahmadian, M.; Suh, J.M.; Hah, N.; Liddle, C.; Atkins, A.R.; Downes, M.; Evans, R.M. PPARgamma signaling and metabolism: The good, the bad and the future. *Nat. Med.* **2013**, *19*, 557–566. [CrossRef]

63. Guignabert, C.; Alvira, C.M.; Alastalo, T.P.; Sawada, H.; Hansmann, G.; Zhao, M.; Wang, L.; El-Bizri, N.; Rabinovitch, M. Tie2-mediated loss of peroxisome proliferator-activated receptor-γ in mice causes PDGF receptor-β-dependent pulmonary arterial muscularization. *Am. J. Physiol. Lung Cell. Mol. Physiol.* **2009**, *297*, L1082–L1090. [CrossRef]

64. Alastalo, T.P.; Li, M.; Perez Vde, J.; Pham, D.; Sawada, H.; Wang, J.K.; Koskenvuo, M.; Wang, L.; Freeman, B.A.; Chang, H.Y.; et al. Disruption of PPARgamma/β-catenin-mediated regulation of apelin impairs BMP-induced mouse and human pulmonary arterial EC survival. *J. Clin. Investig.* **2011**, *121*, 3735–3746. [CrossRef] [PubMed]

65. Li, C.G.; Mahon, C.; Sweeney, N.M.; Verschueren, E.; Kantamani, V.; Li, D.; Hennigs, J.K.; Marciano, D.P.; Diebold, I.; Abu-Halawa, O.; et al. PPARgamma Interaction with UBR5/ATMIN Promotes DNA Repair to Maintain Endothelial Homeostasis. *Cell Rep.* **2019**, *26*, 1333–1343.e7. [CrossRef] [PubMed]

66. Meloche, J.; Le Guen, M.; Potus, F.; Vinck, J.; Ranchoux, B.; Johnson, I.; Antigny, F.; Tremblay, E.; Breuils-Bonnet, S.; Perros, F.; et al. miR-223 reverses experimental pulmonary arterial hypertension. *Am. J. Physiol. Cell Physiol.* **2015**, *309*, C363–C372. [CrossRef] [PubMed]

67. Lampron, M.C.; Vitry, G.; Nadeau, V.; Grobs, Y.; Paradis, R.; Samson, N.; Tremblay, E.; Boucherat, O.; Meloche, J.; Bonnet, S.; et al. PIM1 (Moloney Murine Leukemia Provirus Integration Site) Inhibition Decreases the Nonhomologous End-Joining DNA Damage Repair Signaling Pathway in Pulmonary Hypertension. *Arterioscler. Thromb. Vasc. Biol.* **2020**, *40*, 783–801. [CrossRef]

68. Hsu, J.L.; Leong, P.K.; Ho, Y.F.; Hsu, L.C.; Lu, P.H.; Chen, C.S.; Guh, J.H. Pim-1 knockdown potentiates paclitaxel-induced apoptosis in human hormone-refractory prostate cancers through inhibition of NHEJ DNA repair. *Cancer Lett.* **2012**, *319*, 214–222. [CrossRef]

69. Cook, P.J.; Ju, B.G.; Telese, F.; Wang, X.; Glass, C.K.; Rosenfeld, M.G. Tyrosine dephosphorylation of H2AX modulates apoptosis and survival decisions. *Nature* **2009**, *458*, 591–596. [CrossRef]

70. Wang, Y.; Pandey, R.N.; York, A.J.; Mallela, J.; Nichols, W.C.; Hu, Y.C.; Molkentin, J.D.; Wikenheiser-Brokamp, K.A.; Hegde, R.S. The EYA3 tyrosine phosphatase activity promotes pulmonary vascular remodeling in pulmonary arterial hypertension. *Nat. Commun.* **2019**, *10*, 4143. [CrossRef]

71. Carrassa, L.; Broggini, M.; Erba, E.; Damia, G. Chk1, but not Chk2, is involved in the cellular response to DNA damaging agents: Differential activity in cells expressing or not p53. *Cell Cycle* **2004**, *3*, 1177–1181. [CrossRef]

72. Pilie, P.G.; Tang, C.; Mills, G.B.; Yap, T.A. State-of-the-art strategies for targeting the DNA damage response in cancer. *Nat. Rev. Clin. Oncol.* **2019**, *16*, 81–104. [CrossRef] [PubMed]

73. Bourgeois, A.; Bonnet, S.; Breuils-Bonnet, S.; Habbout, K.; Paradis, R.; Tremblay, E.; Lampron, M.C.; Orcholski, M.E.; Potus, F.; Bertero, T.; et al. Inhibition of CHK 1 (Checkpoint Kinase 1) Elicits Therapeutic Effects in Pulmonary Arterial Hypertension. *Arterioscler. Thromb. Vasc. Biol.* **2019**, *39*, 1667–1681. [CrossRef] [PubMed]

74. Liu, Q.; Guntuku, S.; Cui, X.S.; Matsuoka, S.; Cortez, D.; Tamai, K.; Luo, G.; Carattini-Rivera, S.; DeMayo, F.; Bradley, A.; et al. Chk1 is an essential kinase that is regulated by Atr and required for the G(2)/M DNA damage checkpoint. *Genes Dev.* **2000**, *14*, 1448–1459. [PubMed]

75. Zhao, H.; Piwnica-Worms, H. ATR-mediated checkpoint pathways regulate phosphorylation and activation of human Chk1. *Mol. Cell. Biol.* **2001**, *21*, 4129–4139. [CrossRef] [PubMed]

76. Kim, J.; Kang, Y.; Kojima, Y.; Lighthouse, J.K.; Hu, X.; Aldred, M.A.; McLean, D.L.; Park, H.; Comhair, S.A.; Greif, D.M.; et al. An endothelial apelin-FGF link mediated by miR-424 and miR-503 is disrupted in pulmonary arterial hypertension. *Nat. Med.* **2013**, *19*, 74–82. [CrossRef]

77. Vattulainen-Collanus, S.; Southwood, M.; Yang, X.D.; Moore, S.; Ghatpande, P.; Morrell, N.W.; Lagna, G.; Hata, A. Bone morphogenetic protein signaling is required for RAD51-mediated maintenance of genome integrity in vascular endothelial cells. *Commun. Biol.* **2018**, *1*, 149. [CrossRef]

78. Diebold, I.; Hennigs, J.K.; Miyagawa, K.; Li, C.G.; Nickel, N.P.; Kaschwich, M.; Cao, A.; Wang, L.; Reddy, S.; Chen, P.I.; et al. BMPR2 preserves mitochondrial function and DNA during reoxygenation to promote endothelial cell survival and reverse pulmonary hypertension. *Cell Metab.* **2015**, *21*, 596–608. [CrossRef]

79. Courboulin, A.; Paulin, R.; Giguere, N.J.; Saksouk, N.; Perreault, T.; Meloche, J.; Paquet, E.R.; Biardel, S.; Provencher, S.; Cote, J.; et al. Role for miR-204 in human pulmonary arterial hypertension. *J. Exp. Med.* **2011**, *208*, 535–548. [CrossRef]

80. Marsboom, G.; Toth, P.T.; Ryan, J.J.; Hong, Z.; Wu, X.; Fang, Y.H.; Thenappan, T.; Piao, L.; Zhang, H.J.; Pogoriler, J.; et al. Dynamin-related protein 1-mediated mitochondrial mitotic fission permits hyperproliferation of vascular smooth muscle cells and offers a novel therapeutic target in pulmonary hypertension. *Circ. Res.* **2012**, *110*, 1484–1497. [CrossRef]

81. Paulin, R.; Meloche, J.; Jacob, M.H.; Bisserier, M.; Courboulin, A.; Bonnet, S. Dehydroepiandrosterone inhibits the Src/STAT3 constitutive activation in pulmonary arterial hypertension. *Am. J. Physiol. Heart Circ. Physiol.* **2011**, *301*, H1798–H1809. [CrossRef]

82. Fessel, J.P.; West, J.D. Redox biology in pulmonary arterial hypertension (2013 Grover Conference Series). *Pulm. Circ.* **2015**, *5*, 599–609. [CrossRef] [PubMed]

83. Ryan, J.J.; Archer, S.L. Emerging concepts in the molecular basis of pulmonary arterial hypertension: Part I: Metabolic plasticity and mitochondrial dynamics in the pulmonary circulation and right ventricle in pulmonary arterial hypertension. *Circulation* **2015**, *131*, 1691–1702. [CrossRef] [PubMed]

84. Heikal, A.A. Intracellular coenzymes as natural biomarkers for metabolic activities and mitochondrial anomalies. *Biomark. Med.* **2010**, *4*, 241–263. [CrossRef] [PubMed]

85. Murata, M.M.; Kong, X.; Moncada, E.; Chen, Y.; Imamura, H.; Wang, P.; Berns, M.W.; Yokomori, K.; Digman, M.A. NAD+ consumption by PARP1 in response to DNA damage triggers metabolic shift critical for damaged cell survival. *Mol. Biol. Cell* **2019**, *30*, 2584–2597. [CrossRef]

86. Archer, S.L.; Gomberg-Maitland, M.; Maitland, M.L.; Rich, S.; Garcia, J.G.; Weir, E.K. Mitochondrial metabolism, redox signaling, and fusion: A mitochondria-ROS-HIF-1alpha-Kv1.5 O2-sensing pathway at the intersection of pulmonary hypertension and cancer. *Am. J. Physiol. Heart Circ. Physiol.* **2008**, *294*, H570–H578. [CrossRef]

87. Yakes, F.M.; Van Houten, B. Mitochondrial DNA damage is more extensive and persists longer than nuclear DNA damage in human cells following oxidative stress. *Proc. Natl. Acad. Sci. USA* **1997**, *94*, 514–519. [CrossRef]

88. Boucherat, O.; Peterlini, T.; Bourgeois, A.; Nadeau, V.; Breuils-Bonnet, S.; Boilet-Molez, S.; Potus, F.; Meloche, J.; Chabot, S.; Lambert, C.; et al. Mitochondrial HSP90 Accumulation Promotes Vascular Remodeling in Pulmonary Arterial Hypertension. *Am. J. Respir. Crit. Care Med.* **2018**, *198*, 90–103. [CrossRef]

89. Li, M.; Vattulainen, S.; Aho, J.; Orcholski, M.; Rojas, V.; Yuan, K.; Helenius, M.; Taimen, P.; Myllykangas, S.; De Jesus Perez, V.; et al. Loss of bone morphogenetic protein receptor 2 is associated with abnormal DNA repair in pulmonary arterial hypertension. *Am. J. Respir. Cell Mol. Biol.* **2014**, *50*, 1118–1128. [CrossRef]

90. Wang, Y.; Huang, J.W.; Calses, P.; Kemp, C.J.; Taniguchi, T. MiR-96 downregulates REV1 and RAD51 to promote cellular sensitivity to cisplatin and PARP inhibition. *Cancer Res.* **2012**, *72*, 4037–4046. [CrossRef]

91. Long, L.; Ormiston, M.L.; Yang, X.; Southwood, M.; Graf, S.; Machado, R.D.; Mueller, M.; Kinzel, B.; Yung, L.M.; Wilkinson, J.M.; et al. Selective enhancement of endothelial BMPR-II with BMP9 reverses pulmonary arterial hypertension. *Nat. Med.* **2015**, *21*, 777–785. [CrossRef] [PubMed]

Publisher's Note: MDPI stays neutral with regard to jurisdictional claims in published maps and institutional affiliations.

Review

At the X-Roads of Sex and Genetics in Pulmonary Arterial Hypertension

Meghan M. Cirulis [1,2,*], Mark W. Dodson [1,2], Lynn M. Brown [1,2], Samuel M. Brown [1,2], Tim Lahm [3,4,5] and Greg Elliott [1,2]

1   Division of Pulmonary, Critical Care and Occupational Medicine, University of Utah, Salt Lake City, UT 84132, USA; mark.dodson@imail.org (M.W.D.); Lynn.brown@imail.org (L.M.B.); Samuel.brown@imail.org (S.M.B.); Greg.Elliott_MD@imail.org (G.E.)
2   Division of Pulmonary and Critical Care Medicine, Intermountain Medical Center, Salt Lake City, UT 84107, USA
3   Division of Pulmonary, Critical Care, Sleep and Occupational Medicine, Department of Medicine, Indiana University School of Medicine, Indianapolis, IN 46202, USA; tlahm@iu.edu
4   Richard L. Roudebush Veterans Affairs Medical Center, Indianapolis, IN 46202, USA
5   Department of Anatomy, Cell Biology & Physiology, Indiana University School of Medicine, Indianapolis, IN 46202, USA
*   Correspondence: meghan.cirulis@hsc.utah.edu; Tel.: +1-801-581-7806

Received: 29 September 2020; Accepted: 17 November 2020; Published: 20 November 2020

Abstract: Group 1 pulmonary hypertension (pulmonary arterial hypertension; PAH) is a rare disease characterized by remodeling of the small pulmonary arteries leading to progressive elevation of pulmonary vascular resistance, ultimately leading to right ventricular failure and death. Deleterious mutations in the serine-threonine receptor bone morphogenetic protein receptor 2 (*BMPR2*; a central mediator of bone morphogenetic protein (BMP) signaling) and female sex are known risk factors for the development of PAH in humans. In this narrative review, we explore the complex interplay between the BMP and estrogen signaling pathways, and the potentially synergistic mechanisms by which these signaling cascades increase the risk of developing PAH. A comprehensive understanding of these tangled pathways may reveal therapeutic targets to prevent or slow the progression of PAH.

Keywords: bone morphogenetic protein receptor type 2; heritable; familial; estrogen; estradiol; penetrance; gender; PAH

---

1. Introduction

Group 1 pulmonary hypertension (pulmonary arterial hypertension; PAH) is a rare disease characterized by remodeling of the small pulmonary arteries which leads to the progressive elevation of pulmonary vascular resistance (PVR), and ultimately to right ventricular failure and death. While numerous inciting factors are known (e.g., connective tissue disease, HIV infection, drug and toxin exposure), the pathobiology of PAH generally convenes on a final common pathway of endothelial (EC) and smooth muscle cell (SMC) dysfunction, with an imbalance of apoptotic and proliferative signaling, vasoconstriction, and structural changes in vessel walls [1]. Deficiency of bone morphogenetic protein receptor 2 (BMPR2), a serine threonine kinase "type II" receptor in the transforming growth factor (TGF)-β superfamily, has been inexorably identified as a central mediator in this process [2].

Heterozygous germline mutations in *BMPR2* were first associated with PAH via genetic linkage analysis of families with the disease [3,4]. Since this discovery in 2000, further analysis of up- and down-stream signaling through the receptor, including both canonical and non-canonical pathways, has illuminated several mechanisms by which deficiency in BMPR2 signaling leads to PAH [5]. Not only are deleterious mutations in *BMPR2* associated with both heritable (~80%) and idiopathic (~20%)

PAH [6–8], but decreased BMPR2 expression and signaling has also been demonstrated in other subtypes of PAH and non-PAH pulmonary hypertension (PH) in the absence of mutations [9,10]. Furthermore, the presence of a deleterious *BMPR2* mutation in heritable, idiopathic, and anorexigen-associated PAH portends a more severe clinical phenotype and decreased survival [11–14]. Despite the strong association between *BMPR2* mutations and the development of PAH, and despite the high frequency of *BMPR2* mutations in heritable PAH, having a *BMPR2* mutation alone is not sufficient; heterozygous carriers of deleterious *BMPR2* mutations only have an approximately 20% lifetime risk of disease penetrance [15]. Decades of investigation have revealed that there are likely multiple genetic and environmental "second hits" that may be necessary to spur PAH development in the setting of a deleterious *BMPR2* mutation [2].

2. The "Estrogen Puzzle" of PAH

One piece of the complex pathobiology of PAH is biologic sex and the "estrogen puzzle", as it is referred to in the literature. In various animal models, estrogen and estrogen metabolites have been shown to protect the organism from developing PH in the setting of other provoking factors, whereas in human registry studies, a striking female predominance suggests increased susceptibility to disease [16]. Female carriers of deleterious *BMPR2* variants are more likely to develop PAH compared to males; however, once diagnosed, women are less likely than men to have severe disease [17]. A recently published meta-analysis of clinical trials also suggests that men with *BMPR2* mutations have more severe disease, but interestingly more men with idiopathic or heritable PAH were found to have a pathogenic *BMPR2* variant [18].

Increasingly recognized sex differences in right ventricular (RV) adaptation to chronic PH contribute to the "estrogen puzzle" of PAH and likely play a significant role in disease severity and associated mortality. Studies of PH in several distinct rat models have demonstrated that at baseline female rats have better RV function than males [19], ovariectomy attenuates the beneficial effect of female sex [19], and that the restoration of estrogen signaling (genomic and non-genomic) prevents progression of and can rescue the failing RV phenotype in both male and female rats via alterations in metabolism, inflammation, collagen deposition/fibrosis, and angiogenesis [19–24]. Human studies support these findings in healthy subjects and those with PAH. In the MESA-RV study, higher estradiol (E2) levels in healthy post-menopausal women using hormone replacement therapy were associated with higher RV ejection fraction and lower RV end-systolic volume [25]. In idiopathic PAH at baseline, men have lower RV ejection fraction and stroke volume compared to age-matched females despite similar pulmonary artery pressure (PAP) and PVR [26]. Two independent investigations demonstrated that after initiation of PAH-specific therapy, only women show improvement in RV function despite similar improvement in PVR between the sexes [27,28]. Disparate RV recovery is thought to explain, at least in part, the poorer prognosis seen in men.

Despite strong evidence for a substantial role, female sex and the effects of estrogen signaling do not fully explain the observed sex differences in PAH. Other sex-driven differences in the hormonal milieu (e.g., testosterone, dehydroepiandrosterone (DHEA), and progesterone), as well as non-hormonal sex effects (e.g., the recent finding that a Y-chromosome-encoded transcription factor may mediate *BMPR2* expression [29,30]) likely contribute to the complex and differential effects of biologic sex on pulmonary vascular remodeling and RV adaptation [31].

In this review, we will explore one piece of the intricate "estrogen puzzle", specifically how estrogen and estrogen metabolites interact with the BMPR2 signaling pathway. Readers are directed to two recent reviews for a more comprehensive overview of the interaction between sex and PAH [31,32].

3. *BMPR2* Signaling

BMPR2 is a type II constitutively active serine-threonine kinase receptor integral to canonical bone morphogenetic protein (BMP) signaling. BMPR2 signaling occurs in both the EC and SMC of the pulmonary vasculature. Signal transduction is activated by BMP binding and the formation of a heterotetrameric complex of two dimers of type I and type II receptors. Type I receptors include activin receptor-like kinases (ALKs) 1–7. Complexing of the two receptors allows for phosphorylation of the downstream substrate proteins: receptor-regulated Smads (R-Smads), specifically Smads 1, 5, and 8 in BMP signaling. Activated R-Smads then associate with a co-Smad (Smad4) and translocate to the nucleus where they bind to BMP response element DNA sequences and promote gene expression of transcription factors such as inhibitor of DNA binding factors (ID1, ID2 and ID3) [2]. The inhibitor of DNA binding (also known as inhibitor of differentiation) proteins are known to have important regulatory effects on vascular homeostasis [33]. BMPR2 also activates "non-canonical" signaling pathways such as extracellular signal-related kinase (ERK), p38 mitogen-activated protein kinase (MAPK), Lin11, Isl-1, Mec-3 domain kinase (LIMK), Wingless (Wnt pathway), and NOTCH. The complex regulation of both the canonical and non-canonical pathways occurs at multiple levels, including via co-receptors (endoglin), pseudoreceptors, BMP antagonists, and inhibitory Smad proteins [34]. In addition to mutations in *BMPR2* itself, mutations in a number of components of the BMPR2 signaling pathway are linked to the development of PAH, including *ALK1* [35], *SMAD8* [36], *BMP9* [37], and *CAV1* (encoding caveolin-1) [38]. Although acting at different points in the signaling cascade, all of these mutations cause a deficiency in BMPR2 signaling, which is ultimately thought to drive the vascular remodeling and dysregulation central to PAH pathogenesis [2].

4. Estrogen Signaling

Similar to the BMPR2 signaling cascade, essential components of estrogen signaling pathways are expressed in the ECs, vascular SMCs, and fibroblasts responsible for vascular remodeling and the development of PAH [39,40]. Three primary estrogens (estrone, E1; estradiol, E2; estriol, E3) and their metabolites signal through two classical estrogen receptors (ERα and ERβ) and one newly discovered G-protein-coupled receptor (GPER) [41]. In the absence of pregnancy, E2 is the most abundant estrogen. Estrogens primarily signal via "genomic" and "non-genomic" pathways; the former facilitating the classic role of estrogens as transcription factors in the nucleus, the latter triggering rapid effects such as ion channel, kinase, endothelial nitric oxide synthase (eNOS) and prostacyclin synthase activation in the cytoplasm [42–44]. Via 2-, 4-, or 16-hydroxylation, E1 and E2 are metabolized to active compounds with varying potency and activity, signaling through ER-dependent and independent mechanisms. The 2-hydroxylation metabolites are generally considered weakly or anti-mitogenic, anti-estrogenic, and do not signal through an ER [45]. For example, 2-methoxyestradiol (2-ME) and 2-hydroxyestradiol (2-OHE) are considered to be anti-proliferative. In ECs, 2-ME is a potent modulator of nitric oxide (NO), prostacyclin, and endothelin synthesis [45,46]. On the other hand, 16-hydroxylation produces 16α-OHE$_1$, which is similar in potency to E2 and has potent pro-proliferative, pro-inflammatory, and pro-angiogenic effects [47,48].

5. Estrogen and *BMPR2*

As discussed in the following subsections, multiple lines of evidence, in both health and disease, suggest that baseline BMPR2 expression and signaling may be reduced in females compared to males. A relative deficiency in BMPR2 expression in females may be the "second hit" required to reduce BMPR2 signaling below a critical threshold and allow disease penetrance in *BMPR2* mutation carriers (typically a haplo-insufficient state). However, interactions between estrogen and BMPR2 are complex and context-dependent, and may depend on such factors as patient age, menopausal status, cell type studied, and dose responses and time courses. Figure 1 summarizes key points of interaction between estrogen and *BMPR2* signaling.

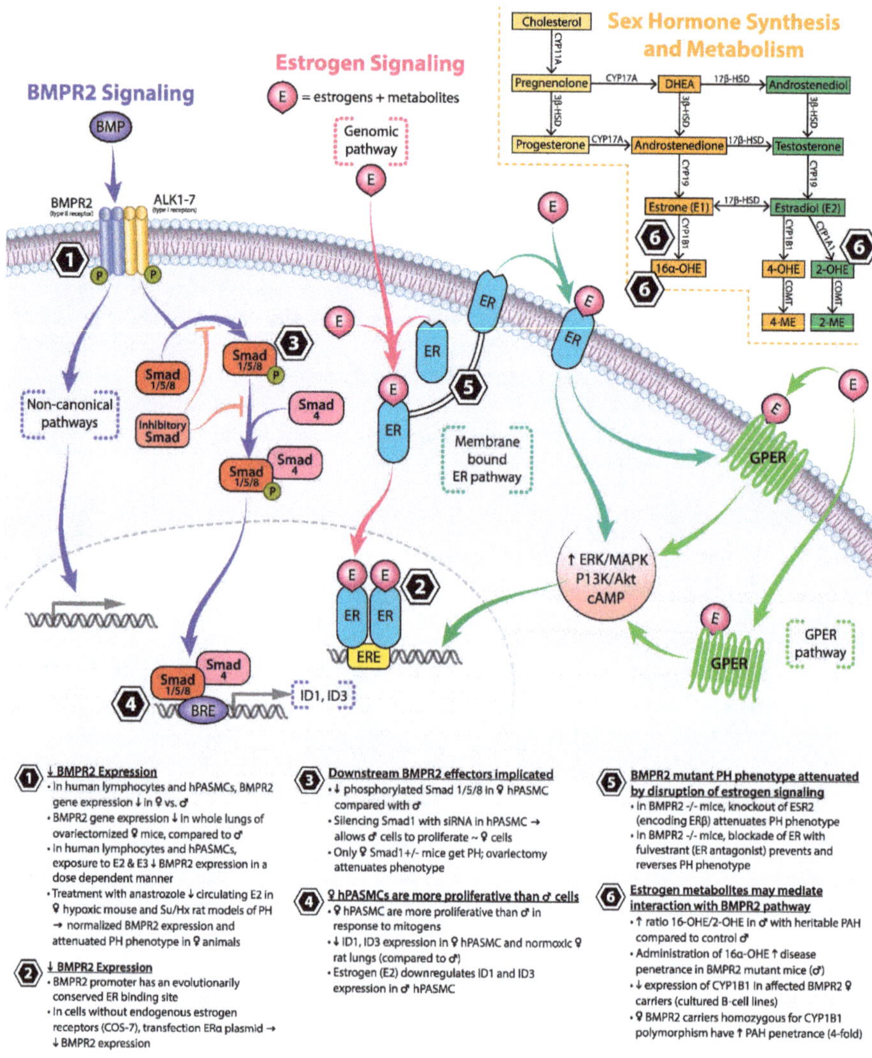

Figure 1. Overview of sex hormone synthesis and metabolism and the interaction between estrogen and *BMPR2* signaling pathways. BMP = bone morphogenetic protein; *BMPR2* = bone morphogenetic protein receptor 2; BRE = BMP response element; DHEA = dehydroepiandrosterone; E3 = estriol; ER = estrogen receptor; ERE = estrogen response element; GPER = G-protein-coupled estrogen receptor; hPASMC = human pulmonary artery smooth muscle cell; PH = pulmonary hypertension.

5.1. Estrogens and Their Receptors Reduce BMPR2 Expression and Downstream Signaling

When assessed in human lymphocytes (commercial cell line) and pulmonary artery smooth muscle cells (hPASMCs; from healthy subjects; cell donor age range: 58–76 years), BMPR2 expression was shown to be lower in female cells compared to males [49,50]. BMPR2 expression in human lymphocytes is suppressed in a dose-dependent manner by the administration of both E2 and E3, with further suppression in the setting of proliferative signals [49]. Additionally, the *BMPR2* promoter has an active and evolutionarily conserved ER binding site [49]. Transfection of a cell line lacking endogenous estrogen receptors with increasing concentrations of ERα plasmid decreased activity at the *BMPR2* promoter (using a luciferase reporter construct), suggesting that direct binding of the ER to the *BMRP2* promoter site may be a mechanism for reduced BMPR2 expression in females [49].

Downstream effectors of BMPR2, including the phosphorylated-Smads 1/5/8 and ID1 and ID3, may also be reduced in female hPASMCs compared to males (healthy subject donors as described above) [50]. In keeping with this finding, female hPASMCs are more proliferative in response to mitogens compared to male cells (suspected to be due to reduced activity of the BMPR2 signaling pathway) and have reduced induction of phosphorylated-Smads when exposed to BMP4, an agonist of BMPR2. Silencing of *SMAD1* using microRNA, thus inhibiting the BMPR2 signaling pathway, allowed male hPASMCs to proliferate in a similar fashion to those of females [50].

Similar findings have been observed in murine models. When examined via whole-lung analysis in normal mice, *BMPR2* gene expression was lower in ovariectomized females compared to males [49]. These findings are corroborated by a second study which found that in normoxic rodents (mice and rats), lung transcript levels of *BMPR2* and downstream effectors ID1 and ID3 were significantly lower in females than males [51].

In two well-established experimental rodent models of PH (hypoxia (mouse) and sugen 5416 + hypoxia (rat; Su/Hx) [52]), Mair and colleagues demonstrated that BMPR2 levels were significantly downregulated in hypoxic male and female rodents compared to normoxic control animals. Treatment with the aromatase inhibitor anastrozole reduced circulating E2 levels in female rodents, with corresponding normalization of BMPR2 expression and attenuation of changes in right ventricular systolic pressure (RVSP) and pulmonary vascular remodeling [51].

5.2. Loss of Estrogen Signaling Attenuates Experimental PH Phenotypes Driven by Mutations in Components of the BMPR2 Signaling Pathway

Further evidence of the interplay between the *BMPR2* and estrogen signaling pathways comes from experimental PH models driven by mutations in *BMPR2* and *SMAD1*. Using anastrozole or the ER antagonist fulvestrant to inhibit estrogen signaling in *BMPR2*$^{-/-}$ mutant mice, Chen et al. [53] demonstrated both prevention and reversal of the typical *BMPR2*-mutation-associated experimental PH phenotype. Knock-out of *ESR2* (the gene encoding estrogen receptor β), and less so *ESR1* (the gene encoding estrogen receptor α), reduced the elevation in RVSP typically seen in *BMPR2*$^{-/-}$ mice, attenuated the muscularization of small pulmonary vessels, and eliminated the presence of vessel occlusion occasionally seen in *BMPR2*$^{-/-}$ mutant mouse lungs [53].

Conditional knock-out of *SMAD1*, a receptor-regulated Smad phosphorylated by BMPR2, in either endothelial cells or smooth muscle cells of mice has been previously shown to cause elevated RVSP and increased muscularization of pulmonary arteries [54]. Interestingly, only female conditional knock-out *SMAD1*$^{+/-}$ mice develop elevated RVSP and pulmonary vascular remodeling, and PASMCs isolated from these mice proliferated faster than those of female wild type mice, suggesting synergism between female sex and the heterozygous loss of *SMAD1* in adult mice. Ovariectomy attenuated the PH phenotype in *SMAD1*$^{+/-}$ females, further suggesting that it is the presence of female sex hormones that drives a difference in penetrance between sexes [50].

5.3. Estrogen Metabolites May Mediate Interaction between BMPR2 and Estrogen Signaling

Recent attention to estrogen metabolites and related enzymes has provided more evidence supporting the interaction between *BMPR2* and estrogen signaling pathways in the development of PAH. CYP1B1 is a p450 enzyme that is highly expressed in lung tissue, and catalyzes 2- and 4-hydroxylation of estrogen. A different p450 enzyme catalyzes the hydroxylation of estrogen at C-16, primarily to 16α-hydroxyestrone (16α-OHE$_1$). A low ratio of urinary 2-hydroxyestrogen (2-OHE)/16α-OHE$_1$ is used as a biologic marker of decreased CYP1B1 activity. In contrast to 2- and 4-hydroxy estrogen metabolites (which are considered weakly or anti-mitogenic), 16α-OHE$_1$ drives cellular proliferation via activation of the estrogen receptor, and preferential hydroxylation to 16α-OHE$_1$ with a low 2-OHE/16α-OHE$_1$ ratio has been associated with an increased risk of diseases resulting from the proliferative effects of estrogen signaling [55].

Comparison of gene array and RT-PCR from cultured B-cell lines of PAH-affected carriers of deleterious *BMPR2* variants, non-affected carriers, and normal control patients identified significantly lower expression of *CYP1B1* in the female PAH-affected *BMPR2* carriers [56]. A second study examined affected and unaffected female *BMPR2* mutation carriers for the presence of *CYP1B1 N453S*, a genetic polymorphism associated with increased protein degradation and previously implicated in hormone-related malignancies (breast, ovarian, prostate, and endometrial cancer) [57,58]. Similar to the results from West et al. [56], the investigation revealed a four-fold higher penetrance of PAH in carriers of deleterious *BMPR2* variants who were homozygous for the polymorphism. Supporting this genetic observation, the urinary 2-OHE/16α-OHE$_1$ ratio was $2.3\times$ lower in affected mutation carriers [58].

Two follow-up studies further examined the role of 16α-OHE$_1$ in mediating the development of *BMPR2*-associated PH. Fessel et al. [59] found a higher ratio of 16α-OHE$_1$/2-OHE (note that this is the inverse ratio from the previous study) in male patients with heritable PAH compared to healthy male controls, although the ratio was less divergent in males compared to the difference previously identified in females [59]. The same study demonstrated that in several genotypes of male *BMPR2* mutant mice (males were used to avoid the complexity of the native estrous cycle in female mice), administration of 16α-OHE$_1$ roughly doubled disease penetrance (as defined by an elevated PVR) and, in one genotype, reduced cardiac output significantly. In a similar experiment, administration of 2-OHE was not found to be protective. Administration of 16α-OHE$_1$ resulted in a relatively lung-specific decrease in Smad 1/5/8 phosphorylation and a decrease in BMPR2 protein level in control animals, but no further reduction in Smad phosphorylation in *BMPR2* mutants, suggesting a differential mechanism of how 16α-OHE$_1$ may act as a second hit in the presence of *BMPR2* mutations. Gene expression data from these experiments demonstrated that in the presence of *BMPR2* mutations, 16α-OHE$_1$ blunts classical cytokine and inflammatory signaling (as previously shown in the literature) but promotes vascular injury through unclear alternative mechanisms (angiogenesis, metabolism, and planar polarity). In a separate study, 16α-OHE$_1$ was shown to upregulate the microRNA-29 family (miR-29) in *BMPR2* mutants, which alters energy metabolism; antagonism of miR-29 improved the in vivo and in vitro features of PH [60].

6. Conclusions

Taken together, scientific investigations demonstrate multifaceted interactions between estrogens, estrogen metabolites, and BMPR2 signaling. The data suggest that abnormalities in BMPR2 signaling pathways may synergize with estrogenic signaling to create a permissive environment for promoting PAH development. Such a paradigm could explain why there is a discrepancy between pro-proliferative and PAH-promoting estrogen effects noted in these studies and protective estrogen effects in other contexts and model systems [31]. In the context of genetic alterations in BMPR2 signaling, estrogens may exert PAH-promoting effects that they do not exert in other contexts. In particular, there may be a shift towards the production of mitogenic metabolites such as 16α-OHE$_1$. Further study may elucidate exact mechanisms and potential therapeutic targets.

Author Contributions: M.M.C., T.L., and G.E. conceptualized the review content; M.M.C., M.W.D., L.M.B., S.M.B., T.L., and G.E. contributed to drafting the manuscript. All authors have read and agreed to the published version of the manuscript.

Funding: M.W.D. receives research funding from Actelion Pharmaceuticals through the Entelligence Young Investigator Program. S.M.B. serves as a DSMB chair for a clinical trial in respiratory failure funded by Hamilton Medical. S.M.B. receives consultancy fees from Faron Pharmaceuticals and Sedana Pharmaceuticals for service on study steering committees for two trials in acute respiratory distress.

Acknowledgments: TL receives research funding via the Department of Veterans Affairs Merit Review Award 1I01BX002042 and NIH 1R01HL144727-01A1. SMB is supported by NHLBI (R01HL144624).

Conflicts of Interest: The authors declare no conflict of interest.

References

1. Humbert, M.; Guignabert, C.; Bonnet, S.; Dorfmuller, P.; Klinger, J.R.; Nicolls, M.R.; Olschewski, A.J.; Pullamsetti, S.S.; Schermuly, R.T.; Stenmark, K.R.; et al. Pathology and pathobiology of pulmonary hypertension: State of the art and research perspectives. *Eur. Respir. J.* **2019**, *53*, 1–14. [CrossRef] [PubMed]
2. Andruska, A.; Spiekerkoetter, E. Consequences of BMPR2 Deficiency in the Pulmonary Vasculature and Beyond: Contributions to Pulmonary Arterial Hypertension. *Int. J. Mol. Sci.* **2018**, *19*, 2499. [CrossRef] [PubMed]
3. Lane, K.B.; Machado, R.D.; Pauciulo, M.W.; Thomson, J.R.; Phillips, J.A., 3rd; Loyd, J.E.; Nichols, W.C.; Trembath, R.C. Heterozygous germline mutations in BMPR2, encoding a TGF-beta receptor, cause familial primary pulmonary hypertension. *Nat. Genet.* **2000**, *26*, 81–84. [CrossRef] [PubMed]
4. Deng, Z.; Morse, J.H.; Slager, S.L.; Cuervo, N.; Moore, K.J.; Venetos, G.; Kalachikov, S.; Cayanis, E.; Fischer, S.G.; Barst, R.J.; et al. Familial primary pulmonary hypertension (gene PPH1) is caused by mutations in the bone morphogenetic protein receptor-II gene. *Am. J. Hum. Genet.* **2000**, *67*, 737–744. [CrossRef] [PubMed]
5. Machado, R.D.; Aldred, M.A.; James, V.; Harrison, R.E.; Patel, B.; Schwalbe, E.C.; Gruenig, E.; Janssen, B.; Koehler, R.; Seeger, W.; et al. Mutations of the TGF-beta type II receptor BMPR2 in pulmonary arterial hypertension. *Hum. Mutat.* **2006**, *27*, 121–132. [CrossRef]
6. Cogan, J.D.; Pauciulo, M.W.; Batchman, A.P.; Prince, M.A.; Robbins, I.M.; Hedges, L.K.; Stanton, K.C.; Wheeler, L.A.; Phillips, J.A., 3rd; Loyd, J.E.; et al. High frequency of BMPR2 exonic deletions/duplications in familial pulmonary arterial hypertension. *Am. J. Respir. Crit. Care Med.* **2006**, *174*, 590–598. [CrossRef]
7. Aldred, M.A.; Vijayakrishnan, J.; James, V.; Soubrier, F.; Gomez-Sanchez, M.A.; Martensson, G.; Galie, N.; Manes, A.; Corris, P.; Simonneau, G.; et al. BMPR2 gene rearrangements account for a significant proportion of mutations in familial and idiopathic pulmonary arterial hypertension. *Hum. Mutat.* **2006**, *27*, 212–213. [CrossRef]
8. Thomson, J.R.; Machado, R.D.; Pauciulo, M.W.; Morgan, N.V.; Humbert, M.; Elliott, G.C.; Ward, K.; Yacoub, M.; Mikhail, G.; Rogers, P.; et al. Sporadic primary pulmonary hypertension is associated with germline mutations of the gene encoding BMPR-II, a receptor member of the TGF-beta family. *J. Med. Genet.* **2000**, *37*, 741–745. [CrossRef]
9. Chen, N.Y.; Collum, S.D.; Luo, F.; Weng, T.; Le, T.T.; Hernandez, A.M.; Philip, K.; Molina, J.G.; Garcia-Morales, L.J.; Cao, Y.; et al. Macrophage bone morphogenic protein receptor 2 depletion in idiopathic pulmonary fibrosis and Group III pulmonary hypertension. *Am. J. Physiol. Lung Cell. Mol. Physiol.* **2016**, *311*, L238–L254. [CrossRef]
10. Atkinson, C.; Stewart, S.; Upton, P.D.; Machado, R.; Thomson, J.R.; Trembath, R.C.; Morrell, N.W. Primary pulmonary hypertension is associated with reduced pulmonary vascular expression of type II bone morphogenetic protein receptor. *Circulation* **2002**, *105*, 1672–1678. [CrossRef]
11. Evans, J.D.; Girerd, B.; Montani, D.; Wang, X.J.; Galie, N.; Austin, E.D.; Elliott, G.; Asano, K.; Grunig, E.; Yan, Y.; et al. BMPR2 mutations and survival in pulmonary arterial hypertension: An individual participant data meta-analysis. *Lancet Respir. Med.* **2016**, *4*, 129–137. [CrossRef]
12. Sztrymf, B.; Coulet, F.; Girerd, B.; Yaici, A.; Jais, X.; Sitbon, O.; Montani, D.; Souza, R.; Simonneau, G.; Soubrier, F.; et al. Clinical outcomes of pulmonary arterial hypertension in carriers of BMPR2 mutation. *Am. J. Respir. Crit. Care Med.* **2008**, *177*, 1377–1383. [CrossRef] [PubMed]
13. Elliott, C.G.; Glissmeyer, E.W.; Havlena, G.T.; Carlquist, J.; McKinney, J.T.; Rich, S.; McGoon, M.D.; Scholand, M.B.; Kim, M.; Jensen, R.L.; et al. Relationship of BMPR2 mutations to vasoreactivity in pulmonary arterial hypertension. *Circulation* **2006**, *113*, 2509–2515. [CrossRef] [PubMed]
14. Rosenzweig, E.B.; Morse, J.H.; Knowles, J.A.; Chada, K.K.; Khan, A.M.; Roberts, K.E.; McElroy, J.J.; Juskiw, N.K.; Mallory, N.C.; Rich, S.; et al. Clinical implications of determining BMPR2 mutation status in a large cohort of children and adults with pulmonary arterial hypertension. *J. Heart Lung Transpl.* **2008**, *27*, 668–674. [CrossRef] [PubMed]
15. Austin, E.D.; Loyd, J.E.; Phillips, J.A., 3rd. Genetics of pulmonary arterial hypertension. *Semin. Respir. Crit. Care Med.* **2009**, *30*, 386–398. [CrossRef] [PubMed]

16. Lahm, T.; Tuder, R.M.; Petrache, I. Progress in solving the sex hormone paradox in pulmonary hypertension. *Am. J. Physiol. Lung Cell. Mol. Physiol.* **2014**, *307*, L7–L26. [CrossRef]

17. Larkin, E.K.; Newman, J.H.; Austin, E.D.; Hemnes, A.R.; Wheeler, L.; Robbins, I.M.; West, J.D.; Phillips, J.A., 3rd; Hamid, R.; Loyd, J.E. Longitudinal analysis casts doubt on the presence of genetic anticipation in heritable pulmonary arterial hypertension. *Am. J. Respir. Crit. Care Med.* **2012**, *186*, 892–896. [CrossRef]

18. Ge, X.; Zhu, T.; Zhang, X.; Liu, Y.; Wang, Y.; Zhang, W. Gender differences in pulmonary arterial hypertension patients with BMPR2 mutation: A meta-analysis. *Respir. Res.* **2020**, *21*, 44. [CrossRef]

19. Frump, A.L.; Goss, K.N.; Vayl, A.; Albrecht, M.; Fisher, A.; Tursunova, R.; Fierst, J.; Whitson, J.; Cucci, A.R.; Brown, M.B.; et al. Estradiol improves right ventricular function in rats with severe angioproliferative pulmonary hypertension: Effects of endogenous and exogenous sex hormones. *Am. J. Physiol. Lung Cell. Mol. Physiol.* **2015**, *308*, L873–L890. [CrossRef]

20. Umar, S.; Iorga, A.; Matori, H.; Nadadur, R.D.; Li, J.; Maltese, F.; Van der Laarse, A.; Eghbali, M. Estrogen rescues preexisting severe pulmonary hypertension in rats. *Am. J. Respir. Crit. Care Med.* **2011**, *184*, 715–723. [CrossRef]

21. Liu, Z.; Duan, Y.L.; Ge, S.L.; Zhang, C.X.; Gong, W.H.; Xu, J.J. Effect of estrogen on right ventricular remodeling of monocrotaline-induced pulmonary arterial hypertension in rats and its mechanism. *Eur. Rev. Med. Pharm. Sci.* **2019**, *23*, 1742–1750. [CrossRef]

22. Lahm, T.; Albrecht, M.; Fisher, A.J.; Selej, M.; Patel, N.G.; Brown, J.A.; Justice, M.J.; Brown, M.B.; Van Demark, M.; Trulock, K.M.; et al. 17beta-Estradiol attenuates hypoxic pulmonary hypertension via estrogen receptor-mediated effects. *Am. J. Respir. Crit. Care Med.* **2012**, *185*, 965–980. [CrossRef] [PubMed]

23. Nadadur, R.D.; Umar, S.; Wong, G.; Eghbali, M.; Iorga, A.; Matori, H.; Partow-Navid, R.; Eghbali, M. Reverse right ventricular structural and extracellular matrix remodeling by estrogen in severe pulmonary hypertension. *J. Appl. Physiol.* **2012**, *113*, 149–158. [CrossRef] [PubMed]

24. Lahm, T.; Frump, A.L.; Albrecht, M.E.; Fisher, A.J.; Cook, T.G.; Jones, T.J.; Yakubov, B.; Whitson, J.; Fuchs, R.K.; Liu, A.; et al. 17beta-Estradiol mediates superior adaptation of right ventricular function to acute strenuous exercise in female rats with severe pulmonary hypertension. *Am. J. Physiol. Lung Cell. Mol. Physiol.* **2016**, *311*, L375–L388. [CrossRef] [PubMed]

25. Ventetuolo, C.E.; Ouyang, P.; Bluemke, D.A.; Tandri, H.; Barr, R.G.; Bagiella, E.; Cappola, A.R.; Bristow, M.R.; Johnson, C.; Kronmal, R.A.; et al. Sex hormones are associated with right ventricular structure and function: The MESA-right ventricle study. *Am. J. Respir. Crit. Care Med.* **2011**, *183*, 659–667. [CrossRef]

26. Swift, A.J.; Capener, D.; Hammerton, C.; Thomas, S.M.; Elliot, C.; Condliffe, R.; Wild, J.M.; Kiely, D.G. Right ventricular sex differences in patients with idiopathic pulmonary arterial hypertension characterised by magnetic resonance imaging: Pair-matched case controlled study. *PLoS ONE* **2015**, *10*, e0127415. [CrossRef]

27. Jacobs, W.; Van de Veerdonk, M.C.; Trip, P.; De Man, F.; Heymans, M.W.; Marcus, J.T.; Kawut, S.M.; Bogaard, H.J.; Boonstra, A.; Vonk Noordegraaf, A. The right ventricle explains sex differences in survival in idiopathic pulmonary arterial hypertension. *Chest* **2014**, *145*, 1230–1236. [CrossRef]

28. Kozu, K.; Sugimura, K.; Aoki, T.; Tatebe, S.; Yamamoto, S.; Yaoita, N.; Shimizu, T.; Nochioka, K.; Sato, H.; Konno, R.; et al. Sex differences in hemodynamic responses and long-term survival to optimal medical therapy in patients with pulmonary arterial hypertension. *Heart Vessel.* **2018**, *33*, 939–947. [CrossRef]

29. Umar, S.; Cunningham, C.M.; Itoh, Y.; Moazeni, S.; Vaillancourt, M.; Sarji, S.; Centala, A.; Arnold, A.P.; Eghbali, M. The Y Chromosome Plays a Protective Role in Experimental Hypoxic Pulmonary Hypertension. *Am. J. Respir. Crit. Care Med.* **2018**, *197*, 952–955. [CrossRef]

30. Yan, L.; Cogan, J.D.; Hedges, L.K.; Nunley, B.; Hamid, R.; Austin, E.D. The Y Chromosome Regulates BMPR2 Expression via SRY: A Possible Reason "Why" Fewer Males Develop Pulmonary Arterial Hypertension. *Am. J. Respir. Crit. Care Med.* **2018**, *198*, 1581–1583. [CrossRef]

31. Hester, J.; Ventetuolo, C.; Lahm, T. Sex, Gender, and Sex Hormones in Pulmonary Hypertension and Right Ventricular Failure. *Compr. Physiol.* **2019**, *10*, 125–170. [CrossRef] [PubMed]

32. Tofovic, S.P.; Jackson, E.K. Estradiol Metabolism: Crossroads in Pulmonary Arterial Hypertension. *Int. J. Mol. Sci.* **2019**, *21*, 116. [CrossRef] [PubMed]

33. Yang, J.; Li, X.; Morrell, N.W. Id proteins in the vasculature: From molecular biology to cardiopulmonary medicine. *Cardiovasc Res.* **2014**, *104*, 388–398. [CrossRef] [PubMed]

34. De Vinuesa, A.G.; Abdelilah-Seyfried, S.; Knaus, P.; Zwijsen, A.; Bailly, S. BMP signaling in vascular biology and dysfunction. *Cytokine Growth Factor Rev.* **2016**, *27*, 65–79. [CrossRef]

35. Girerd, B.; Montani, D.; Coulet, F.; Sztrymf, B.; Yaici, A.; Jais, X.; Tregouet, D.; Reis, A.; Drouin-Garraud, V.; Fraisse, A.; et al. Clinical outcomes of pulmonary arterial hypertension in patients carrying an ACVRL1 (ALK1) mutation. *Am. J. Respir. Crit. Care Med.* **2010**, *181*, 851–861. [CrossRef]

36. Shintani, M.; Yagi, H.; Nakayama, T.; Saji, T.; Matsuoka, R. A new nonsense mutation of SMAD8 associated with pulmonary arterial hypertension. *J. Med. Genet.* **2009**, *46*, 331–337. [CrossRef] [PubMed]

37. Wang, G.; Fan, R.; Ji, R.; Zou, W.; Penny, D.J.; Varghese, N.P.; Fan, Y. Novel homozygous BMP9 nonsense mutation causes pulmonary arterial hypertension: A case report. *BMC Pulm. Med.* **2016**, *16*, 17. [CrossRef]

38. Austin, E.D.; Ma, L.; LeDuc, C.; Berman Rosenzweig, E.; Borczuk, A.; Phillips, J.A., 3rd; Palomero, T.; Sumazin, P.; Kim, H.R.; Talati, M.H.; et al. Whole exome sequencing to identify a novel gene (caveolin-1) associated with human pulmonary arterial hypertension. *Circ. Cardiovasc. Genet.* **2012**, *5*, 336–343. [CrossRef]

39. Hamidi, S.A.; Dickman, K.G.; Berisha, H.; Said, S.I. 17beta-estradiol protects the lung against acute injury: Possible mediation by vasoactive intestinal polypeptide. *Endocrinology* **2011**, *152*, 4729–4737. [CrossRef]

40. Dougherty, S.M.; Mazhawidza, W.; Bohn, A.R.; Robinson, K.A.; Mattingly, K.A.; Blankenship, K.A.; Huff, M.O.; McGregor, W.G.; Klinge, C.M. Gender difference in the activity but not expression of estrogen receptors alpha and beta in human lung adenocarcinoma cells. *Endocr. Relat. Cancer* **2006**, *13*, 113–134. [CrossRef]

41. Carmeci, C.; Thompson, D.A.; Ring, H.Z.; Francke, U.; Weigel, R.J. Identification of a gene (GPR30) with homology to the G-protein-coupled receptor superfamily associated with estrogen receptor expression in breast cancer. *Genomics* **1997**, *45*, 607–617. [CrossRef] [PubMed]

42. Simoncini, T.; Hafezi-Moghadam, A.; Brazil, D.P.; Ley, K.; Chin, W.W.; Liao, J.K. Interaction of oestrogen receptor with the regulatory subunit of phosphatidylinositol-3-OH kinase. *Nature* **2000**, *407*, 538–541. [CrossRef] [PubMed]

43. Shaul, P.W. Rapid activation of endothelial nitric oxide synthase by estrogen. *Steroids* **1999**, *64*, 28–34. [CrossRef]

44. Heldring, N.; Pike, A.; Andersson, S.; Matthews, J.; Cheng, G.; Hartman, J.; Tujague, M.; Strom, A.; Treuter, E.; Warner, M.; et al. Estrogen receptors: How do they signal and what are their targets. *Physiol. Rev.* **2007**, *87*, 905–931. [CrossRef] [PubMed]

45. Dubey, R.K.; Gillespie, D.G.; Zacharia, L.C.; Rosselli, M.; Korzekwa, K.R.; Fingerle, J.; Jackson, E.K. Methoxyestradiols mediate the antimitogenic effects of estradiol on vascular smooth muscle cells via estrogen receptor-independent mechanisms. *Biochem. Biophys. Res. Commun.* **2000**, *278*, 27–33. [CrossRef]

46. Fenoy, F.J.; Hernandez, M.E.; Hernandez, M.; Quesada, T.; Salom, M.G.; Hernandez, I. Acute effects of 2-methoxyestradiol on endothelial aortic No release in male and ovariectomized female rats. *Nitric Oxide* **2010**, *23*, 12–19. [CrossRef]

47. Tofovic, S.P. Estrogens and development of pulmonary hypertension: Interaction of estradiol metabolism and pulmonary vascular disease. *J. Cardiovasc. Pharm.* **2010**, *56*, 696–708. [CrossRef]

48. Fishman, J.; Martucci, C. Biological properties of 16 alpha-hydroxyestrone: Implications in estrogen physiology and pathophysiology. *J. Clin. Endocrinol. Metab.* **1980**, *51*, 611–615. [CrossRef]

49. Austin, E.D.; Hamid, R.; Hemnes, A.R.; Loyd, J.E.; Blackwell, T.; Yu, C.; Phillips Iii, J.A.; Gaddipati, R.; Gladson, S.; Gu, E.; et al. BMPR2 expression is suppressed by signaling through the estrogen receptor. *Biol. Sex Differ.* **2012**, *3*, 6. [CrossRef]

50. Mair, K.M.; Yang, X.D.; Long, L.; White, K.; Wallace, E.; Ewart, M.A.; Docherty, C.K.; Morrell, N.W.; MacLean, M.R. Sex affects bone morphogenetic protein type II receptor signaling in pulmonary artery smooth muscle cells. *Am. J. Respir. Crit. Care Med.* **2015**, *191*, 693–703. [CrossRef]

51. Mair, K.M.; Wright, A.F.; Duggan, N.; Rowlands, D.J.; Hussey, M.J.; Roberts, S.; Fullerton, J.; Nilsen, M.; Loughlin, L.; Thomas, M.; et al. Sex-dependent influence of endogenous estrogen in pulmonary hypertension. *Am. J. Respir. Crit. Care Med.* **2014**, *190*, 456–467. [CrossRef] [PubMed]

52. Gomez-Arroyo, J.; Saleem, S.J.; Mizuno, S.; Syed, A.A.; Bogaard, H.J.; Abbate, A.; Taraseviciene-Stewart, L.; Sung, Y.; Kraskauskas, D.; Farkas, D.; et al. A brief overview of mouse models of pulmonary arterial hypertension: Problems and prospects. *Am. J. Physiol. Lung Cell. Mol. Physiol.* **2012**, *302*, L977–L991. [CrossRef] [PubMed]

53. Chen, X.; Austin, E.D.; Talati, M.; Fessel, J.P.; Farber-Eger, E.H.; Brittain, E.L.; Hemnes, A.R.; Loyd, J.E.; West, J. Oestrogen inhibition reverses pulmonary arterial hypertension and associated metabolic defects. *Eur. Respir. J.* **2017**, *50*, 1–14. [CrossRef] [PubMed]

54. Han, C.; Hong, K.H.; Kim, Y.H.; Kim, M.J.; Song, C.; Kim, M.J.; Kim, S.J.; Raizada, M.K.; Oh, S.P. SMAD1 deficiency in either endothelial or smooth muscle cells can predispose mice to pulmonary hypertension. *Hypertension* **2013**, *61*, 1044–1052. [CrossRef]

55. Muti, P.; Bradlow, H.L.; Micheli, A.; Krogh, V.; Freudenheim, J.L.; Schunemann, H.J.; Stanulla, M.; Yang, J.; Sepkovic, D.W.; Trevisan, M.; et al. Estrogen metabolism and risk of breast cancer: A prospective study of the 2:16alpha-hydroxyestrone ratio in premenopausal and postmenopausal women. *Epidemiology* **2000**, *11*, 635–640. [CrossRef]

56. West, J.; Cogan, J.; Geraci, M.; Robinson, L.; Newman, J.; Phillips, J.A.; Lane, K.; Meyrick, B.; Loyd, J. Gene expression in BMPR2 mutation carriers with and without evidence of pulmonary arterial hypertension suggests pathways relevant to disease penetrance. *BMC Med. Genom.* **2008**, *1*, 45. [CrossRef]

57. Gajjar, K.; Martin-Hirsch, P.L.; Martin, F.L. CYP1B1 and hormone-induced cancer. *Cancer Lett.* **2012**, *324*, 13–30. [CrossRef]

58. Austin, E.D.; Cogan, J.D.; West, J.D.; Hedges, L.K.; Hamid, R.; Dawson, E.P.; Wheeler, L.A.; Parl, F.F.; Loyd, J.E.; Phillips, J.A., 3rd. Alterations in oestrogen metabolism: Implications for higher penetrance of familial pulmonary arterial hypertension in females. *Eur. Respir. J.* **2009**, *34*, 1093–1099. [CrossRef]

59. Fessel, J.P.; Chen, X.; Frump, A.; Gladson, S.; Blackwell, T.; Kang, C.; Johnson, J.; Loyd, J.E.; Hemnes, A.; Austin, E.; et al. Interaction between bone morphogenetic protein receptor type 2 and estrogenic compounds in pulmonary arterial hypertension. *Pulm. Circ.* **2013**, *3*, 564–577. [CrossRef]

60. Chen, X.; Talati, M.; Fessel, J.P.; Hemnes, A.R.; Gladson, S.; French, J.; Shay, S.; Trammell, A.; Phillips, J.A.; Hamid, R.; et al. Estrogen Metabolite 16alpha-Hydroxyestrone Exacerbates Bone Morphogenetic Protein Receptor Type II-Associated Pulmonary Arterial Hypertension Through MicroRNA-29-Mediated Modulation of Cellular Metabolism. *Circulation* **2016**, *133*, 82–97. [CrossRef]

Publisher's Note: MDPI stays neutral with regard to jurisdictional claims in published maps and institutional affiliations.

eview

Novel Advances in Modifying BMPR2 Signaling in PAH

venja Dannewitz Prosseda [1,2,3], Md Khadem Ali [1,2] and Edda Spiekerkoetter [1,2,*]

1   Division Pulmonary, Allergy and Critical Care Medicine, Department of Medicine, Stanford University, Stanford, CA 94305, USA; svenja.dannewitz@gmail.com (S.D.P.); mdali@stanford.edu (M.K.A.)
2   Vera Moulton Wall Center for Pulmonary Vascular Diseases, Stanford, CA 94305, USA
3   Institute for Experimental and Clinical Pharmacology and Toxicology, Albert-Ludwigs University Freiburg, 79104 Freiburg, Germany
*   Correspondence: eddas@stanford.edu

Abstract: Pulmonary Arterial Hypertension (PAH) is a disease of the pulmonary arteries, that is characterized by progressive narrowing of the pulmonary arterial lumen and increased pulmonary vascular resistance, ultimately leading to right ventricular dysfunction, heart failure and premature death. Current treatments mainly target pulmonary vasodilation and leave the progressive vascular remodeling unchecked resulting in persistent high morbidity and mortality in PAH even with treatment. Therefore, novel therapeutic strategies are urgently needed. Loss of function mutations of the Bone Morphogenetic Protein Receptor 2 (BMPR2) are the most common genetic factor in hereditary forms of PAH, suggesting that the BMPR2 pathway is fundamentally important in the pathogenesis. Dysfunctional BMPR2 signaling recapitulates the cellular abnormalities in PAH as well as the pathobiology in experimental pulmonary hypertension (PH). Approaches to restore BMPR2 signaling by increasing the expression of BMPR2 or its downstream signaling targets are currently actively explored as novel ways to prevent and improve experimental PH as well as PAH in patients. Here, we summarize existing as well as novel potential treatment strategies for PAH that activate the BMPR2 receptor pharmaceutically or genetically, increase the receptor availability at the cell surface, or reconstitute downstream BMPR2 signaling.

Keywords: PAH; pulmonary hypertension; bone morphogenetic protein receptor 2; signaling; repurposed drugs; pharmaceuticals; miRNA; clinical trials

Citation: Dannewitz Prosseda, S.; Ali, .K.; Spiekerkoetter, E. Novel Advances Modifying BMPR2 Signaling in PAH. *enes* **2021**, *12*, 8. https://dx.doi.org/ .3390/genes12010008

eceived: 30 November 2020
ccepted: 21 December 2020
ublished: 23 December 2020

ublisher's Note: MDPI stays neu- al with regard to jurisdictional claims published maps and institutional filiations.

1. Introduction

Pulmonary Arterial Hypertension (PAH) is a cardio-pulmonary-vascular condition, where a progressive occlusion of the distal pulmonary vasculature leads to an increase in pulmonary vascular resistance and right ventricular (RV) afterload, resulting in RV failure and premature death [1,2]. Histopathological analysis suggests that dysfunction of key cellular components of the pulmonary vasculature, namely endothelial and smooth muscle cells, pericytes, inflammatory cells, and adventitial fibroblasts, induce pulmonary vascular remodeling [3,4]. This results in narrowing of the vessel lumen and formation of complex vascular lesions, which together raise pulmonary vascular resistance, increasing pulmonary arterial pressure as well as the afterload for the right ventricle.

Although PAH is a rare disease affecting only about 1–2 of every 1 million individuals annually, the mortality and morbidity rate is high and, if untreated, PAH quickly leads to right ventricle failure and death after 2–3 years [5,6]. PAH may be heritable (with a family history of PAH), idiopathic (without a family history, unknown cause), or associated (linked to interstitial lung disease, congenital heart disease, autoimmune disease, etc.) [7]. Whilst the exact cause of PAH is not known, genetic factors (mutations or epigenetic changes), environmental factors (e.g., hypoxia, viral infections, anorectic agents, stimulants, etc.) and immune or inflammatory triggers may contribute to the cause or progression of the disease [4]. Importantly, there is no cure for PAH. Existing drugs target pulmonary vasodilation, proliferation and endothelial function by increasing nitric oxide (NO), inhibiting

endothelin and voltage-gated calcium channels and by augmenting prostacyclin signalin
pathways [8]. However, these drugs only partially increase survival and improve quality c
life, while the majority of patients ultimately become resistant to medication and succum
to the disease [9]. With current treatments, the 5-year survival of PAH patients has bee
improved from 34% to 60%, yet these drugs are not capable of reducing the extent an
progression of vascular and cardiac remodeling, resulting in eventual clinical deterioratio
of PAH patients over time [10].

Thus, new, effective and disease modifying therapies are urgently needed [11], ther
pies that target the underlying molecular mechanisms responsible for pulmonary vascula
remodeling, which is the hallmark of PAH. Over the past two decades, many cellular an
molecular mechanisms have been described as playing key roles in the pathogenesis c
disease in preclinical and clinical settings [4,12]. Here, we focus on modulation of bon
morphogenic protein receptor 2 (BMPR2) signaling [7] as a key mechanistic pathway an
potential master switch in the pathogenesis of PAH.

2. The BMPR2 Signaling Pathway

In 2000 two independent groups identified mutations in BMPR2 as causative for th
familial form of PAH [13,14]. BMPR2 carriers with PAH have an earlier disease onse
than idiopathic PAH patients [15]. Interestingly, male patients were more likely to posses
a BMPR2 mutation than women and develop severe disease in presence of a BMPR
mutation [16].

Meanwhile, researchers have identified mutations in over 16 genes in patient wit
hereditary PAH (HPAH) that may predispose to PAH, including BMPR2 of course, but als
receptors that are part of or are interacting with the BMPR2 pathway such as activin *
receptor type II-like 1 (ACVRL1), endoglin (ENG), caveolin-1 (CAV1), SMAD1, SMAD
SMAD9, bone morphogenetic protein receptor type 1B (BMPR1B), eukaryotic translatio
initiation factor 2 α kinase 4 (EIF2AK4), and growth differentiation factor 2 (GDF2) [17
While most identified gene mutations are relatively rare (1–3% cases), heterozygous los
of-function mutations in the BMPR2 gene are the most common and occur in 53–86%
of HPAH and 14%–35% of idiopathic PAH (IPAH) patients [18]. To date, more than 30
mutations, predominantly nonsense and frameshift types, have been identified in th
BMPR2 gene in PAH patients. BMPR2, encoded by the *BMPR2* gene, is a member of th
serine/threonine kinase transmembrane proteins belonging to the TGFβ receptor supe
family. BMPR2 binds BMP ligands such as BMP2, BMP4, BMP6, BMP7 and BMP9. BMP
typically play a role in a wide range of signal pathways involved in cellular differentiatio
growth, and apoptosis and in embryogenesis, development, and tissue homeostasis. In th
canonical BMP signaling pathway, upon binding of BMP ligands, BMP type 2 receptor
(e.g., BMPR2 (ActRIIA) and ActRIIB)) recruit, complex and phosphorylate BMP type
receptors (e.g., Activin receptor-like kinase 1(ALK1), BMPR-1A (ALK3), BMPR-1B (ALK6
and ActR-1A (ALK2)), which then phosphorylate receptor-regulated SMADs (R-SMADs
These R-SMADs form a complex with co-SMADs (e.g., SMAD4) and translocate to th
nucleus where the complex binds to a BMP response element DNA sequence. As a resu
the complex acts as transcriptional regulator of target gene expression including Inhibito
of DNA Binding 1, 2, and 3 (ID1, ID2, ID3) or cyclin-dependent kinase inhibitor 1A and 2
(CDKN1A and CDKN2B) by binding to the BMP responsive element (BRE), which play
a critical role in cell proliferation, apoptosis and migration. In addition to the canonica
SMAD mediated signaling pathway, several non-canonical BMP signaling pathways ar
also activated by BMPR2, including p38 Mitogen-Activated Protein Kinase (MAPK), Extra
cellular Signal-Regulated Kinase (ERK), Phosphoinositide 3-kinase (PI3K)/Akt signalin
Peroxisome proliferator-activated receptor γ (PPAR γ)/Apolipoprotein E (ApoE)/ Hig
–density lipoprotein cholesterol (HDLC), Wingless (Wnt), Caveolin, Rho-GTPases, Protei
Kinase C (PKC) signaling and NOTCH signaling [19].

3. Regulation of BMPR2 Signaling

A tight regulation of BMPR2 signaling is exerted by extracellular agonists and antagonists, such as the inhibitory molecule Noggin, Chordin and gremlin1 [20], which is upregulated by endothelin1 [21]. Intracellularly, a feedback loop controls BMPR2 signaling through the activity of inhibitory SMADs (iSMADs) SMAD6 and SMAD7, which inhibit the phosphorylation of SMAD2 and SMAD3, signaling molecules that function as counterparts to SMAD 1/5 signaling. SMAD1 degradation is initiated through SMURF1 and SMURF2 targeting, which downregulates further downstream gene expression [22]. BMPR2 downstream signaling is further regulated by FK binding protein 12 (FKBP12), which prevents the activation and phosphorylation of type 1 receptors in absence of a ligand [23] FKBP12 furthermore maintains the balance of rSMAD and iSMAD signaling, by regulating SMAD2/3 activity and recruiting SMAD7 [24].

The availability of BMPR2 receptors at the cell surface is provided by the balance of receptor expression and degradation, as well as receptor shuttling to the cell surface [25]. While upregulation of BMPR2 receptor expression has recently been explored as a therapeutic strategy in PAH [26], little is known about intracellular signaling molecules that target BMPR2 expression. We recently explored upstream modulators of BMPR2 expression and described two novel players in BMPR2 signaling that can increase BMPR2 expression, namely Fragile Histidine Triad (FHIT) and lymphocyte-specific protein tyrosine kinase (LCK) [27].

In contrast to the lack of data on the positive regulation of BMPR2 expression, the mechanisms of its downregulation are well-described, whereas regulation via micro RNAs (miRs) and receptor degradation play major roles. miR-20a and miR17 have both been connected to the downregulation of BMPR2 expression [28,29] whereas the miR17-92 cluster downregulated BMPR2 by engaging the inflammatory cytokine IL-6 via STAT3 [30]. Hypoxia downregulates BMPR2 signaling through miR-21 and miR-125a [31]. miR-302 targets BMPR2 signaling in PASMCs, thereby reducing their proliferation [32] miR21 is connected to a feedback inhibition of BMPR2 signaling, as its expression is induced by BMPR2 signaling on the one hand, but also reduces BMPR2 expression in PAECs. Therefore, the lack of its expression in vivo induces PH, while the use of miR-21 inhibitors in a rodent model of PH supports vascular regeneration in the hypoxia-remodeled pulmonary vasculature [33,34]. In addition to BMPR2 regulation by micro RNAs, it was recently described that 17-estradiol-induced binding of the estrogen receptor to the BMPR2 gene promotor, inhibited BMPR2 transcription, a finding that might explain the sex-based differences in PAH pathogenesis [35,36]. A reduction of BMPR2 receptor presence on the cell surface can be achieved by its premature degradation in connection to infection and inflammation. The inflammatory cytokine Tumor necrosis factor α (TNFalpha) activates metalloproteases that can cleave the receptor, and viral particles (i.e., Kaposi sarcoma-associated herpesvirus KSHV) can ubiquinate BMPR2, leading to its lysosomal degradation [37]. Furthermore, in the absence of BMPR2, SMAD signaling can shift from rSMAD-dominated signals of BMPR2 to the activation of the rSMADs SMAD2, SMAD3 and SMAD4, which are controlled by TGFβ [38] activating EC ITGB1 transcription, leading to EndMT, stress fiber production and actomyosin contractility.

Defective BMPR2 signaling caused by a mutational change in the BMPR2 gene can be rescued, as shown in unaffected BMPR2 mutant carriers through an effective feedback loop. When BMPR2 is functionally inactivated or reduced, the expression of receptor antagonists such as FKBP1A or Gremlin1 is reduced, while, similarly, cellular receptor activators are being upregulated [39].

4. BMPR2 Deficiency and Pulmonary Hypertension

Despite the high frequency of BMPR2 mutations in PAH patients, the disease penetrance rate is ~20% of the mutation carriers, suggesting that, in addition to BMPR2 mutations, other unidentified genetic, epigenetic, or environmental factors are involved in the development of the disease, potentially by decreasing BMPR2 expression and signaling activity below a specific threshold required to cause disease.

Furthermore, in PAH patients with and without BMPR2 mutations, BMPR2 expression and signaling activity is impaired in the pulmonary vasculature [40,41], suggesting that dysfunction of BMPR2 signaling is a key common feature in PAH patients.

Pulmonary endothelial-specific deletion of BMPR2 in mice recapitulates human PAH features [42]. PAH manifestations are also observed in mice expressing a dominant-negative BMPR2 gene in pulmonary smooth muscle cells [43,44]. Similarly, haplo-insufficient BMPR2 mutant rats developed severe dysfunction of the cardio-pulmonary-vascular system, such as distal vessel muscularization, loss of microvascular vessels, inflammation, RV and endothelial dysfunction as well as intrinsic cardiomyocyte dysfunction [45].

Impaired BMPR2 signaling is associated with aberrant vascular cell phenotypes, including pulmonary arterial endothelial cells (PAEC) apoptosis, hyperproliferation and apoptosis resistance of pulmonary arterial smooth muscle cells (PASMC), and inflammation [3,12]. These findings suggest that targeting and thereby increasing BMPR2 expression and signaling could be an effective therapeutic approach for treating PAH.

5. Therapeutic Strategies to Modify BMPR2 Signaling

As outlined above, the mechanistic causes of BMPR2 deficiency in PAH can be defined as either receptor inactivation, decreased receptor expression, or an impairment of the receptor's downstream signaling pathway [19]. In recent years, many novel approaches have emerged that target the BMPR2 pathway and are promising for clinical translation. Here, we have grouped and classified pharmacological and genetic interventions as follows: (a) targeting the BMPR2 receptor to increase its activity by pharmacological activators or gene-directed modulation, (b) increasing receptor availability at the cell surface by increasing signaling upstream of BMPR2, preventing receptor degradation or increasing the receptor shuttling to the cell surface, and (c) reconstituting BMPR2 downstream signaling by targeting interacting signaling pathways.

5.1. Targeting BMPR2 Receptor Activity

5.1.1. Receptor Activation

The activity of the BMPR2 receptor can be pharmaceutically increased by pharmacological activators, as long as a small quantity of BMPR2 exists and/or the potential of the BMPR2 protein to be activated is not prevented by the presence of a mutation in the BMPR gene. The most direct activation of BMPR2 signaling can be achieved pharmaceutically through the administration of recombinant BMP-9 ligand which has been proposed as therapeutic strategy for use in PAH [26].

5.1.2. Relieving Receptor Inhibition

The inhibition of the BMPR2 receptor can be pharmacological or genetical in nature. The functional activity of the BMPR2 receptor complex can be repressed by the intracellular binding of FKBP12 to the intracellular domain of the type 1 transmembrane receptor activin receptor-like kinase 1 (ALK1), ALK2, and ALK3 and presence of the phosphatase Calcineurin, which binds to FKBP12. The release of FKBP12 from the receptor complex by BMPR2 ligands activates downstream (intracellular) BMPR2 signaling [23].

The activation of the receptor complex could therefore be induced by inhibitors of Calcineurin and compounds that bind FKBP12 themselves and prevent interaction with the type 1 receptors. Cyclosporine is an inhibitor of Calcineurin that decreased pulmonary arterial smooth muscle cell proliferation in vivo and apoptosis in vitro, while partially reversing the severity of experimental PH in monocrotaline treated rats [46]. Rapamycin, an FKBP12 ligand, has been shown to ameliorate the extend of artery smooth muscle cell proliferation [47]. The dual inhibition of both FKBP12 and Calcineurin was achieved by FK506(Tacrolimus) [23], which facilitates the release of the FKBP12/Calcineurin complex from the type 1 receptor by binding to both molecules, and thereby activating downstream canonical and non-canonical BMPR2 signaling even in presence of a BMPR2 mutation. FK506 was identified as the best BMPR2 activator in a high-throughput luciferase reporter assay of 3756 FDA-approved drugs using ID1 expression as the assay readout [23], superior to Rapamycin and Cyclosporine. In vitro, FK506 activated downstream BMPR2 signaling via SMAD1/5, MAPK and ID1 signaling in healthy PA endothelial cells (PAECS), while normalizing endothelial dysfunction in PAH PAECs. FK506 prevented experimental PH in BMPR2+/- mice and reversed PH in both the rat monocrotaline induced and Sugen-Hypoxia induced PH, whereby it reduced right ventricular systolic pressure (RVSP), right ventricular hypertrophy (RVH), pulmonary vascular medial hypertrophy and neointima formation. FK506 was found to be safe and well tolerated in a Phase 2a proof-of concept safety and tolerability study [48] and has shown promise as compassionate use in three end-stage PAH patients [49].

5.1.3. Gene-Directed Modulation of the BMPR2 Receptor

The promise of genetic interventions to correct a specific BMPR2 mutation in familial and idiopathic PAH patients is on the advent. Gene-directed modulation of the BMPR2 receptor showed promise in experimental PH models. However, the use of CRISPR modulation as a pharmaceutical strategy, while a powerful tool, is not yet available for PAH patients.

In 2007: Reynolds et al. used an adenoviral vector for the targeted delivery of the BMPR2 gene to prevent BMPR2 inhibition in a rat model of hypoxia-induced PH. This treatment strategy significantly reduced the RVSP, RVH, and distal pulmonary vasculature muscularization [50]. However, the concern about a neutralizing immune response mounted after adenoviral transduction for horizontal gene transfer posed a concern for the efficacy of the method [51]. Building on these promising results, Harper et al. [52] improved established experimental PH in Monocrotaline (MCT) induced PH in rats using genetic modifications. Endothelial-like progenitor cells (ELPC) from the femoral bone-marrow of rats were transduced with a BMPR2 adenoviral vector (AdCMVBMPR2myc) and were injected into the tail-vein of experimental PH rats. While the injected cells were short lived in the lungs (<24 h), the injected animals showed an immediate increase in BMPR2 in their lungs, which was thought to be exosome mediated, as well as an improvement in muscularized vessels over time [52]. Another avenue to overcome the obstacle of a neutralizing immune response would be the use of Adeno-associated virus (AAV) for gene delivery, which elicits only a neglectable immune response [53,54]. The use of the adenovirus for BMPR2 AAV1.SERCA2a reduced RV hypertrophy, RVSP, mPAP and vascular remodeling, thereby overall reducing experimental PH [55]. Currently, this strategy is investigated in translational studies for heart failure and PH [56]. The effect of the correction of a BMPR2 mutation by CRISPR was investigated by Gu et al. [39], where induced pluripotent stem cell-derived endothelial cells (iPSC-ECs) from different individuals amongst three single families were examined for their characteristics. Endothelial cells derived from FPAH patients were defective in cell survival, adhesion, migration, tube formation and BMPR2 signaling, whilst unaffected mutation carriers as well as CRISPR corrected iPSC-ECs were not.

As the presence of a BMPR2 mutation can reduce not only the functionality, but also the expression of the BMPR2 receptor, the induction of readthrough of nonsense mutations by ataluren has been employed to increase BMPR2 signaling in several lung and blood cell types [57]. Similarly, gentamycin was used to treat premature stop codons and readthrough in PAH [58,59].

5.2. Modulating the Availability of the BMPR2 Receptor at the Cell Surface

The increased availability of the BMPR2 receptor at the cell surface is a potent strategy to increase downstream BMPR2 signaling. We have shown that BMPR2 signaling can be modulated by upstream modifiers that we have targeted by repurposed pharmaceuticals [27]. In an siRNA high-throughput screen of over 20,000 genes of potential BMPR2 modulators, the tumor suppressor gene Fragile Histidine Triad (FHIT) was identified an thereafter pharmaceutically targeted by the repurposed drug Enzastaurin, which reverse established experimental PH in Sugen Hypoxia rats. Despite the established pharmacolog cal role of Enzastaurin in PKC-inhibition, the authors showed that the action of Enzastauri on BMPR2 signaling may likely be an unspecific effect of the drug, as other PKC inhibitor were unable to achieve the same effect on BMPR2 gene expression and PAEC function.

Similarly, treatment with the elastase-specific inhibitor elafin stabilized the BMPR receptor at the membrane in animal PH models by enhancing its interaction with Caveolin- and thus reversed established PH in Sugen-hypoxia rats and patient PAECs [60].

Likewise, the prevention of receptor degradation at the cell surface by chloroquin and hydro-chloroquine by inhibition of autophagy and lysosomal degradation prevente the development of experimental PH [61,62]. BMPR2 receptor degradation and recepto shedding is also targeted by the TNF-α antagonist Etanercept [37], which prevented an reversed experimental PH in rats [63] and endotoxic pigs [64].

Increasing BMPR2 shuttling to the membrane using the chaperone 4-phenylbutyri acid (4PBA) [65] led to a mild improvement in BMPR2 downstream signaling in patier fibroblasts that contained a specific inactivating mutation C118W, which served as a proc of concept for the applicability of this method in patients and an important step toward precision medicine in PAH [66].

5.3. Increasing Downstream Gene Transcription by Targeting BMPR2 Signaling or Interacting Pathways

Classical activation of BMPR2 signaling is achieved through ligand binding to BMPR receptor complexes. BMPR2/Alk1 heterocomplexes are mainly targeted by BMP9 [26 whereas other BMP ligands, such as BMP2 and BMP4, can activate multiple BMPR2 hetero complexes (i.e., BMPR2/BMPR1A-B, BMPR2/Alk3), resulting in a higher probability fo off-target effects in gene expression of bone formation signaling [67]. In PAECs, the admi istration of BMP9 prevents EC apoptosis consistent with the desired therapeutic outcom of preventing early vessel loss. Injection of the BMP9 ligand in mice and rats reversed estal lished experimental PH (MCT-induced and SuHx) even in the presence of a heterozygou inactivating BMPR2 mutation.

A different approach would be to interfere with pathways or molecules that i hibit the BMPR2 receptor or its pathway such as TGF-β signaling or the binding protei FKBP12. FKBP12 can be pharmaceutically targeted by FK506 and also FKVP, a non immunosuppressive FK506 analog, which both activate BMPR2 signaling by FKBP1 antagonism [23,68]. The drug etanercept likewise increased BMPR2 signaling by inhibitin the BMP inhibitory pathway TGF-β, an effect that can also be observed with other TGF- inhibitory substances, such as Paclitaxel [69].

Lastly, many pathways converge to induce BMPR2 signaling, opening the pharmaceu tical potential of combined use of BMPR2-potentiating medication to achieve the synergisti or additive activation of BMPR2 signaling. As a proof-of-concept, FK506 and Enzastauri showed additive effects on BMPR2 signaling activation in vitro [27]. Moreover, the los of BMPR2 leads to changes in several of its downstream signaling pathways, such a p38/MAPK/ERK [70], PI3K/Akt [71] and Wnt [72] signaling, which have thus also bee investigated as therapeutic targets at a molecular level. Exploring the potential additiv effects of targeting the BMPR2 receptor, as well as its downstream signaling, may be o therapeutical value.

6. Conclusions

PAH is a progressive and ultimately fatal disease, while current treatments are insuff cient to substantially prolong patient survival. Targeting BMPR2 signaling and interactin signaling pathways has emerged as a promising approach to identify disease modifyin therapies that address fundamental, genetically based molecular pathways importar in PAH pathogenesis. Additive and synergistic effects of a combination treatment wit several BMPR2 enhancing drugs have been shown to increase the therapeutic effect. Hov ever, the off-target effects of existing BMPR2-targeting pharmaceuticals hinder the precise assessment of the full potential of BMPR2 targeting in PAH therapy.

In summary, the use of BMPR2 targeted treatments in addition to conventional vasodilatory drugs in PAH is a promising avenue to explore in the search for novel PAH treatments, but the development of novel compounds to target BMPR2 signaling with increased specificity is of utmost importance.

Author Contributions: S.D.P., Manuscript preparation and illustrations; M.K.A., Addition to BMPR2 signaling chapter and revision; E.S., Manuscript discussion and revision. All authors have read and agreed to the published version of the manuscript.

Funding: This research received no external funding.

Conflicts of Interest: The authors declare the following conflicts of interest: E.S. has served as scientific adviser for Selten Pharma, Inc., Vivus (modest). E.S. is listed as inventor on patent applications Use of FK506 for the Treatment of Pulmonary Arterial Hypertension (Serial No 61/481317). E.S. and S.D.P. are listed as inventors on patent application Enzastaurin and Fragile Histidine Trial (FHIT) Increasing Agents for the Treatment of Pulmonary Hypertension (PCT/US2018/033533).

References

Frost, A.E.; Badesch, D.; Gibbs, J.S.R.; Gopalan, D.; Khanna, D.; Manes, A.; Oudiz, R.; Satoh, T.; Torres, F.; Torbicki, A. Diagnosis of pulmonary hypertension. *Eur. Respir. J.* **2019**, *53*, 1801904. [CrossRef]

Noordegraaf, A.V.; Chin, K.M.; Haddad, F.; Hassoun, P.M.; Hemnes, A.R.; Hopkins, S.R.; Kawut, S.M.; Langleben, D.; Lumens, J.; Naeije, R. Pathophysiology of the right ventricle and of the pulmonary circulation in pulmonary hypertension: An update. *Eur. Respir. J.* **2019**, *53*, 1801900. [CrossRef]

Rabinovitch, M. Molecular pathogenesis of pulmonary arterial hypertension. *J. Clin. Investig.* **2012**, *122*, 4306–4313. [CrossRef] [PubMed]

Humbert, M.; Guignabert, C.; Bonnet, S.; Dorfmüller, P.; Klinger, J.R.; Nicolls, M.R.; Olschewski, A.J.; Pullamsetti, S.S.; Schermuly, R.T.; Stenmark, K.R.; et al. Pathology and pathobiology of pulmonary hypertension: State of the art and research perspectives. *Eur. Respir. J.* **2019**, *53*, 1801887. [CrossRef] [PubMed]

Benza, R.L.; Miller, D.P.; Barst, R.J.; Badesch, D.B.; Frost, A.E.; McGoon, M.D. An Evaluation of Long-term Survival from Time of Diagnosis in Pulmonary Arterial Hypertension from the REVEAL Registry. *Chest* **2012**, *142*, 448–456. [CrossRef]

PPeacock, A.J.; Murphy, N.F.; McMurray, J.J.V.; Caballero, L.; Stewart, S. An epidemiological study of pulmonary arterial hypertension. *Eur. Respir. J.* **2007**, *30*, 104–109. [CrossRef] [PubMed]

Morrell, N.W.; Aldred, M.A.; Chung, W.K.; Elliott, C.G.; Nichols, W.C.; Soubrier, F.; Trembath, R.C.; Loyd, J.E. Genetics and genomics of pulmonary arterial hypertension. *Eur. Respir. J.* **2019**, *53*, 1801899. [CrossRef] [PubMed]

Galie, N.; Channick, R.N.; Frantz, R.P.; Grünig, E.; Jing, Z.C.; Moiseeva, O.; Preston, I.R.; Pulido, T.; Safdar, Z.; Tamura, Y.; et al. Risk stratification and medical therapy of pulmonary arterial hypertension. *Eur. Respir. J.* **2019**, *53*, 1801889. [CrossRef]

Mayeux, J.D.; Pan, I.Z.; Dechand, J.; Jacobs, J.A.; Jones, T.L.; McKellar, S.H.; Beck, E.; Hatton, N.D.; Ryan, J.J. Management of Pulmonary Arterial Hypertension. *Curr. Cardiovasc. Risk Rep.* **2021**, *15*, 1–24. [CrossRef]

Humbert, M.; Sitbon, O.; Yaici, A.; Montani, D.; O'Callaghan, D.S.; Jais, X.; Parent, F.; Savale, L.; Natali, D.; Gunther, S.; et al. Survival in incident and prevalent cohorts of patients with pulmonary arterial hypertension. *Eur. Respir. J.* **2010**, *36*, 549–555. [CrossRef]

Woodcock, C.-S.C.; Chan, S.Y. The Search for Disease-Modifying Therapies in Pulmonary Hypertension. *J. Cardiovasc. Pharmacol. Ther.* **2019**, *24*, 334–354. [CrossRef]

Spiekerkoetter, A.; Kawut, S.M.; Perez, V.A.D.J. New and Emerging Therapies for Pulmonary Arterial Hypertension. *Annu. Rev. Med.* **2019**, *70*, 45–59. [CrossRef]

Deng, Z.; Morse, J.H.; Slager, S.L.; Cuervo, N.; Moore, K.J.; Venetos, G.; Kalachikov, S.; Cayanis, E.; Fischer, S.G.; Barst, R.J.; et al. Familial Primary Pulmonary Hypertension (Gene PPH1) Is Caused by Mutations in the Bone Morphogenetic Protein Receptor–II Gene. *Am. J. Hum. Genet.* **2000**, *67*, 737–744. [CrossRef] [PubMed]

Lane, K.B.; Machado, R.D.; Pauciulo, M.W.; Thomson, J.R.; Phillips, J.A.; Loyd, J.E.; Nichols, W.C.; Trembath, R.C. Heterozygous germline mutations in BMPR2, encoding a TGF-β receptor, cause familial primary pulmonary hypertension. *Nat. Genet.* **2000**, *26*, 81–84. [CrossRef]

Austin, E.D.; Loyd, J.E. The Genetics of Pulmonary Arterial Hypertension. *Circ. Res.* **2014**, *115*, 189–202. [CrossRef] [PubMed]

Ge, X.; Zhu, T.; Zhang, X.; Liu, Y.; Wang, Y.; Zhang, W. Gender differences in pulmonary arterial hypertension patients with BMPR2 mutation: A meta-analysis. *Respir. Res.* **2020**, *21*, 1–10. [CrossRef]

Southgate, L.; Machado, R.D.; Gräf, S.; Morrell, N.W. Molecular genetic framework underlying pulmonary arterial hypertension. *Nat. Rev. Cardiol.* **2019**, *17*, 85–95. [CrossRef] [PubMed]

18. Machado, R.D.; Southgate, L.; Eichstaedt, C.A.; Aldred, M.A.; Austin, E.D.; Best, D.H.; Chung, W.K.; Benjamin, N.; Elliott, C.(Eyries, M.; et al. Pulmonary Arterial Hypertension: A Current Perspective on Established and Emerging Molecular Gene(Defects. *Hum. Mutat.* **2015**, *36*, 1113–1127. [CrossRef] [PubMed]

19. Andruska, A.; Spiekerkoetter, E. Consequences of BMPR2 Deficiency in the Pulmonary Vasculature and Beyond: Contributio: to Pulmonary Arterial Hypertension. *Int. J. Mol. Sci.* **2018**, *19*, 2499. [CrossRef]

20. Chalazonitis, A.; D'Autreaux, F.; Guha, U.; Pham, T.D.; Faure, C.; Chen, J.J.; Roman, D.; Kan, L.; Rothman, T.P.; Kessler, J.A.; et Bone morphogenetic protein-2 and -4 limit the number of enteric neurons but promote development of a TrkC-expressir neurotrophin-3-dependent subset. *J. Neurosci.* **2004**, *24*, 4266–4282. [CrossRef]

21. Maruyama, H.; Dewachter, C.; Belhaj, A.; Rondelet, B.; Sakai, S.; Remmelink, M.; Vachiéry, J.-L.; Naeije, R.; Dewachter, Endothelin-Bone morphogenetic protein type 2 receptor interaction induces pulmonary artery smooth muscle cell hyperplasia pulmonary arterial hypertension. *J. Heart Lung Transplant.* **2015**, *34*, 468–478. [CrossRef] [PubMed]

22. Murakami, G.; Watabe, T.; Takaoka, K.; Miyazono, K.; Imamura, T. Cooperative Inhibition of Bone Morphogenetic Prote Signaling by Smurf1 and Inhibitory Smads. *Mol. Biol. Cell* **2003**, *14*, 2809–2817. [CrossRef] [PubMed]

23. Spiekerkoetter, E.; Tian, X.; Cai, J.; Hopper, R.K.; Sudheendra, D.; Li, C.G.; El-Bizri, N.; Sawada, H.; Haghighat, I Chan, R.; et al. FK506 activates BMPR2, rescues endothelial dysfunction, and reverses pulmonary hypertension. *J. Clin. Invest* **2013**, *123*, 3600–3613. [CrossRef]

24. Yamaguchi, T.; Kurisaki, A.; Yamakawa, N.; Minakuchi, K.; Sugino, H. FKBP12 functions as an adaptor of the Smad7-Smur complex on activin type I receptor. *J. Mol. Endocrinol.* **2006**, *36*, 569–579. [CrossRef]

25. Sanchez-Duffhues, G.; Williams, E.; Goumans, M.-J.; Heldin, C.-H.; Dijke, P.T. Bone morphogenetic protein receptors: Structu: function and targeting by selective small molecule kinase inhibitors. *Bone* **2020**, *138*, 115472. [CrossRef]

26. Long, L.; Ormiston, M.L.; Yang, X.; Southwood, M.; Gräf, S.; Machado, R.D.; Mueller, M.; Kinzel, B.; Yung, L.N Wilkinson, J.M.; et al. Selective enhancement of endothelial BMPR-II with BMP9 reverses pulmonary arterial hypertension. *N: Med.* **2015**, *21*, 777–785. [CrossRef] [PubMed]

27. Prosseda, S.D.; Tian, X.; Kuramoto, K.; Boehm, M.; Sudheendra, D.; Miyagawa, K.; Zhang, F.; Solow-Cordero, I Saldivar, J.C.; Austin, E.D.; et al. FHIT, a Novel Modifier Gene in Pulmonary Arterial Hypertension. *Am. J. Respir. C: Care Med.* **2019**, *199*, 83–98. [CrossRef] [PubMed]

28. Larabee, S.M.; Michaloski, H.; Jones, S.; Cheung, E.; Gallicano, G.I. miRNA-17 Members that Target Bmpr2 Influence Sign; ing Mechanisms Important for Embryonic Stem Cell Differentiation In Vitro and Gastrulation in Embryos. *Stem Cells D: **2015**, *24*, 354–371. [CrossRef] [PubMed]

29. Pullamsetti, S.S.; Doebele, C.; Fischer, A.; Savai, R.; Kojonazarov, B.; Dahal, B.K.; Ghofrani, H.A.; Weissmann, N.; Grimminger, Bonauer, A.; et al. Inhibition Of MicroRNA-17 Improves Lung And Heart Function In Experimental Pulmonary Hypertensic *Am. J. Respir. Crit. Care Med.* **2012**, *185*, 409–419. [CrossRef]

30. Brock, M.; Trenkmann, M.; Gay, R.E.; Michel, B.A.; Gay, S.; Fischler, M.; Ulrich, S.; Speich, R.; Huber, L.C. Interleukin-6 modulat: the expression of the bone morphogenic protein receptor type II through a novel STAT3-microRNA cluster 17/92 pathway. Ci Res. **2009**, *104*, 1184–1191. [CrossRef]

31. Yang, S.; Banerjee, S.; De Freitas, A.; Cui, H.; Xie, N.; Abraham, E.; Liu, G. miR-21 regulates chronic hypoxia-induced pulmona vascular remodeling. *Am. J. Physiol. Lung Cell. Mol. Physiol.* **2012**, *302*, L521–L529. [CrossRef] [PubMed]

32. Kang, H.; Louie, J.; Weisman, A.; Sheu-Gruttadauria, J.; Davis-Dusenbery, B.; Lagna, G.; Hata, A. Inhibition of MicroRNA-3((miR-302) by Bone Morphogenetic Protein 4 (BMP4) Facilitates the BMP Signaling Pathway. *J. Biol. Chem.* **2012**, *287*, 38656–386([CrossRef] [PubMed]

33. Negi, V.; Chan, S.Y. Discerning functional hierarchies of microRNAs in pulmonary hypertension. *JCI Insight* **2017**, *2*, e913: [CrossRef] [PubMed]

34. Parikh, V.N.; Jin, R.C.; Rabello, S.; Gulbahce, N.; White, K.; Hale, A.; Cottrill, K.A.; Shaik, R.S.; Waxman, A.B.; Zhang, Y.-Y.; et MicroRNA-21 Integrates Pathogenic Signaling to Control Pulmonary Hypertension: Results of a Network Bioinformati Approach. *Circulation* **2012**, *125*, 1520–1532. [CrossRef] [PubMed]

35. Fessel, J.P.; Chen, X.; Frump, A.; Gladson, S.; Blackwell, T.; Kang, C.; Johnson, J.; Loyd, J.E.; Hemnes, A.; Austin, E.; et al. Interactic between Bone Morphogenetic Protein Receptor Type 2 and Estrogenic Compounds in Pulmonary Arterial Hypertension. *Pul Circ.* **2013**, *3*, 564–577. [CrossRef] [PubMed]

36. Austin, E.D.; Hamid, R.; Hemnes, A.R.; Loyd, J.E.; Blackwell, T.; Yu, C.; Iii, J.A.P.; Gaddipati, R.; Gladson, S.; Gu, E.; et al. BMPI expression is suppressed by signaling through the estrogen receptor. *Biol. Sex Differ.* **2012**, *3*, 6. [CrossRef]

37. Hurst, L.A.; Dunmore, B.J.; Long, L.; Crosby, A.; Al-Lamki, R.; Deighton, J.; Southwood, M.; Yang, X.; Nikolic, M.; Herrera, B.; et al. TNFalpha drives pulmonary arterial hypertension by suppressing the BMP type-II receptor and alterir NOTCH signalling. *Nat. Commun.* **2017**, *8*, 14079. [CrossRef]

38. Hiepen, C.; Jatzlau, J.; Hildebrandt, S.; Kampfrath, B.; Goktas, M.; Murgai, A.; Camacho, J.L.C.; Haag, R.; Ruppert, (Sengle, G.; et al. BMPR2 acts as a gatekeeper to protect endothelial cells from increased TGFbeta responses and altered c(mechanics. *PLoS Biol.* **2019**, *17*, e3000557. [CrossRef]

39. Gu, M.; Shao, N.Y.; Sa, S.; Li, D.; Termglinchan, V.; Ameen, M.; Karakikes, I.; Sosa, G.; Grubert, F.; Lee, J.; et al. Patient-Speci iPSC-Derived Endothelial Cells Uncover Pathways that Protect against Pulmonary Hypertension in BMPR2 Mutation Carrie *Cell Stem Cell* **2017**, *20*, 490–504.e5. [CrossRef]

Atkinson, C.; Stewart, S.; Upton, P.D.; Machado, R.D.; Thomson, J.R.; Trembath, R.C.; Morrell, N.W. Primary Pulmonary Hypertension Is Associated With Reduced Pulmonary Vascular Expression of Type II Bone Morphogenetic Protein Receptor. *Circulation* **2002**, *105*, 1672–1678. [CrossRef]

Chen, N.-Y.; Collum, S.D.; Luo, F.; Weng, T.; Le, T.-T.; Hernandez, A.M.; Philip, K.; Molina, J.G.; Garcia-Morales, L.J.; Cao, Y.; et al. Macrophage bone morphogenic protein receptor 2 depletion in idiopathic pulmonary fibrosis and Group III pulmonary hypertension. *Am. J. Physiol. Lung Cell. Mol. Physiol.* **2016**, *311*, L238–L254. [CrossRef] [PubMed]

Hong, K.; Lee, Y.J.; Lee, E.; Park, S.O.; Han, C.; Beppu, H.; Li, E.; Raizada, M.K.; Bloch, K.D.; Oh, S.P. Genetic Ablation of the Bmpr2 Gene in Pulmonary Endothelium Is Sufficient to Predispose to Pulmonary Arterial Hypertension. *Circulation* **2008**, *118*, 722–730. [CrossRef] [PubMed]

West, J.; Fagan, K.; Steudel, W.; Fouty, B.; Lane, K.; Harral, J.; Hoedt-Miller, M.; Tada, Y.; Ozimek, J.; Tuder, R.; et al. Pulmonary Hypertension in Transgenic Mice Expressing a Dominant-Negative BMPRII Gene in Smooth Muscle. *Circ. Res.* **2004**, *94*, 1109–1114. [CrossRef] [PubMed]

West, J.; Harral, J.; Lane, K.; Deng, Y.; Ickes, B.; Crona, D.J.; Albu, S.; Stewart, D.; Fagan, K. Mice expressing BMPR2R899X transgene in smooth muscle develop pulmonary vascular lesions. *Am. J. Physiol. Lung Cell. Mol. Physiol.* **2008**, *295*, L744–L755. [CrossRef] [PubMed]

Hautefort, A.; Mendes-Ferreira, P.; Sabourin, J.; Manaud, G.; Bertero, T.; Rucker-Martin, C.; Riou, M.; Adão, R.; Manoury, B.; Lambert, M.; et al. Bmpr2 Mutant Rats Develop Pulmonary and Cardiac Characteristics of Pulmonary Arterial Hypertension. *Circulation* **2019**, *139*, 932–948. [CrossRef]

Bonnet, S.; Rochefort, G.; Sutendra, G.; Archer, S.L.; Haromy, A.; Webster, L.; Hashimoto, K.; Bonnet, S.N.; Michelakis, E.D. The nuclear factor of activated T cells in pulmonary arterial hypertension can be therapeutically targeted. *Proc. Natl. Acad. Sci. USA* **2007**, *104*, 11418–11423. [CrossRef]

Houssaïni, A.; Abid, S.; Mouraret, N.; Wan, F.; Rideau, D.; Saker, M.; Marcos, E.; Tissot, C.-M.; Dubois-Randé, J.-L.; Amsellem, V.; et al. Rapamycin reverses pulmonary artery smooth muscle cell proliferation in pulmonary hypertension. *Am. J. Respir. Cell Mol. Biol.* **2013**, *48*, 568–577. [CrossRef]

Spiekerkoetter, E.; Sung, Y.K.; Sudheendra, D.; Scott, V.; Del Rosario, P.; Bill, M.; Haddad, F.; Long-Boyle, J.; Hedlin, H.; Zamanian, R.T. Randomised placebo-controlled safety and tolerability trial of FK506 (tacrolimus) for pulmonary arterial hypertension. *Eur. Respir. J.* **2017**, *50*, 1602449. [CrossRef]

Spiekerkoetter, A.; Sung, Y.K.; Sudheendra, D.; Bill, M.; Aldred, M.A.; Van De Veerdonk, M.C.; Noordegraaf, A.V.; Long-Boyle, J.; Dash, R.; Yang, P.C.; et al. Low-Dose FK506 (Tacrolimus) in End-Stage Pulmonary Arterial Hypertension. *Am. J. Respir. Crit. Care Med.* **2015**, *192*, 254–257. [CrossRef]

Reynolds, A.M.; Xia, W.; Holmes, M.D.; Hodge, S.; Danilov, S.; Curiel, D.T.; Morrell, N.W.; Reynolds, P.N. Bone morphogenetic protein type 2 receptor gene therapy attenuates hypoxic pulmonary hypertension. *Am. J. Physiol. Lung Cell. Mol. Physiol.* **2007**, *292*, L1182–L1192. [CrossRef]

Lee, C.S.; Bishop, E.S.; Zhang, R.; Yu, X.; Farina, E.M.; Yan, S.; Zhao, C.; Zeng, Z.; Shu, Y.; Wu, X.; et al. Adenovirus-mediated gene delivery: Potential applications for gene and cell-based therapies in the new era of personalized medicine. *Genes Dis.* **2017**, *4*, 43–63. [CrossRef] [PubMed]

Harper, R.L.; Maiolo, S.; Ward, R.J.; Seyfang, J.; Cockshell, M.P.; Bonder, C.S.; Reynolds, P.N. BMPR2-expressing bone marrow-derived endothelial-like progenitor cells alleviate pulmonary arterial hypertension in vivo. *Respirology* **2019**, *24*, 1095–1103. [CrossRef] [PubMed]

Daya, S.; Berns, K.I. Gene Therapy Using Adeno-Associated Virus Vectors. *Clin. Microbiol. Rev.* **2008**, *21*, 583–593. [CrossRef] [PubMed]

Watanabe, S.; Ishikawa, K.; Plataki, M.; Bikou, O.; Kohlbrenner, E.; Aguero, J.; Hadri, L.; Zarragoikoetxea, I.; Fish, K.; Leopold, J.A.; et al. Safety and long-term efficacy of AAV1.SERCA2a using nebulizer delivery in a pig model of pulmonary hypertension. *Pulm. Circ.* **2018**, *8*. [CrossRef]

Hadri, L.; Kratlian, R.G.; Benard, L.; Maron, B.A.; Dorfmüller, P.; Ladage, D.; Guignabert, C.; Ishikawa, K.; Aguero, J.; Ibanez, B.; et al. Therapeutic Efficacy of AAV1.SERCA2a in Monocrotaline-Induced Pulmonary Arterial Hypertension. *Circulation* **2013**, *128*, 512–523. [CrossRef]

Lyon, A.R.; Babalis, D.; Morley-Smith, A.C.; Hedger, M.; Barrientos, A.S.; Foldes, G.; Couch, L.S.; Chowdhury, R.A.; Tzortzis, K.N.; Peters, N.S.; et al. Investigation of the safety and feasibility of AAV1/SERCA2a gene transfer in patients with chronic heart failure supported with a left ventricular assist device—The SERCA-LVAD TRIAL. *Gene Ther.* **2020**, 1–12. [CrossRef]

Drake, K.M.; Dunmore, B.J.; McNelly, L.N.; Morrell, N.W.; Aldred, M.A. Correction of NonsenseBMPR2andSMAD9Mutations by Ataluren in Pulmonary Arterial Hypertension. *Am. J. Respir. Cell Mol. Biol.* **2013**, *49*, 403–409. [CrossRef]

Nasim, M.T.; Ghouri, A.; Patel, B.; James, V.; Rudarakanchana, N.; Morrell, N.W.; Trembath, R.C. Stoichiometric imbalance in the receptor complex contributes to dysfunctional BMPR-II mediated signalling in pulmonary arterial hypertension. *Hum. Mol. Genet.* **2008**, *17*, 1683–1694. [CrossRef]

Hamiid, R.; Hedges, L.K.; Austin, E.; Phillips, J.A., III; Loyd, J.E.; Cogan, J.D. Transcripts from a novel BMPR2 termination mutation escape nonsense mediated decay by downstream translation re-initiation: Implications for treating pulmonary hypertension. *Clin. Genet.* **2010**, *77*, 280–286. [CrossRef]

60. Perez, V.A.D.J. Faculty Opinions recommendation of Elafin Reverses Pulmonary Hypertension via Caveolin-1-Dependent Bone Morphogenetic Protein Signaling. *Am. J. Respir. Crit. Care Med.* **2015**, *191*, 1273–1286.

61. Dunmore, B.J.; Drake, K.M.; Upton, P.D.; Toshner, M.R.; Aldred, M.A.; Morrell, N.W. The lysosomal inhibitor, chloroquine, increases cell surface BMPR-II levels and restores BMP9 signalling in endothelial cells harbouring BMPR-II mutations. *Hum. Mol. Genet.* **2013**, *22*, 3667–3679. [CrossRef] [PubMed]

62. Long, L.; Yang, X.; Southwood, M.; Lu, J.; Marciniak, S.J.; Dunmore, B.J.; Morrell, N.W. Chloroquine Prevents Progression of Experimental Pulmonary Hypertension via Inhibition of Autophagy and Lysosomal Bone Morphogenetic Protein Type II Receptor Degradation. *Circ. Res.* **2013**, *112*, 1159–1170. [CrossRef] [PubMed]

63. Zhang, L.-L.; Lu, J.; Li, M.; Wang, Q.; Zeng, X. Preventive and remedial application of etanercept attenuate monocrotaline-induced pulmonary arterial hypertension. *Int. J. Rheum. Dis.* **2014**, *19*, 192–198. [CrossRef] [PubMed]

64. Mutschler, D.; Wikström, G.; Lind, L.; Larsson, A.; Lagrange, A.; Eriksson, M. Etanercept Reduces Late Endotoxin-Induced Pulmonary Hypertension in the Pig. *J. Interf. Cytokine Res.* **2006**, *26*, 661–667. [CrossRef] [PubMed]

65. Dunmore, B.J.; Yang, X.; Crosby, A.; Moore, S.; Long, L.; Huang, C.; Southwood, M.; Austin, E.D.; Rana, A.; Upton, P.D.; et al. 4PBA Restores Signaling of a Cysteine-substituted Mutant BMPR2 Receptor Found in Patients with Pulmonary Arterial Hypertension. *Am. J. Respir. Cell Mol. Biol.* **2020**, *63*, 160–171. [CrossRef]

66. Andruska, A.; Ali, M.K.; Spiekerkoetter, E. Targeting BMPR2 Trafficking with Chaperones: An Important Step toward Precision Medicine in Pulmonary Arterial Hypertension. *Am. J. Respir. Cell Mol. Biol.* **2020**, *63*, 137–138. [CrossRef]

67. Ou, M.; Zhao, Y.; Zhang, F.; Huang, X. Bmp2 and Bmp4 accelerate alveolar bone development. *Connect. Tissue Res.* **2014**, *56*, 204–211. [CrossRef]

68. Peiffer, B.J.; Qi, L.; Ahmadi, A.R.; Wang, Y.; Guo, Z.; Peng, H.; Sun, Z.; Liu, J.O. Activation of BMP Signaling by FKBP12 Ligands Synergizes with Inhibition of CXCR4 to Accelerate Wound Healing. *Cell Chem. Biol.* **2019**, *26*, 652–661.e4. [CrossRef]

69. Kassa, B.; Mickael, C.; Kumar, R.; Sanders, L.; Koyanagi, D.; Hernandez-Saavedra, D.; Tuder, R.M.; Graham, B.B. Paclitaxel blocks Th2-mediated TGF-β activation in Schistosoma mansoni-induced pulmonary hypertension. *Pulm. Circ.* **2019**, *9*, 204589401882081. [CrossRef]

70. Church, C.; Martin, D.H.; Wadsworth, R.; Bryson, G.; Fisher, A.J.; Welsh, D.J.; Peacock, A.J. The reversal of pulmonary vascular remodeling through inhibition of p38 MAPK-α: A potential novel anti-inflammatory strategy in pulmonary hypertension. *Am. J. Physiol. Lung Cell. Mol. Physiol.* **2015**, *309*, L333–L347. [CrossRef]

71. Wang, Y.-Y.; Cheng, X.-D.; Jiang, H. Effect of atorvastatin on pulmonary arterial hypertension in rats through PI3K/AKT signaling pathway. *Eur. Rev. Med. Pharmacol. Sci.* **2019**, *23*, 10549–10556. [PubMed]

72. Perez, V.D.J.; Yuan, K.; Alastalo, T.-P.; Spiekerkoetter, E.; Rabinovitch, M. Targeting the Wnt signaling pathways in pulmonary arterial hypertension. *Drug Discov. Today* **2014**, *19*, 1270–1276. [CrossRef] [PubMed]

MDPI

St. Alban-Anlage 66

4052 Basel

Switzerland

Tel. +41 61 683 77 34

Fax +41 61 302 89 18

www.mdpi.com

Genes Editorial Office

E-mail: genes@mdpi.com

www.mdpi.com/journal/genes